A MATTER OF DENSITY

A MATTER OF DENSITY

Exploring the Electron Density Concept in the Chemical, Biological, and Materials Sciences

Edited by

N. SUKUMAR
Department of Chemistry
Shiv Nadar University
Dadri, UP, India

A JOHN WILEY & SONS, INC., PUBLICATION

Published by John Wiley & Sons, Inc., Hoboken, New Jersey
Published simultaneously in Canada

Library of Congress Cataloging-in-Publication Data:

A matter of density : exploring the electron density concept in the chemical, biological, and materials sciences / edited by N. Sukumar.
pages cm
Includes index.
ISBN 978-0-470-76900-3 (hardback)
1. Electron distribution. I. Sukumar, N., editor of compilation.
QC793.5.E626M38 2013
539.7′2112—dc23

2012023635

Printed in the United States of America

ISBN: 9780470769003

10 9 8 7 6 5 4 3 2 1

CONTENTS

PREFACE

Electron density is one of the fundamental concepts underpinning modern chemistry. Introduced through Max Born's probability interpretation of the wave function, it is an enigma that bridges the classical concepts of particles and fluids. The electronic structure of matter is intimately related to the quantum laws of composition of probabilities and the Born–Oppenheimer separation of electronic and nuclear motions in molecules. The topology of the electron density determines the details of molecular structure and stability. The electron density is a quantity that is directly accessible to experimental determination through diffraction experiments. It is the basic variable of density functional theory, which has enabled practical applications of the mathematical theory of quantum physics to chemical and biological systems in recent years. The importance of density functional theory was recognized by the 1998 Nobel Prize in chemistry to Walter Kohn and John Pople.

In the first part (Chapters 1–6) of this book, we aim to present the reader with a coherent and logically connected treatment of theoretical foundations of the electron density concept, beginning with its statistical underpinnings: the use of probabilities in statistical physics (Chapter 1) and the origins of quantum mechanics. We delve into the philosophical questions at the heart of the quantum theory such as quantum entanglement (Chapter 2), and also describe methods for the experimental determination of electron density distributions (Chapter 3). The conceptual and statistical framework developed in earlier chapters is then employed to treat electron exchange and correlation, the partitioning of molecules into atoms (Chapter 4), density functional theory, and the theory of the insulating state of matter (Chapter 5). Chapter 6 concludes with an in-depth treatment of density-functional approximations for exchange and correlation by Viktor Staroverov.

The second part (Chapters 7–11) deals with applications of the electron density concept in chemical, biological, and materials sciences. In Chapter 7, Chakraborty, Duley, Giri, and Chattaraj describe how a deep understanding of the origins of chemical reactivity can be gleaned from the concepts of density functional theory. Applications of electron density in molecular similarity analysis and of electron-density-derived molecular descriptors form the subject matter of Chapter 8. In Chapter 9, Politzer, Bulat, Burgess, Baldwin, and Murray, elaborate on two of the most important such descriptors, namely, electrostatic potentials and local ionization energies, with particular reference to nanomaterial applications. All the applications discussed thus far have dealt with electron density in position space. A complementary perspective is obtained by considering the electron density in momentum space. MacDougall and Levit illustrate this in Chapter 10, by employing the Laplacian of the electron momentum density as a probe of electron dynamics. Pilania, Zhu, and Ramprasad conclude the discussion in Chapter 11 with some applications of modern density functional theory to surfaces and interfaces. The book is addressed to senior undergraduate and graduate students in chemistry and philosophers of science, as well as to current and aspiring practitioners of computational quantum chemistry, and anyone interested in exploring the applications of the electron density concept in chemistry, biology, and materials sciences.

I would like to express my sincere thanks and appreciation to the numerous friends and colleagues who helped to make this book a reality by graciously contributing their precious time and diligent efforts in reviewing various chapters or otherwise offering their valuable suggestions, namely, Drs. Felipe Bulat and A. K. Rajagopal (Naval Research Laboratory, Washington, DC), Prof. Shridhar Gadre (University of Pune and Indian Institute of Technology, Kanpur, India), Dr. Michael Krein (Lockheed Martin Advanced Technology Laboratories, Cherry Hill, NJ and Rensselaer Polytechnic Institute, Troy, NY), Prof. Preston MacDougall (Middle Tennessee State University, Murfreesboro, TN), Prof. Cherif Matta (Mount Saint Vincent University and Dalhousie University, Halifax, Nova Scotia, Canada), Dr. Salilesh Mukhopadhyay (Feasible Solutions, NJ), Profs. Peter Politzer and Jane Murray (CleveTheoComp LLC, Cleveland, OH), Prof. Sunanda Sukumar (Albany College of Pharmacy, Albany, NY and Shiv Nadar University, Dadri, India), Prof. Ajit Thakkar (University of New Brunswick, Fredericton, Canada), and Prof. Viktor Staroverov (University of Western Ontario, Canada). I also owe a deep debt of gratitude to the institutions and individuals who hosted me at various times during the last couple of years and provided me with the facilities to complete this book, namely, Rensselaer Polytechnic Institute in Troy, NY, and my host there Prof. Curt Breneman; the Institute of Mathematical Sciences in Chennai, India, and my host there Prof. G. Baskaran; and Shiv Nadar University in Dadri, India. The patient assistance of Senior Acquisitions Editor, Anita Lekhwani, and her very capable and efficient team at John Wiley & Sons has also been invaluable in this process.

N. Sukumar

Department of Chemistry
Shiv Nadar University
Dadri, UP, India

CONTRIBUTORS

Jeffrey W. Baldwin, Acoustics Division, Naval Research Laboratory, Washington, DC

Felipe A. Bulat, Acoustics Division, Naval Research Laboratory, Washington, DC

James Burgess, Acoustics Division, Naval Research Laboratory, Washington, DC

Arindam Chakraborty, Department of Chemistry and Center for Theoretical Studies, Indian Institute of Technology, Kharagpur, India

Pratim Kumar Chattaraj, Department of Chemistry and Center for Theoretical Studies, Indian Institute of Technology, Kharagpur, India

Soma Duley, Department of Chemistry and Center for Theoretical Studies, Indian Institute of Technology, Kharagpur, India

Santanab Giri, Department of Chemistry and Center for Theoretical Studies, Indian Institute of Technology, Kharagpur, India

M. Creon Levit, NASA, Advanced Supercomputing Division, Ames Research Center, Moffett Field, CA

Preston J. MacDougall, Department of Chemistry and Center for Computational Science, Middle Tennessee State University, Murfreesboro, TN

Jane S. Murray, CleveTheoComp LLC, Cleveland, OH

G. Pilania, Department of Chemical, Materials and Biomolecular Engineering, Institute of Materials Science, University of Connecticut, Storrs, CT

Peter Politzer, CleveTheoComp LLC, Cleveland, OH

R. Ramprasad, Department of Chemical, Materials and Biomolecular Engineering, Institute of Materials Science, University of Connecticut, Storrs, CT

Viktor N. Staroverov, Department of Chemistry, The University of Western Ontario, London, Ontario, Canada

N. Sukumar, Department of Chemistry, Shiv Nadar University, India; Rensselaer Exploratory Center for Cheminformatics Research, Troy, NY

Sunanda Sukumar, Department of Chemistry, Shiv Nadar University, India

H. Zhu, Department of Chemical, Materials and Biomolecular Engineering, Institute of Materials Science, University of Connecticut, Storrs, CT

1

INTRODUCTION OF PROBABILITY CONCEPTS IN PHYSICS—THE PATH TO STATISTICAL MECHANICS

N. SUKUMAR

It was an Italian gambler who gave us the first scientific study of probability theory. But Girolamo Cardano, also known as Hieronymus Cardanus or Jerome Cardan (1501–1576), was no ordinary gambler. He was also an accomplished mathematician, a reputed physician, and author. Born in Pavia, Italy, Cardan was the illegitimate son of Fazio Cardano, a Milan lawyer and mathematician, and Chiara Micheria. In addition to his law practice, Fazio lectured on geometry at the University of Pavia and at the Piatti Foundation and was consulted by the likes of Leonardo da Vinci on matters of geometry. Fazio taught his son mathematics and Girolamo started out as his father's legal assistant, but then went on to study medicine at Pavia University, earning his doctorate in medicine in 1525. But on account of his confrontational personality, he had a difficult time finding work after completing his studies. In 1525, he applied to the College of Physicians in Milan, but was not admitted owing to his illegitimate birth. Upon his father's death, Cardan squandered his bequest and turned to gambling, using his understanding of probability to make a living off card games, dice, and chess. Cardan's book on games of chance, *Liber de ludo aleae* (*On Casting the Die*, written in the 1560s, but not published until 1663), contains the first ever exploration of the laws of probability, as well as a section on effective cheating methods! In this book, he considered the fundamental scientific principles

A Matter of Density: Exploring the Electron Density Concept in the Chemical, Biological, and Materials Sciences, First Edition. Edited by N. Sukumar.

governing the likelihood of achieving double sixes in the rolling of dice and how to divide the stakes if a game of dice is incomplete.

First of all, note that each die has six faces, each of which is equally likely (assuming that the dice are unloaded). As the six different outcomes of a single die toss are mutually exclusive (only one face can be up at any time), their probabilities have to add up to 1 (a certainty). In other words, the probabilities of mutually exclusive events are additive. Thus, $P(\mathrm{A} = 6) = 1/6$ is the probability of die A coming up a six; likewise $P(\mathrm{B} = 6) = 1/6$ is the probability of die B coming up a six. Then, according to Cardan, the probability of achieving double sixes is the simple product:

$$P(\mathrm{A} = 6; \mathrm{B} = 6) = P(\mathrm{A} = 6) \times P(\mathrm{B} = 6) = 1/36.$$

The fundamental assumption here is that the act of rolling (or not rolling) die A does not affect the outcome of the roll of die B. In other words, the two dice are independent of each other, and their probabilities are found to compound in a multiplicative manner. Of course, the same conclusion holds for the probability of two fives or two ones or indeed that of die A coming up a one and die B coming up a five. So we can generalize this law to read

$$P(\mathrm{A}; \mathrm{B}) = P(\mathrm{A}) \times P(\mathrm{B}), \tag{1.1}$$

provided A and B are independent events. Notice, however, that the probability of obtaining a five and a one when rolling two dice is 1/18, since there are two equally likely ways of achieving this result: $\mathrm{A} = 1$; $\mathrm{B} = 5$ and $\mathrm{A} = 5$; $\mathrm{B} = 1$. Thus

$$P(\mathrm{A} = 1; \mathrm{B} = 5) + P(\mathrm{A} = 5; \mathrm{B} = 1) = \frac{1}{6} \times \frac{1}{6} + \frac{1}{6} \times \frac{1}{6} = \frac{1}{18}.$$

Likewise, the probability of obtaining a head and a tail in a two-coin toss is $1/2 \times 1/2 + 1/2 \times 1/2 = 1/2$, while that of two heads is $1/2 \times 1/2$ (and the same for two tails) because the two-coin tosses, whether performed simultaneously or sequentially, are independent of each other.

Eventually, Cardan developed a great reputation as a physician, successfully treating popes and archbishops, and was highly sought after by many wealthy patients. He was appointed Professor of Medicine at Pavia University, and was the first to provide a (clinical) description of typhus fever and (what we now know as) imaginary numbers. Cardan's book *Arts Magna* (*The Great Art or The Rules of Algebra*) is one of the classics in algebra. Cardan did, however, pass on his gambling addiction to his younger son Aldo; he was also unlucky in his eldest son Giambatista. Giambatista poisoned his wife, whom he suspected of infidelity, and was then executed in 1560. Publishing the horoscope of Jesus and writing a book in praise of Nero (tormentor of Christian martyrs) earned Girolamo Cardan a conviction for heresy in 1570 and a jail term. Forced to give up his professorship, he lived the remainder of his days in Rome off a pension from the Pope.

The foundations of probability theory were thereafter further developed by Blaise Pascal (1623–1662) in correspondence with Pierre de Fermat (1601–1665). Following Cardan, they studied the dice problem and solved the problem of points, considered by Cardan and others, for a two player game, as also the “gambler’s ruin”: the problem of finding the probability that when two men are gambling together, one will ruin the other. Blaise Pascal was the third child and only son of Étienne Pascal, a French lawyer, judge, and amateur mathematician. Blaise’s mother died when he was three years old. Étienne had unorthodox educational views and decided to homeschool his son, directing that his education should be confined at first to the study of languages, and should not include any mathematics. This aroused the boy’s curiosity and, at the age of 12, Blaise started to work on geometry on his own, giving up his playtime to this new study. He soon discovered for himself many properties of figures, and, in particular, the proposition that the sum of the angles of a triangle is equal to two right angles. When Étienne realized his son’s dedication to mathematics, he relented and gave him a copy of Euclid’s elements.

In 1639, Étienne was appointed tax collector for Upper Normandy and the family went to live in Rouen. To help his father with his work collecting taxes, Blaise invented a mechanical calculating machine, the Pascaline, which could do the work of six accountants, but the Pascaline never became a commercial success. Blaise Pascal also repeated Torricelli’s experiments on atmospheric pressure (*New Experiments Concerning Vacuums*, October 1647), and showed that a vacuum could and did exist above the mercury in a barometer, contradicting Aristotle’s and Descartes’ contentions that nature abhors vacuum. In August 1648, he observed that the pressure of the atmosphere decreases with height, confirming his theory of the cause of barometric variations by obtaining simultaneous readings at different altitudes on a nearby hill, and thereby deduced the existence of a vacuum above the atmosphere. Pascal also worked on conic sections and derived important theorems in projective geometry. These studies culminated in his *Treatise on the Equilibrium of Liquids* (1653) and *The Generation of Conic Sections* (1654 and reworked on 1653–1658). Following his father’s death in 1651 and a road accident in 1654 where he himself had a narrow escape, Blaise turned increasingly to religion and mysticism. Pascal’s philosophical treatise *Pensées* contains his statistical cost-benefit argument (known as Pascal’s wager) for the rationality of belief in God:

> If God does not exist, one will lose nothing by believing in him, while if he does exist, one will lose everything by not believing.

In his later years, he completely renounced his interest in science and mathematics, devoting the rest of his life to God and charitable acts. Pascal died of a brain hemorrhage at the age of 39, after a malignant growth in his stomach spread to the brain.

In the following century, several physicists and mathematicians drew upon the ideas of Pascal and Fermat, in advancing the science of probability and statistics. Christiaan Huygens (1629–1694), mathematician and physicist, wrote a book on probability, *Van Rekeningh in Spelen van Geluck* (*The Value of all Chances in Games of Fortune*), outlining the calculation of the expectation in a game of chance. Jakob Bernoulli (1654–1705), professor of mathematics at the University of Basel, originated the term *permutation* and introduced the terms *a priori* and *a posteriori* to distinguish two ways of deriving probabilities. Daniel Bernoulli (1700–1782), mathematician, physicist, and a nephew of Jakob Bernoulli, working in St. Petersburg and at the University of Basel, wrote nine papers on probability, statistics, and demography, but is best remembered for his *Exposition of a New Theory on the Measurement of Risk* (1737). Thomas Bayes (1702–1761), clergyman and mathematician, wrote only one paper on probability, but one of great significance: *An Essay towards Solving a Problem in the Doctrine of Chances* published posthumously in 1763. Bayes' theorem is a simple mathematical formula for calculating conditional probabilities. In its simplest form, Bayes' theorem relates the conditional probability (also called the *likelihood*) of event A given B to its converse, the conditional probability of B given A:

$$P(\mathrm{A}|\mathrm{B}) = \frac{P(\mathrm{B}|\mathrm{A})P(\mathrm{A})}{P(\mathrm{B})}, \tag{1.2}$$

where $P(\mathrm{A})$ and $P(\mathrm{B})$ are the prior or marginal probabilities of A ("prior" in the sense that it does not take into account any information about B) and B, respectively; $P(\mathrm{A}|\mathrm{B})$ is the conditional probability of A, given B (also called the *posterior probability* because it is derived from or depends on the specified value of B); and $P(\mathrm{B}|\mathrm{A})$ is the conditional probability of B given A. To derive the theorem, we note that from the product rule, we have

$$P(\mathrm{A}|\mathrm{B})\,P(\mathrm{B}) = P(\mathrm{A};\mathrm{B}) = P(\mathrm{B}|\mathrm{A})\,P(\mathrm{A}). \tag{1.3}$$

Dividing by $P(\mathrm{B})$, we obtain Bayes' theorem (Eq. 1.2), provided that neither $P(\mathrm{B})$ nor $P(\mathrm{A})$ is zero.

To see the wide-ranging applications of this theorem, let us consider a couple of examples (given by David Dufty). If a patient exhibits fever and chills, a doctor might suspect tuberculosis, but would like to know the conditional probability $P(\mathrm{TB}|\text{fever \& chills})$ that the patient has tuberculosis given the present symptoms. Some half of all TB sufferers exhibit these symptoms at any point in time. Thus, $P(\text{fever \& chills}|\mathrm{TB}) = 0.5$. While tuberculosis is now rare in the United States and affects some 0.01% of the population, $P(\mathrm{TB}) = 0.0001$; fever is a common symptom, generated by hundreds of diseases, and affecting 3% of Americans every year, and hence $P(\text{fever \& chills}) = 0.03$. Thus the conditional probability of TB given the symptoms of fever and chills is

$$P(\mathrm{TB}|\text{fever \& chills}) = 0.5 \times 0.0001/0.03 = 0.001667$$

or about 1.6 in a thousand. Another common situation is when a patient has a blood test done for lupus. If the test result is positive, it can be a concern, but the test is known to give a false positive result in 2% of cases: $P(\text{test}\oplus|\text{no lupus}) = 0.02$. In patients with lupus, 99% of the time the test result is positive, that is, $P(\text{test}\oplus|\text{lupus}) = 0.99$. A doctor would like to know the conditional probability $P(\text{lupus}|\text{test}\oplus)$ that the patient has lupus, given the positive test result. Lupus occurs in 0.5% of the US population, so that $P(\text{lupus}) = 0.005$. The probability of a positive result in general is

$$\begin{aligned}
P(\text{test}\oplus) &= P(\text{test}\oplus;\ \text{lupus}) + P(\text{test}\oplus;\ \text{no lupus}) \\
&= P(\text{test}\oplus;\ \text{lupus}) \times P(\text{lupus}) + P(\text{test}\oplus;\ \text{no lupus}) \times P(\text{no lupus}) \\
&= 0.99 \times 0.005 + 0.02 \times 0.995 \\
&= 0.02485,
\end{aligned}$$

where we have used the sum rule for mutually exclusive events in the first step, and Equation 1.3 in the next step. The probability of lupus, given the positive test result, is then $P(\text{lupus}|\text{test}\oplus) = 0.99 \times 0.005/0.02485 = 0.199$. So, in spite of the 99% accuracy of the test, there is only a 20% chance that a patient testing positive actually has lupus. This seemingly nonintuitive result is due to the fact that lupus is a very rare disease, while the test gives a large number of false positives, so that there are more false positives in any random population than actual cases of the disease.

The next actor in our story is Pierre-Simon de Laplace (1749–1827), a mathematician and a physicist, who worked on probability and calculus over a period of more than 50 years. His father, Pierre Laplace, was in the cider trade and expected his son to make a career in the church. However, at Caen University, Pierre-Simon discovered his love and talent for mathematics and, at the age of 19, went to Paris without taking his degree, but with a letter of introduction to d'Alembert, from his teacher at Caen. With d'Alembert's help, Pierre-Simon was appointed professor of mathematics at École Militaire, from where he started producing a series of papers on differential equations and integral calculus, the first of which was read to the Académie des Sciences in Paris in 1770. His first paper to appear in print was on integral calculus in *Nova Acta Eruditorum*, Leipzig, in 1771. He also read papers on mathematical astronomy to the Académie, including the work on the inclination of planetary orbits and a study of the perturbation of planetary orbits by their moons. Within 3 years Pierre-Simon had read 13 papers to the Académie, and, in 1773, he was elected as an adjoint in the Académie des Sciences. His' 1774 *Mémoire sur la Probabilité des Causes par les Évènemens* gave a Bayesian analysis of errors of measurement. Laplace has many other notable contributions to his credit, such as the central limit theorem, the probability generating function, and the characteristic function. He also applied his probability theory to compare the mortality rates at several hospitals in France.

Working with the chemist Antoine Lavoisier in 1780, Laplace embarked on a new field of study, applying quantitative methods to a comparison of living and inanimate systems. Using an ice calorimeter that they devised, Lavoisier and Laplace showed respiration to be a form of combustion. In 1784, Laplace was appointed examiner at the Royal Artillery Corps, where he examined and passed the young Napoleon Bonaparte. As a member of a committee of the Académie des Sciences to standardize weights and measures in 1790, he advocated a decimal base, which led to the creation of the metric system. He married in May 1788; he and his wife went on to have two children. While Pierre-Simon was not modest about his abilities and achievements, he was at least cautious, perhaps even politically opportunistic, but certainly a survivor. Thus, he managed to avoid the fate of his colleague Lavoisier, who was guillotined during the French Revolution in 1794. He was a founding member of the Bureau des Longitudes and went on to lead the Bureau and the Paris Observatory. In this position, Laplace published his *Exposition du Systeme du Monde* as a series of five books, the last of which propounded his nebular hypothesis for the formation of the solar system in 1796, according to which the solar system originated from the contraction and cooling of a large, oblate, rotating cloud of gas.

During Napoleon's reign, Laplace was a member, then chancellor of the Senate, receiving the Legion of Honor in 1805 and becoming Count of the Empire the following year. In *Mécanique Céleste* (4th edition, 1805), he propounded an approach to physics that influenced thinking for generations, wherein he "*sought to establish that the phenomena of nature can be reduced in the last analysis to actions at a distance between molecule and molecule, and that the consideration of these actions must serve as the basis of the mathematical theory of these phenomena*." Laplace's *Théorie Analytique des Probabilités* (1812) is a classic of probability and statistics, containing Laplace's definition of probability; the Bayes rule; methods for determining probabilities of compound events; a discussion of the method of least squares; and applications of probability to mortality, life expectancy, and legal affairs. Later editions contained supplements applying probability theory to measurement errors; to the determination of the masses of Jupiter, Saturn, and Uranus; and to problems in surveying and geodesy. On restoration of the Bourbon monarchy, which he supported by casting his vote against Napoleon, Pierre-Simon became Marquis de Laplace in 1817. He died on March 5, 1827.

Another important figure in probability theory was Carl Friedrich Gauss (1777–1855). Starting elementary school at the age of seven, he amazed his teachers by summing the integers from 1 to 100 instantly (the sum equals 5050, being the sum of 50 pairs of numbers, each pair summing to 101). At the Brunswick Collegium Carolinum, Gauss independently discovered the binomial theorem, as well as the law of quadratic reciprocity and the prime number theorem. Gauss' first book *Disquisitiones Arithmeticae* published in 1801 was devoted to algebra and number theory. His second book, *Theoria Motus Corporum Coelestium in Sectionibus Conicis Solem Ambientium* (1809), was a two-volume treatise on the motion of celestial bodies. Gauss also used the method of least squares approximation (published in *Theoria Combinationis Observationum Erroribus Minimis Obnoxiae*, 1823, supplement 1828) to successfully predict the orbit of Ceres in 1801. In 1807, he was appointed director of the Göttingen observatory. As the story goes, Gauss' assistants were unable to exactly reproduce the results of their astronomical measurements. Gauss got angry and stormed into the lab, claiming he would show them how to do the measurements properly. But, Gauss was not able to repeat his measurements exactly either! On plotting a histogram of the results of a particular measurement, Gauss discovered the famous bell-shaped curve that now bears his name, the Gaussian function:

$$G(x) = A\,\mathrm{e}^{-x^2/2\sigma^2}, \tag{1.4}$$

where σ is the spread, standard deviation, or variance and A is a normalization constant. $A = (2\pi)^{-1/2}/\sigma$ if the function is normalized such that $\int_{-\infty}^{\infty} G(x) = 1$. The error function of x is twice the integral of a normalized Gaussian function between 0 and x:

$$\mathrm{erf}(x) = \frac{2}{\sqrt{\pi}} \int_0^x \mathrm{e}^{-t^2}\,\mathrm{d}t. \tag{1.5}$$

It is of a sigmoid shape and has wide applications in probability and statistics. In the field of statistics, Gauss is best known for his theory of errors, but this represents only one of Gauss' many remarkable contributions to science. He published over 70 papers between 1820 and 1830 and in 1822, won the Copenhagen University Prize for *Theoria Attractioniscorporum Sphaeroidicorum Ellipticorum Momogeneorum Methodus Nova Tractata*, dealing with geodesic problems and potential theory. In *Allgemeine Theorie des Erdmagnetismus* (1839), Gauss showed that there can only be two poles in the globe and went on to specify a location for the magnetic South pole, establish a worldwide net of magnetic observation points, and publish a geomagnetic atlas. In electromagnetic theory, Gauss discovered the relationship between the charge density and the electric field. In the absence of time-dependent magnetic fields, Gauss's law relates the divergence of the electric field **E** to the charge density $\rho(\mathbf{r})$:

$$\nabla \cdot \mathbf{E} = \rho(\mathbf{r}), \tag{1.6}$$

which now forms one of Maxwell's equations.

The stage is now set for the formal entry of probability concepts into physics, and the credit for this goes to the Scottish physicist James Clerk Maxwell and the Austrian physicist Ludwig Boltzmann. James Clerk Maxwell (1831–1879) was born in Edinburgh on June 13, 1831, to John Clerk Maxwell, an advocate, and his wife Frances. Maxwell's father, a man of comfortable means, had been born John Clerk, and added the surname Maxwell to his own after he inherited a country estate in Middlebie, Kirkcudbrightshire, from the Maxwell family. The family moved when James was young to "Glenlair," a house his parents had built on the

1500-acre Middlebie estate. Growing up in the Scottish countryside in Glenlair, James displayed an unquenchable curiosity from an early age. By the age of three, everything that moved, shone, or made a noise drew the question: "what's the go o' that?" He was fascinated by geometry at an early age, rediscovering the regular polyhedron before any formal instruction. However, his talent went largely unnoticed until he won the school's mathematical medal at the age of 13, and first prizes for English and poetry. He then attended Edinburgh Academy and, at the age of 14, wrote a paper *On the Description of Oval Curves, and Those Having a Plurality of Foci* describing the mechanical means of drawing mathematical curves with a piece of twine and generalizing the definition of an ellipse, which was read to the Royal Society of Edinburgh on April 6, 1846. Thereafter, in 1850, James went to Cambridge, where (according to Peter Guthrie Tait) he displayed a wealth of knowledge, but in a state of disorganization unsuited to mastering the cramming methods required to succeed in the Tripos. Nevertheless, he obtained the position of Second Wrangler, graduating with a degree in mathematics from Trinity College in 1854, and was awarded a fellowship by Trinity to continue his work. It was during this time that he extended Michael Faraday's theories of electricity and magnetism. His paper *On Faraday's Lines of Force*, read to the Cambridge Philosophical Society in 1855 and 1856, reformulated the behavior of and relation between electric and magnetic fields as a set of four partial differential equations (now known as *Maxwell's equations*, published in a fully developed form in Maxwell's *Electricity and Magnetism* 1873).

In 1856, Maxwell was appointed professor of natural philosophy at Marischal College in Aberdeen, Scotland, where he became engaged to Katherine Mary Dewar. They were married in 1859. At 25, Maxwell was a decade and a half younger than any other professors at Marischal, and lectured 15 hours a week, including a weekly pro bono lecture to the local working men's college. During this time, he worked on the perception of color and on the kinetic theory of gases. In 1860, Maxwell was appointed to the chair of natural philosophy at King's College in London. This was probably the most productive time of his career. He was awarded the Royal Society's Rumford Medal in 1860 for his work on color, and elected to the Society in 1861. Maxwell is credited with the discovery that color photographs could be formed using red, green, and blue filters. In 1861, he presented the world's first color photograph during a lecture at the Royal Institution. It was also here that he came into regular contact with Michael Faraday, some 40 years his senior, whose theories of electricity and magnetism would be refined and perfected by Maxwell. Around 1862, Maxwell calculated that the speed of propagation of an electromagnetic field is approximately the speed of light and concluded, "*We can scarcely avoid the conclusion that light consists in the transverse undulations of the same medium which is the cause of electric and magnetic phenomena*." Maxwell then showed that the equations predict the existence of waves of oscillating electric and magnetic fields that travel through an empty space at a speed of 310,740,000 m/s. In his 1864 paper *A Dynamical Theory of the Electromagnetic Field*, Maxwell wrote, "*The agreement of the results seems to show that light and magnetism are affections of the same substance, and that light is an electromagnetic disturbance propagated through the field according to electromagnetic laws.*"

In 1865, Maxwell left London and returned to his Scottish estate in Glenlair. There he continued his work on the kinetic theory of gases and, using a statistical treatment, showed in 1866 that temperature and heat involved only molecular movement. Maxwell's statistical picture explained heat transport in terms of molecules at higher temperature having a high probability of moving toward those at lower temperature. In his 1867 paper, he also derived (independently of Boltzmann) what is known today as the *Maxwell–Boltzmann velocity distribution*:

$$f_v\left(v_x, v_y, v_z\right) = \left(\frac{m}{2kT}\right)^{3/2} \exp\left[-\frac{m\left({v_x}^2 + {v_y}^2 + {v_z}^2\right)}{2kT}\right], \qquad (1.7)$$

where $f_v(v_x, v_y, v_z)\, dv_x\, dv_y\, dv_z$ is the probability of finding a particle with velocity in the infinitesimal element $[dv_x, dv_y, dv_z]$ about velocity $v = [v_x, v_y, v_z]$, k is a constant now known as the *Boltzmann constant* (1.38062×10^{-23} J/K), and T is the temperature. This distribution is the product of three independent Gaussian distributions of the variables v_x, v_y, and v_z, with variance kT/m.

Maxwell's work on thermodynamics also led him to devise the *Gedankenexperiment* (thought experiment) that came to be known as *Maxwell's demon*. In 1871, Maxwell accepted an offer from Cambridge to be the first Cavendish Professor of Physics. He designed the Cavendish Laboratory, which was formally opened on June 16, 1874. His four famous equations of electrodynamics first appeared in their modern form of partial differential equations in his 1873 textbook *A Treatise on Electricity and Magnetism*:

$$\nabla \cdot \mathbf{E} = \rho(\mathbf{r}) \qquad \text{Gauss's law,} \qquad (1.8)$$

$$\nabla \cdot \mathbf{B} = 0 \qquad \text{Gauss's law for magnetism,} \qquad (1.9)$$

$$\nabla \times \mathbf{E} = -\partial \mathbf{B}/\partial t \qquad \text{Faraday's law of induction,} \qquad (1.10)$$

$$\nabla \times \mathbf{B} = \mathbf{J} + \partial \mathbf{E}/\partial t \qquad \text{Ampère's law with Maxwell's correction,} \qquad (1.11)$$

where **E** is the electric field, **B** the magnetic field, **J** the current density, and we have suppressed the universal constants, the permittivity, and permeability of free space. Maxwell delivered his last lecture at Cambridge in May 1879 and passed away on November 5, 1879, in Glenlair.

The story goes that Einstein was once asked whom he would most like to meet if he could go back in time and meet any physicist of the past. Without hesitation, Einstein gave the name of Newton and then Boltzmann. Ludwig Eduard Boltzmann was born on February 20, 1844, in Vienna, the son of a tax official. Ludwig attended high school in Linz and subsequently studied physics at the University of Vienna, receiving his doctorate in 1866 for a thesis on the kinetic theory of gases, under the supervision of Josef Stefan. Boltzmann's greatest contribution to science is, of course, the invention of statistical mechanics, relating the behavior and motions of atoms and molecules with the mechanical and thermodynamic properties of bulk matter. We owe to the American physicist Josiah Willard Gibbs the first use of the term *statistical mechanics*. In his 1866 paper, entitled *Über die Mechanische Bedeutung des Zweiten Hauptsatzes der Warmetheorie*, Boltzmann set out to seek a mechanical analog of the second law of thermodynamics, noting that while the first law of thermodynamics corresponded exactly with the principle of conservation of energy, no such correspondence existed for the second law. Already in this 1866 paper, Boltzmann used a $\rho \log \rho$ formula, interpreting ρ as density in phase space. To obtain a mechanical formulation of the second law, he started out by providing a mechanical interpretation of temperature by means of the concept of thermal equilibrium, showing that at equilibrium both temperature and the average kinetic energy exchanged are zero.

To establish this result, Boltzmann considered a subsystem consisting of two molecules and studied their behavior assuming that they are in equilibrium with the rest of the gas. The condition of equilibrium requires that this subsystem and the rest of the molecules exchange kinetic energy and change their state in such a way that the average value of the kinetic energy exchanged in a finite time interval is zero, so that the time average of the kinetic energy is stable. However, one cannot apply the laws of elastic collision to this subsystem, as it is in equilibrium with, and exchanging energy and momentum with, the rest of the gas. To overcome this obstacle, Boltzmann proposed a remarkable argument: he argued that, at equilibrium, the evolution of the two-particle subsystem is such that, sooner or later, it would pass through two states having the same total energy and momentum. But, this is just the same outcome as if these states had resulted from an elastic collision. Herein, we can find the germ of the ergodic hypothesis. Boltzmann regarded the irregularity of the system evolution as a sort of spreading out or diffusion of the system trajectory among the possible states and thus reasoned that if such states are able to occur, they will occur. It is only the existence of such states that is of importance and no assumption was made regarding the time interval required for the system to return to a state with the same energy and momentum. In particular, Boltzmann made no assumption of periodicity for the trajectory. Only the fact of closure of the trajectory matters to the argument, not when such closure occurs.

Next, Boltzmann derived the kinetic energy exchanged by the two molecules in passing from one state to the other, and then generalized the results to other states, assuming the equiprobability of the direction of motion (again based on the irregularity and complexity of molecular motion) and averaging the results for the elastic collision over all collision angles. He thus obtained a condition for the average kinetic energy to reach equilibrium, which was analogous to that for thermal equilibrium. Concluding from this that temperature is a function of the average kinetic energy, Boltzmann then proceeded to derive a mechanical analog of the second law of thermodynamics in the form of a least action principle. He showed that, if a mechanical system obeys the principle of least action, the kinetic energy (in analogy to heat $\mathrm{d}Q$) supplied to a periodic system is given by

$2\mathrm{d}(\tau\overline{\mathrm{E}})/\tau$, where $\overline{\mathrm{E}} = (1/\tau)\int_0^\tau E\,\mathrm{d}t$ is the average kinetic energy. Hence, if no energy is supplied (corresponding to the adiabatic condition $\mathrm{d}Q = 0$), the ratio $\overline{\mathrm{E}}/v$ is an invariant. The concept of probability enters into this formulation in a most fundamental way. We can distinguish at least two interpretations of the concept of probability in Boltzmann's writings. He used the term probability in the sense of relative frequency or sojourn time of a trajectory in order to interpret thermodynamic parameters as average mechanical quantities. Elsewhere, he had defined probability of a trajectory with certain constraints as the ratio between the number of trajectories satisfying those constraints and the number of all possible trajectories. The equivalence between these two probability measures leads directly to ergodic hypothesis.

On completing his Privatdozenten (lectureship) in 1867, Boltzmann was appointed professor of mathematical physics at the University of Graz. The next year he set out to create a general theory of the equilibrium state. Boltzmann argued on probabilistic grounds that the average energy of motion of a molecule in an ideal gas is the same in each direction (an assumption also made by Maxwell) and thus derived the Maxwell–Boltzmann velocity distribution (Eq. 1.7). Since for an ideal gas, all energy is in the form of kinetic energy, $E = \frac{1}{2}mv^2$, the Boltzmann distribution for the fractional number of molecules N_i/N occupying a set of states i and possessing energy E_i is thus proportional to the probability density function (Eq. 1.7):

$$\frac{N_i}{N} = \frac{g_i \exp(-E_i/k_\mathrm{B}T)}{\sum_j g_j \exp(-E_j/k_\mathrm{B}T)}, \tag{1.12}$$

where g_i is the degeneracy (the number of states having energy E_i), N_i the number of molecules at equilibrium temperature T in a state i with energy E_i and degeneracy g_i, and $N = \sum_i N_i$ the total number of molecules. The denominator in Equation 1.12 is the canonical partition function:

$$Z(T) = \sum_i g_i\,\mathrm{e}^{-E_i/k_\mathrm{B}T}. \tag{1.13}$$

He applied the distribution to increasingly complex cases, treating external forces, potential energy, and motion in three dimensions. In his 1868 paper, he elaborated on his concept of diffuse motion of the trajectory among possible states, generalizing his earlier results to the whole available phase space consistent with the conservation of total energy. In 1879, Maxwell pointed out that this generalization rested on the assumption that the system, if left to itself, will sooner or later pass through every phase consistent with the conservation of energy—namely, the ergodic hypothesis. In his 1868 paper, Boltzmann also pioneered the use of combinatorial arguments, showed the invariance of the phase volume during the motion, and interpreted the phase space density as the probability attributed to a region traversed by a trajectory. Here, we see the precursor to Max Born's statistical interpretation of the quantum wave function.

Boltzmann was also the first one to recognize the importance of Maxwell's electromagnetic theory. He spent several months in Heidelberg with Robert Bunsen and Leo Konigsberg in 1869 and then in Berlin with Gustav Kirchoff and Herman von Helmholtz in 1871, working on problems of electrodynamics. During this time, he continued developing and refining his ideas on statistical mechanics. Boltzmann's nonequilibrium theory was first presented in 1872 and used many ideas from his equilibrium theory of 1866–1871. His famous 95-page article, *Weitere Studien über das Wärmegleichgewicht unter Gasmolecülen* (*Further Studies on the Thermal Equilibrium of Gas Molecules*), published in October 1872, contains what he called his *minimum theorem*, now known as the *H-theorem*, the first explicit probabilistic expression for the entropy of an ideal gas. Boltzmann's probability equation relates the entropy S of an ideal gas to the number of ways W (*Wahrscheinlichkeit*) in which the constituent atoms or molecules can be arranged, that is, the number of microstates corresponding to a given macrostate:

$$S = k \log W. \tag{1.14}$$

Here, log refers to natural logarithms. The H-theorem is an equation based on Newtonian mechanics that quantifies the heat content of an ideal gas by a numerical quantity H (short for heat). Defined in terms of the velocity distributions of the atoms and molecules of the gas, H assumes its minimum value when the velocities of the particle are distributed according to the Maxwell–Boltzmann (or Gaussian) distribution. Any gas system not at its minimal value of H will tend toward the minimum value through molecular collisions that move the system toward the Maxwell–Boltzmann distribution of velocities.

After a stint as professor of mathematics at the University of Vienna from 1873 to 1876, Boltzmann returned to Graz to take the chair of experimental physics. In 1884, Boltzmann initiated a theoretical study of radiation in a cavity (black body radiation) and used the principles of thermodynamics to derive Stefan's law:

$$E \propto T^4, \tag{1.15}$$

according to which the total energy density E radiated by a black body is proportional to the fourth power of its temperature T. Study of this radiation led Wilhelm Wien and Max Planck to their famous scaling law, whereby the energy density in the cavity is given by ν times a function of ν/T, ν being the frequency of radiation and T the temperature of the cavity. Planck's main interest in these studies was the question of the origin of irreversibility in thermodynamics. Following Boltzmann's procedure, he was able to show that, for radiation in the cavity, the entropy S is a function of E/ν. Thus, for a reversible adiabatic process ($dS = 0$), E/ν is an invariant.

In 1890, Boltzmann was appointed to the chair of theoretical physics at the University of Munich in Bavaria, Germany, and succeeded Stefan as professor of theoretical physics in his native Vienna after the latter's death in 1893. In

1900, at the invitation of Wilhelm Ostwald, Boltzmann moved to the University of Leipzig. Although the two were on good personal terms, Ostwald was one of Boltzmann's foremost scientific critics and the latter struggled to gain acceptance for his ideas among his peers. Ostwald argued, for instance, that the actual irreversibility of natural phenomena proved the existence of processes that cannot be described by mechanical equations. Unlike Boltzmann, most chemists at that time did not ascribe a real existence to molecules as mechanical entities; the molecular formula was treated as no more than a combinatorial formula. The Vienna Circle was strongly influenced at that time by the positivist–empiricist philosophy of the Austrian physicist and philosopher Ernst Mach (1838–1916), who occupied the chair for the philosophy of the inductive sciences at the University of Vienna. As an experimental physicist, Mach also held that scientific theories were only provisional and had no lasting place in physics. He advanced the concept that all knowledge is derived from sensation; his philosophy was thus characterized by an antimetaphysical attitude that recognized only sensations as real. According to this view, phenomena investigated by science can be understood only in terms of experiences or the "sensations" experienced in the observation of the phenomena; thus, no statement in science is admissible unless it is empirically verifiable. This led him to reject concepts such as absolute time and space as metaphysical. Mach's views thus stood in stark opposition to the atomism of Boltzmann. Mach's reluctance to acknowledge the reality of atoms and molecules as external, mind-independent objects was criticized by Boltzmann and later by Planck as being incompatible with physics. Mach's main contribution to physics involved his description and photographs of spark shock-waves and ballistic shock-waves. He was the first to systematically study supersonic motion, and describe how passing the sound barrier caused the compression of air in front of bullets and shells; the speed of sound bears his name today. After Mach's retirement following a cardiac arrest, Boltzmann returned to his former position as professor of theoretical physics in Vienna in 1902, where he remained for the rest of his life.

On April 30, 1897, Joseph John Thomson announced the discovery of "the carriers of negative electricity"—the electron—to the Royal Institution in England. He was to be awarded the Nobel Prize in 1906 for his determination of its charge to mass ratio. Meanwhile, in November 1900, Max Planck came to the realization that the Wien law is not exact. In an attempt to define an entropy of radiation conforming with Stefan's empirical result (Eq. 15), Planck was led to postulate the quantum of action:

$$E/\nu = nh. \tag{1.16}$$

This result was first publicly communicated to a small audience of the German Physical Society on December 14, 1900. In his 1901 paper *On the Law of Distribution of Energy in the Normal Spectrum*, Planck used Boltzmann's H-function to explain that the entropy S of a system is proportional to the logarithm of its probability W, to within an arbitrary additive constant. He later called this the *general definition of entropy*. The great Boltzmann, however, took little notice of these revolutionary developments in theoretical and experimental physics that would soon confirm his theories. Growing increasingly isolated and despondent, Boltzmann hanged himself on September 5, 1906, while on vacation in Duino, near Trieste. On Boltzmann's tombstone is inscribed his formula $S = \mathrm{k} \log W$.

FURTHER READING

Boltzmann L. Wien Ber 1866;53:195–220.

Boltzmann L. Wien Ber 1872;66:275–370.

Campbell L, Garnett W. *The Life of James Clerk Maxwell*. London: Macmillan; 1882.

Cohen EGD, Thirring W. The Boltzmann Equation: Theory and Applications: Proceedings of the International Symposium '100 Years Boltzmann Equation' Vienna, 4th-8th September 1972 (Few-Body Systems). New York: Springer-Verlag; 1973.

Gibbs JW. *Elementary Principles in Statistical Mechanics, Developed with Especial Reference to the Rational Foundation of Thermodynamics* [reprint]. New York: Dover; 1960.

Jammer M. *The Conceptual Development of Quantum Mechanics*. New York: McGraw Hill; 1966.

Klein MJ. *The Making of a Theoretical Physicist. Biography of Paul Ehrenfest*. Amsterdam: Elsevier; 1970.

Maxwell JC. *Theory of Heat, 1871* [reprint]. Westport (CT): Greenwood Press; 1970.

Maxwell JC. *A Treatise on Electricity and Magnetism*. Oxford: Clarendon Press; 1873.

Pascal B. Pensees (Penguin Classics). Penguin Books; 1995. Krailsheimer AJ, Translator.

Planck M. Ann Phys 1901;4:553.

Tait PG. Proceedings of the Royal Society of Edinburgh, 1879–1880. Quoted in Everitt CWF. *James Clerk Maxwell: Physicist and Natural Philosopher*. New York: Charles Scribner; 1975.

The MacTutor History of Mathematics archive Index of Biographies. Available at http://www-groups.dcs.st-and.ac.uk/~ history/BiogIndex.html (School of Mathematics and Statistics, University of St Andrews, Scotland). Accessed 2011.

2

DOES GOD PLAY DICE?

N. SUKUMAR

2.1 QUANTA OF RADIATION

We saw in the previous chapter how the study of black-body radiation initiated by Boltzmann led Wien to his famous scaling law. Paul Ehrenfest (PhD, 1904), one of the foremost students of Ludwig Boltzman at Vienna, was the first to point out the divergence of the energy density of black-body radiation predicted by the classical wave theory of light at low frequencies; he called this the "*Rayleigh–Jeans catastrophe in the ultraviolet*" (nowadays known as the *ultraviolet catastrophe*). Realizing that the classical wave theory is not valid at low frequencies, Planck postulated the quantum of action (Eq. 1.16) to obtain an entropy of radiation that would conform with Stefan's empirical fourth power dependence on temperature (Eq. 1.15) and eliminate the ultraviolet catastrophe. Planck's law was essentially an interpolation formula between the Wien radiation law, valid for high frequencies (ν) and low temperatures (T), and the Rayleigh–Jeans formula, valid at low ν and high T. The former is derived from the assumption $\partial^2 S/\partial U^2 = \text{constant}/U$, where S is the entropy of black-body radiation and U its energy, whereas the latter can be obtained from $\partial^2 S/\partial U^2 = \text{constant}/U^2$. Interpolating between these formulae, Planck assumed instead

$$\partial^2 S/\partial U^2 = \frac{\text{constant}}{U(b+U)} \tag{2.1}$$

A Matter of Density: Exploring the Electron Density Concept in the Chemical, Biological, and Materials Sciences, First Edition. Edited by N. Sukumar.

(where b is a constant), from which he derived his famous radiation law for the energy density of black-body radiation:

$$u_\nu = \frac{8\pi h\nu^3}{c^3\left[\exp\left(\frac{h\nu}{kT}\right) - 1\right]} \quad \text{[Planck]}. \tag{2.2}$$

This equation reduces to the Wien law at high ν and low T:

$$u_\nu = \frac{8\pi h\nu^3}{c^3}\exp\left(-\frac{h\nu}{kT}\right) \quad \text{[Wein]} \tag{2.3}$$

and to the Rayleigh–Jeans radiation law at low ν and high T:

$$u_\nu = \frac{8\pi\nu^2 kT}{c^2} \quad \text{[Rayleigh–Jeans radiation]}. \tag{2.4}$$

Planck adopted Boltzmann's statistical concept of entropy in order to elevate Equation 2.1 from a "lucky guess" to a "statement of real physical significance": he assumed that the entropy S of a system of oscillators of frequency ν is given by $S = k \log W$, where W is the number of distributions compatible with the energy of the system and k is Boltzmann's constant. These results were obtained during a period of 8 weeks, which Planck later described thus [1, 2]: "*After a few weeks of the most strenuous work of my life, the darkness lifted and an unexpected vista began to appear*." Planck communicated these results at a meeting of the German Physical Society on December 14, 1900 [3], by reading his paper *On the Theory of the Energy Distribution Law of the Normal Spectrum*, which introduced for the first time his "universal constant h."

Ehrenfest realized that Planck's hypothesis challenged Boltzmann's assumption of equal *a priori* probabilities of volume elements in phase space. It was Albert Einstein, a technical expert at the Swiss patent office in Bern, who recognized the logical inconsistency in combining an electrodynamical description based on Maxwell's equations, which assume that energy can vary continuously, with a statistical description, where the oscillator energy is restricted to assume only discrete values that are integral multiples of $h\nu$. Equation 2.1, in fact, combines the wave and particle aspects of radiation, although this was not Planck's explicit intention. Planck's conception of energy quantization was of oscillators of frequency ν that could only absorb or emit energy in integral multiples of $h\nu$: the quantization only applied to the interaction between matter and radiation. Einstein's 1905 paper in *Annalen der Physik* on the photoelectric effect and the light quantum hypothesis [4], wherein Einstein proposed that radiant energy itself is quantized, launched the quantum revolution in our physical conceptions of matter and radiation, winning him the Nobel Prize in 1921.

Albert Einstein was born on March 14, 1879, in the German town of Ulm, the first child of Pauline and Hermann Einstein. Pauline was a talented pianist and Hermann was a merchant. In 1880, the family moved to Munich, where Hermann started a business with his brother Jacob. A daughter Maria (also called *Maja*) was born in 1881. Albert was a good student and excelled at school, but generally kept to himself and detested sports and gymnastics [5]. Both Albert and Maja learned to play the piano; Albert would play Mozart and Beethoven sonatas, accompanied by his mother, and he delighted in piano improvisations. Their uncle Jacob would pose mathematical problems, which Albert derived great satisfaction from solving. When a family friend gave him a book on Euclidean geometry when Albert was 12 years old, it was like a revelation. Albert also taught himself calculus. After the family moved to Italy to start a new business, Albert moved to Zurich in 1895, and the following year he gave up his German citizenship and enrolled at the Swiss Federal Institute of Technology (Eidgenössische Technische Hochschule, ETH). At ETH, his proposal for an experiment to test the earth's movement relative to the ether was rebuffed by Professor Heinrich Weber. After graduation from ETH, Einstein failed to secure a university position and was unemployed for nearly a year before obtaining a series of temporary teaching positions. He was granted Swiss citizenship in 1901, moved to Bern the following year, and finally secured an appointment at the Swiss federal patent office. Here he found the time to work on his own on scientific problems of interest to him. Hermann Einstein died of a heart attack in Milan in 1902. Albert married Mileva Maric, a former classmate from ETH, the following year. Their son, Hans Albert, was born in 1904.

That brings us to Einstein's *annus mirabilis* or miracle year, 1905: the 8-month period during which he published in the *Annalen der Physik* four of the most important papers of his life, in addition to his PhD thesis [6] for the University of Zurich: (i) *On a Heuristic Viewpoint Concerning the Production and Transformation of Light* dealing with the photoelectric effect, received in March [4]; (ii) *On the Motion—Required by the Molecular Kinetic Theory of Heat—of Small Particles Suspended in a Stationary Liquid* on Brownian motion and the determination of Avogadro's number, which helped to resolve lingering doubts on the reality of molecules, received in May [7], (iii) *On the Electrodynamics of Moving Bodies* on special relativity, received in June [8], (iv) *Does the Inertia of a Body Depend Upon Its Energy Content?* on mass–energy equivalence, received in September [9]. Einstein was then 26. The revolution in physics was under way!

At the turn of the century, the wave theory of light, based on Maxwell's equations and continuous functions in space, was firmly established. The existence of electromagnetic waves had been confirmed by Heinrich Hertz in a series of experiments beginning in 1886, but these same experiments also produced the first evidence for the photoelectric effect [10]. The photoelectric effect occurs when ultraviolet or visible light illuminates the surface of an electropositive metal subjected to a negative potential: there is then a flow of electrons from the cathode (cathode rays). This discovery spurred several others to investigate the phenomenon. It was established that radiation from an electric arc discharges the cathode, without affecting the anode [11], that red and infrared radiation are ineffective in inducing a photoelectric current [12], that the photoelectric current is proportional to the intensity of light absorbed [13], that the emission occurs only at frequencies exceeding a minimum threshold value ν_0 [14], with the more electropositive the metal comprising the electrode, the lower the threshold frequency ν_0 [15] and that the energy of the ejected photoelectrons is independent of the intensity of the incident light but proportional to its frequency above the threshold, that is, to $\nu - \nu_0$ [14, 16]. These observations seemed incompatible with Maxwell's electromagnetic theory, but were explained by Einstein's hypothesis of discrete light corpuscles or quanta, each of energy $h\nu$ and each capable of interacting with a single electron. This electron can then absorb the single light quantum (photon); part of its energy is used to overcome the attraction of the electron to the metal and the rest would appear as the kinetic energy of the photoelectron.

From the law of conservation of energy, the maximum kinetic energy of the photoelectrons is given by $h\nu - h\nu_0$, where $h\nu_0$ is the energy necessary to remove an electron from the metal, a constant for a given material. This was verified by Hughes [17] who measured the maximum velocity of photoelectrons ejected from a number of metals. In a series of painstaking experiments over 10 years culminating in 1916, Robert Millikan [18] confirmed Einstein's prediction that the stopping potential for the electrons would be a linear function of the frequency of the incident light, thereby providing irrefutable evidence for the existence of photons and the first direct photoelectric determination of Planck's constant h. Millikan obtained the value $h = 6.57 \times 10^{-27}$ erg/s for the constant of proportionality between the kinetic energy of photoelectrons and the frequency of absorbed light, and showed that this value is independent of the surface, work for which he received the Nobel Prize in 1923.

2.2 ADIABATIC INVARIANTS

Paul Ehrenfest had long realized the importance of the variable E/ν and he sought a conceptual foundation for the quantum hypothesis in terms of generalized adiabatic invariants:

> If you contract a reflecting cavity infinitely slowly, then the frequency ν and the energy E of each proper vibration increase simultaneously in such a way that E/ν remains invariant under this "adiabatic" influence. The *a priori* probability must always depend on only those quantities which remain invariant under adiabatic influencing, or else the quantity ln W will fail to satisfy the condition, imposed by the second law on the entropy, of remaining invariant under adiabatic changes. [19]

In December 1912, Ehrenfest found the result he was seeking: "*Then my theorem reads... The average kinetic energy of our system increases in the same proportion as the frequency under an adiabatic influencing*." According to Ehrenfest's adiabatic hypothesis [20], quantum admissible motions transform to other admissible motions under adiabatic influences. Ehrenfest thus launched on a program for finding quantum states by quantizing the adiabatic invariants. Ehrenfest's adiabatic principle determined the formal applicability of the formalism of classical mechanics to the quantum theory, and also enabled the determination of stationary states of systems that were adiabatically related to other known systems.

In 1913, the Danish physicist Niels Bohr [21] applied the quantization of energy and angular momentum to explain the spectral lines of hydrogen. He proposed that electrons move without loss of energy in stable, stationary orbits around the nucleus in an atom, with fixed energy and with angular momentum being equal to an integral multiple of $h/2\pi$, and furthermore that absorption or emission of radiation by atoms occurs only when an electron jumps between two different stationary orbits. The frequency of the radiation emitted or absorbed during the transition between stationary orbits of energy E_{m} and E_{n} is given by the energy quantization rule:

$$E_{\mathrm{m}} - E_{\mathrm{n}} = h\nu. \tag{2.5}$$

Wilson [22] and Sommerfeld [23] extended Bohr's model of the atom to include elliptical, in addition to circular, electron orbits, and postulated that stationary states are characterized by constant and quantized action integral of the angular momentum:

$$\oint pdq = nh, \tag{2.6}$$

where p is the momentum corresponding to the generalized coordinate and q and n are integer quantum numbers. Ehrenfest [24] then showed the equivalence of Sommerfeld's quantization condition (Eq 2.6) with the adiabatic principle of mechanics.

2.3 PROBABILITY LAWS

Einstein's light quanta of 1905 were energy quanta satisfying Equation 1.16; there was no concept of momentum associated with them yet. Einstein continued brooding on light quanta in the following years, but increasingly became preoccupied with the development of general relativity. His principle of the equivalence of inertial and gravitational mass was first formulated in 1907. Einstein resigned from the patent office in 1909 and was appointed to the newly created position of associate professor of theoretical physics at the University of Zurich. In 1911, he moved to Prague as full professor, and began work on deriving the equivalence principle from a new theory of gravitation. By 1916 Einstein, now in Berlin, had completed his formulation of the general theory of relativity and returned to the problem of light quanta. Analysis of statistical fluctuations of black-body radiation now led him to associate a definite momentum with a light quantum. Einstein [25] provided a new derivation for Planck's radiation law by assuming that the transitions follow probability laws similar to those known to govern radioactivity [26]. Einstein considered a gas interacting with electromagnetic radiation, the entire system being in thermal equilibrium. Let the probability per unit time of spontaneous emission from a higher energy level E_m to a lower energy level E_n be $A_{m\to n}$, and the probability per unit time of emission from E_m to E_n induced or stimulated by radiation of frequency ν be $B_{m\to n}u_\nu$ and that for stimulated absorption from E_n to E_m be $B_{n\to m}u_\nu$. The equilibrium condition (microscopic reversibility) requires that

$$B_{n\to m}u_\nu \exp\left(-\frac{E_n}{kT}\right) = (B_{m\to n}u_\nu + A_{m\to n}) \exp\left(-\frac{E_m}{kT}\right). \tag{2.7}$$

Assuming that $B_{m\to n} = B_{n\to m}$, as required by the high temperature Rayleigh–Jeans limit, yields

$$u_\nu = \frac{A_{m\to n}/B_{m\to n}}{\exp\left(\dfrac{E_m - E_n}{kT}\right) - 1}. \tag{2.8}$$

Comparing Equation 2.8 with the Wien radiation law (Eq. 2.3) for high frequencies and with the Rayleigh–Jeans result (Eq. 2.4) for low frequencies, Einstein [25] obtained Bohr's quantization condition (Eq. 2.5) for transitions between atomic stationary states, and also

$$A_{m\to n} = \frac{8\pi h\nu^3}{c^3} B_{m\to n}, \tag{2.9}$$

which then immediately gives Planck's radiation law (Eq. 2.2). Note that without the stimulated emission term in Equation 2.7, one obtains only the Wien law.

In order to obtain Planck's law (Eq. 2.2), Einstein had to associate a definite momentum

$$p = h\nu/c, \tag{2.10}$$

with a light quantum: "*if a bundle of radiation causes a molecule to emit or absorb an energy amount $h\nu$, then a momentum $h\nu/c$ is transferred to the molecule, directed along the bundle for absorption and opposite the bundle for emission.*" Consider now the term involving $A_{m\to n}$, the spontaneous emission probability, and recall that radiation can be viewed as consisting of discrete quanta or photons, each with definite energy and momentum. This term has the atom emitting photons at random times in random directions (and suffering a recoil $h\nu/c$) governed purely by the laws of probability, a feature of the quantum theory which many would find deeply disturbing in the years to follow. Einstein himself was the first to realize the deep conceptual crisis caused by spontaneous emission [5]: "*it is a weakness of the theory . . . that it leaves the time and direction of elementary processes to chance*" [27]. Others would consider all of quantum mechanics to be but a revision of statistical mechanics [28]; the laws of quantum mechanics place restrictions on the simultaneous probability distributions of complementary observables. But the probabilistic nature of spontaneous quantum processes and its irreconcilability with classical notions of causality would continue to trouble Einstein for the rest of his life: "*That business about causality causes me a lot of trouble, too. Can the quantum absorption and emission of light ever be understood in the sense of the complete causality requirement, or would a statistical residue remain? ... I would be very unhappy to renounce complete causality*" [29]. It is indeed ironic that the person who was perhaps more responsible than any other for the introduction of statistical ideas into quantum mechanics would be so reluctant to reconcile himself to the statistical nature of the theory!

The final bit of experimental confirmation for the quantum corpuscular view of radiation came from Arthur Holly Compton's experiments on the scattering of X-rays in 1923, but the basic facts underlying the phenomenon had long been known. That the scattered radiation had lower penetrating power than the primary radiation was shown for γ-rays by Eve [30] and by Kleeman [31] and for X-rays by Sadler and Mesham [32]. Madsen [33] and Florance [34] further showed that the wavelength of the scattered radiation depends on the angle of scattering. Gray [35] recognized that these observations could not be explained by the prevailing classical theory on the basis of Maxwell's electrodynamics. To explain his experimental observations, Compton assumed instead that each quantum of X-ray energy is concentrated in a single quantum and acts as a unit on a single electron, that "*each electron which is effective in the scattering scatters a complete quantum*" and that "*the quanta of radiation are received from definite directions and scattered into definite directions*" [36]. Applying the principles of conservation of energy and momentum, he then derived his well-known formula for the change in wavelength on scattering:

$$\Delta\lambda = \left(\frac{h}{mc}\right)(1 - \cos\theta), \tag{2.11}$$

where θ is the scattering angle, m the electron mass, and c the velocity of light. He presented these results at a meeting of the American Physical Society in Washington in April 1923 and in a paper entitled "A quantum theory of the scattering of x-rays by light elements" published in the *Physical Review* [36], wherein he concluded, "*The experimental support of the theory indicates very convincingly that a radiation quantum carries with it directed momentum as well as energy*." These results were also independently derived by Debye [37].

Thus, Einstein's light quanta now acquired physical significance and a firm experimental foundation. However, phenomena such as interference and diffraction could still be understood only by recourse to the wave theory of light. The stabilities of many-electron atoms and molecules also defied explanation in the older version of the quantum theory. The best guide to bridging the classical and quantum theories at that time seemed to be the correspondence principle, championed by Bohr; this principle declared that quantum theory reduces to classical mechanics in the limiting case where the quantum of action becomes infinitely small $h \to 0$. The translation of classical formulae into quantum theory generally required a lot of guess work and differed from problem to problem.

2.4 MATTER WAVES

Before the advent of quantum mechanics, the physics of matter (based on the concepts of particles and atoms obeying the laws of mechanics) and the physics of radiation (based on the idea of wave propagation) seemed difficult to reconcile with each other. Yet, the interaction of matter with radiation called for a unified theory of matter and radiation. This was what attracted Louis Victor de Broglie (1892–1987) to theoretical physics: "*the mystery in which the structure of matter and of radiation was becoming more and more enveloped as the strange concept of the quantum, introduced by Planck in 1900 in his researches into black-body radiation, daily penetrated further unto the whole of physics*." Louis de Broglie was the younger son of Victor, 5th Duc de Broglie, and Pauline d'Amaillé. Their elder son, Maurice, was an experimental physicist who worked on X-rays and the photoelectric effect, and was one of Arthur Compton's early supporters. Their father died in 1906 and Maurice, then 31, assumed responsibility for Louis's education. After completing secondary school in 1909, Louis entered the Sorbonne in Paris, first studying history, then law, and graduated with an arts degree at the age of 18. By then he was already becoming more interested in mathematics and physics, and he decided to study theoretical physics instead. Louis was awarded the Licence ès Sciences in 1913, but then World War I broke out, and he served in the army, as a telegraph operator at the Eiffel Tower.

Resuming his research in theoretical physics after the war, de Broglie tried to reconcile Einstein's light quantum hypothesis with the phenomena of interference and diffraction, and suggested the need to associate the quanta with a periodicity [38]. He had always been impressed by the formal analogy between the principle of least action in mechanics and Fermat's principle of least time in optics, on which Hamilton had published nearly a century ago. Louis also continued his interest in experimental physics, working with his brother at Maurice's laboratory. The concept of matter waves now began to take shape:

> As in my conversations with my brother we always arrived at the conclusion that in the case of X-rays one had both waves and corpuscles, thus suddenly... in the course of summer 1923, I got the idea that one had to extend this duality to material particles, especially to electrons.

Thus, $E = h\nu$ should apply not only for photons but also for electrons. Using Einstein's mass–energy relation and Planck's law, de Broglie now proposed [39] that any particle of rest mass m_0 and velocity v is associated with a wave whose frequency ν_0 is given by

$$\nu_0 = \frac{m_0 c^2}{h} \tag{2.12}$$

or

$$\lambda = \frac{h}{p} = \frac{h}{m_0 v}\sqrt{1 - \frac{v^2}{c^2}}, \tag{2.13}$$

where h is Planck's constant, λ the wavelength of the wave, c the speed of light in vacuum, p the momentum, and m_0c^2 the rest energy of the particle. He then showed that velocity v is just the group velocity of the phase waves [40] and asserted that "*a stream of electrons passing through a sufficiently narrow hole should also exhibit diffraction phenomena*." He further suggested that "*It is in this direction where one has probably to look for experimental confirmation of our ideas*" [40]. In November 1924, de Broglie presented his doctoral thesis "Reserches sur la Théorie des Quanta" containing these ideas to the Faculty of Sciences at the University of Paris [41, 42]. He was awarded the Nobel Prize for this work in 1929. Einstein [43] regarded de Broglie's discovery "*a first feeble ray of light on this worst of our physics enigmas*" and was convinced that the concept of matter waves "*involves more than merely an analogy*" [44].

James Franck, professor of experimental physics at the University of Göttingen, realized that the electron diffraction experiment suggested by de Broglie had already been performed by Davisson and Kunssman [45]. These authors had studied the scattering of electrons from nickel, palladium, and magnesium, interpreting the angular distribution of scattered electrons as due to the variation of the electron density in the atomic shells. Franck explained this scattering as a diffraction phenomenon, and found that the wavelength agreed with de Broglie's formula (Eq. 2.13). Walter Elsasser, a student of Professor Max Born at Göttingen, showed that the Ramsauer effect [46–48] could be explained as an interference effect of matter waves [49]. Davisson was not convinced by Elsasser's interpretation, but his subsequent work with nickel targets [50] confirmed de Broglie's formula (Eq. 2.13), as did the experiments by George Paget Thomson, son of J. J. Thomson, and his student Andrew Reid [51] on electron diffraction by thin films. Davisson and George Thomson shared the Nobel Prize for these discoveries in 1937.

2.5 QUANTUM STATISTICS

In 1924, an Indian physicist, Satyendra Nath Bose at Dacca University, derived Planck's radiation law independent of classical electrodynamics, on the basis of a new method of counting [52], now known as *Bose–Einstein statistics*. Bose replaced the counting of wave frequencies with the counting of cells in phase space. He considered a cell in one-particle phase space with momentum between p and $p + \mathrm{d}p$. Because of the relation (Eq. 2.13) for momentum, he associated a volume h^3 with each such cell, which then led him to Planck's law (Eq. 2.2). This scheme implicitly assumes—and these are the key assumptions explicitly stated later by Einstein in 1925—statistical independence of cells in phase space and the indistinguishability of particles within a cell. His paper having been rejected by the *Philosophical Magazine*, Bose forwarded the manuscript to Einstein, who recognized its importance, translated it into German and sent it for publication to the *Zeitschrift für Physik*, with his recommendation. Einstein [53] also showed that the new statistics, applied to an ideal monatomic gas, gave rise to an interference term in the energy fluctuations, in addition to the term arising from the Maxwell–Boltzmann statistics of noninteracting particles. In a second paper [44], he discussed the interference term in the energy fluctuation formula with reference to de Broglie's work. Einstein also considered the consequences of extremely low temperature and realized that

> in this case, a number of molecules steadily growing with increasing density goes over in the first quantum state (which has zero kinetic energy) while the remaining molecules distribute themselves ... A separation effected; one part condenses, the rest remains a 'saturated ideal gas' [44]

A phenomenon now known as *Bose–Einstein condensation* [54]. The 2001 Nobel Prize was awarded to Cornell, Ketterle, and Wieman for their experimental realization of a gaseous Bose–Einstein condensate consisting of thousands to millions of atoms [55, 56].

In 1926 [57], Paul Adrien Maurice Dirac (1902–1984) showed that, for an ideal gas of $N = \sum_i n_i$ particles with total energy $E = \sum_i n_i \varepsilon_i$, such that n_i particles have energy ε_i; Bose statistics implies that the number of microstates w is

$$w = 1 \qquad \text{[Bose]}, \tag{2.14}$$

whereas Boltzmann statistics requires

$$w = \frac{N!}{\Pi_i n_i!} \qquad \text{[Boltzmann]}. \tag{2.15}$$

Thus the single microstate symmetric in the N particles is the only one allowed in Bose statistics. Dirac also showed that the statistics implied by the condition (Eq. 2.14) led directly to Planck's radiation law (Eq. 2.2), thereby finally lifting the veil of mystery shrouding that equation. Dirac [57] and separately Fermi [58] realized that there is a third possibility, where the state is antisymmetric in the particles. This leads to Fermi–Dirac statistics, the statistics obeyed by electrons.

2.6 MATRIX MECHANICS AND COMMUTATION RELATIONS

Meanwhile, there were momentous developments from the laboratory of Max Born (1882–1970) at the University of Göttingen. Max was the son of the anatomist and embryologist Gustav Born and Margarethe Kauffmann. His mother died when he was four years old. Max studied at Breslau, Heidelberg, and Zurich. After his Habilitation (license to teach) in 1909, he settled as a Privatdozent at Göttingen. Here, he formed close ties with the leading mathematicians David Hilbert and Herman Minkowski. Hilbert was soon impressed with Born's exceptional abilities and took him under his wing. Between 1915 and 1919, Born was professor of physics at the University of Berlin. Here he came into contact with Einstein and the two formed a life-long friendship. In 1913, Max married Hedwig and converted from Judaism to his wife's Lutheran faith the following year. Born moved back to Göttingen in 1921 as professor of theoretical physics and director of the institute for theoretical physics, after negotiating a position for his long-time friend and experimental colleague James Franck. The two would develop a close collaboration over the following dozen years. This collaboration and cross-fertilization between experimental and theoretical physics—and pure mathematics, as represented by Hilbert—would prove crucial to the development of quantum mechanics. Yet another of Born's singular qualities was his extraordinary ability to recognize and nurture talent, even when they might outshine him. Seven of Born's students and assistants at Göttingen—Max Delbrück, Enrico Fermi, Werner Heisenberg, Maria Goeppert-Mayer, Gerhard Herzberg, Wolfgang Pauli, and Eugene Wigner—would eventually go on to win Nobel Prizes!

Among the most brilliant of Born's assistants was Werner Heisenberg (1901–1976), who employed the correspondence principle in an entirely new way: to guess, not the solution to a specific quantum problem, but the mathematical scheme for a new mechanics [2]. Expanding time-dependent variables by Fourier expansions, as in classical mechanics, Heisenberg went on to represent physical quantities in his quantum theory by sets of complex Fourier amplitudes and found that the Fourier amplitudes obeyed a curious multiplication rule:

$$X^2{}_{n,n'} = \sum_{n''} X_{n,n''} X_{n'',n'}. \tag{2.16}$$

He then applied this theory to the problems of the anharmonic oscillator and the rigid rotor, finding results in agreement with observations. These results were published in the *Zeitschrift für Physik* in the summer of 1925 [59]. One of the first to hail the significance of this paper was Neils Bohr in Copenhagen.

On studying Heisenberg's manuscript, Max Born recalled his college algebra lectures at Breslau, and realized that the multiplication rule obeyed by Heisenberg's sets of Fourier amplitudes was precisely the rule for multiplication of matrices, propounded 70 years earlier by Arthur Cayley [60]. Born then employed Pascual Jordan, an assistant to Professor Courant at Göttingen and an expert on matrices, and together they wrote up their paper "On quantum mechanics" in the fall of 1925 [61]. This remarkable paper laid out the foundations of quantum dynamics for nondegenerate systems with one degree of freedom. Representing coordinates and momenta by matrices $\mathbf{q}$ and $\mathbf{p}$, respectively, they derived the canonical equations of motion:

$$\dot{q} = \partial \mathbf{H}/\partial \mathbf{p} \qquad \text{and} \qquad \dot{p} = -\partial \mathbf{H}/\partial \mathbf{q} \tag{2.17}$$

by finding the extrema of the Lagrangian matrix

$$\mathbf{L} = \mathbf{p}\dot{q} - \mathbf{H}, \tag{2.18}$$

where $\mathbf{H}$ is the Hamiltonian matrix, and $\dot{p}$ and $\dot{q}$ are the time derivatives of $\mathbf{p}$ and $\mathbf{q}$, respectively. Employing the adiabatic principle or the equivalent Sommerfeld quantization condition (Eq. 2.6), and using the correspondence principle to expand $\mathbf{p}$ and $\mathbf{q}$ into Fourier series, Born and Jordan also derived the basic commutation relation between coordinates and momenta:

$$[\mathbf{p}, \mathbf{q}] = \mathbf{pq} - \mathbf{qp} = \left(\frac{h}{2\pi i}\right)\mathbf{1}, \tag{2.19}$$

where $\mathbf{1}$ is the unit matrix. Born called this commutation relation the *exact quantum condition* and derived great satisfaction from having condensed Heisenberg's quantum conditions into this one equation [62]. The fact that all diagonal elements in Equation 2.19 are equal to $h/2\pi i$ can be seen as a consequence of the correspondence principle. At that time, however, Heisenberg found "*the fact that xy was not equal to yx was very disagreeable to me*." Although he "*had written down the ... quantization rule*" (Eq. 2.16), he "*did not realize that this was just pq-q*p" [63].

In November 1925, Born et al. [64] generalized these results to systems with an arbitrary, finite number of degrees of freedom in their sequel *On Quantum Mechanics II*. Here, they postulated the commutation relation (Eq. 2.19) and derived the canonical equations of motion (Eq. 2.17), as well as Bohr's quantization condition (Eq. 2.5). This paper also introduced canonical transformations, which are transformations of the variables $\mathbf{p},\mathbf{q} \rightarrow \mathbf{P},\mathbf{Q}$ that preserve the commutation relation (Eq. 2.19). Transformations of the form

$$\mathbf{P} = \mathbf{U}^{-1}\mathbf{pU}; \qquad \mathbf{Q} = \mathbf{U}^{-1}\mathbf{qU} \tag{2.20}$$

are canonical transformations, for any arbitrary matrix $\mathbf{U}$. The canonical equations of motion (Eq. 2.17) are invariant under such canonical transformations. Solving the equations of motion then reduces to finding a canonical transformation such that

$$\mathbf{H}(\mathbf{P},\mathbf{Q}) = \mathbf{H}(\mathbf{U}^{-1}\mathbf{pU}, \mathbf{U}^{-1}\mathbf{qU}) = \mathbf{U}^{-1}\mathbf{H}(\mathbf{p},\mathbf{q})\mathbf{U} = \mathbf{E} \tag{2.21}$$

is diagonal. This is equivalent to the Hamilton–Jacobi equation of classical mechanics and implies that

$$\mathbf{HU} = \mathbf{UE}, \tag{2.22}$$

the solutions to which are the roots of the secular equation:

$$|\mathbf{H} - \lambda\mathbf{1}| = 0. \tag{2.23}$$

In a series of applications, Wolfgang Pauli [65] demonstrated the superiority of Heisenberg's matrix mechanics to the older quantum theory. Heisenberg won the

Nobel Prize in physics for 1932 (actually awarded in 1933) for his "creation of quantum mechanics."

Meanwhile at Cambridge, the British theoretical physicist Paul Dirac [66–68] derived an alternative, but equivalent, formulation of Heisenberg's mechanics, without explicit use of matrices, by starting with the multiplication rule (Eq. 2.16) and seeking the classical analog of the commutator between two functions u and v. This he found in the Poisson bracket:

$$\{u, v\} = \sum_r \{(\partial u/\partial q_r)(\partial v/\partial p_r) - (\partial u/\partial p_r)(\partial v/\partial q_r)\} \tag{2.24}$$

introduced over a century earlier by Simeon Denis Poisson [69]. Using the algebraic properties of the Poisson bracket, Dirac deduced its quantum analog, and by analogy with the corresponding values for the Poisson bracket, he derived the commutation relations between coordinates $\mathbf{q}_r$ and momenta $\mathbf{p}_r$:

$$[\mathbf{q}_r, \mathbf{q}_s] = 0, \quad [\mathbf{p}_r, \mathbf{p}_s] = 0, \quad \text{and} \quad [\mathbf{q}_r, \mathbf{p}_s] = \mathrm{i}\left(\frac{h}{2\pi}\right)\delta_{rs}, \tag{2.25}$$

where the Dirac delta is defined as

$$\begin{aligned} \delta_{rs} &= 0 \text{ for } r \neq s, \\ \delta_{rs} &= 1 \text{ for } r = s. \end{aligned} \tag{2.26}$$

By this means Dirac showed the formal equivalence between quantum mechanics and the Hamilton–Jacobi formulation of classical mechanics. By all accounts, Dirac was a brilliant mind and his methods baffled many of his peers. He would reportedly often be found sitting alone in an empty classroom, staring fixedly at the blank blackboard. After a few hours he would get up and write down a single equation. At various times Einstein wrote about him "*I have trouble with Dirac. This balancing on the dizzying path between genius and madness is awful*" [70] and later on "*Dirac, to whom, in my opinion, we owe the most logically perfect presentation of quantum mechanics*" [71].

2.7 WAVE FUNCTIONS

Erwin Schrödinger (1887–1961) was born in Vienna to Rudolf Schrödinger and Georgine Emilia Brenda. He received his PhD from the University of Vienna in 1910, and his Habilitation in 1914. He served as an assistant to Franz Exner at Vienna and after the First World War, to Max Wien in Jena. Schrödinger developed a deep and lasting interest in Vedanta philosophy. He married Annemarie Bertel in 1920 and, after brief academic stints at Stuttgart and Breslau, moved to Zurich in 1921. Here, he published the famous Schrödinger equation in a paper on "Quantization as an Eigenvalue Problem" in the *Annalen der Physik*. Schrödinger realized that if de Broglie's matter waves are real, there had to be a corresponding wave equation. From Sommerfeld's quantization condition (Eq. 2.6) and de Broglie's equation (Eq. 2.13), he obtained

$$\oint \frac{1}{\lambda} \mathrm{d}q = n, \tag{2.27}$$

which suggests an eigenvalue problem. Generalizing de Broglie's waves to the case of a bound particle, Schrödinger obtained the energy levels as eigenvalues of an operator. When applied to the electron in the hydrogen atom, with a relativistic treatment of the electron, he found results that were not in accord with the experimental observations. Disappointed, Schrödinger abandoned this line of investigation. The discrepancy, however, was not because Schrödinger's approach was incorrect, but because electron "spin" was unknown at that time! When he returned to the problem several months later, Schrödinger found that a nonrelativistic treatment of the electron gave results in agreement with the experiment [72] for hydrogen-like atoms. By analogy with the classical Hamilton's equation, Schrödinger obtained

$$H\left(q, \frac{1}{\psi}\frac{\delta\psi}{\partial q}\right) = E, \tag{2.28}$$

where ψ is a wave function, a continuous differentiable function of q. Applying his theory to the linear harmonic oscillator, the rigid rotor and the diatomic molecule, Schrödinger obtained results in full agreement with Heisenberg's matrix mechanics results. In subsequent papers [73–75], starting from the classical wave equation and the assumption

$$\psi(q,t) = \psi(q)\mathrm{e}^{\frac{-2\pi i Et}{h}}, \tag{2.29}$$

where E/h is the frequency ν, from Planck's law, he deduced

$$\nabla^2\psi + \frac{8\pi^2 m}{h^2}(E - V)\psi = 0, \tag{2.30}$$

for motion in a potential V, and what is now known as *Schrödinger's time-dependent equation*:

$$-\frac{h^2}{8\pi^2 m}\nabla^2\psi + V\psi = \frac{h}{2\pi i}\frac{\partial\psi}{\partial t}. \tag{2.31}$$

Schrödinger interpreted $\psi\psi^*$ as a weight function of the electron charge distribution in configuration space, so that $\rho = \mathrm{e}\psi\psi^*$ is the electron charge density, e being the charge of the electron. He emphasized that ψ is, in general, a function in configuration space and not (except for the one-electron case) in real space. Schrödinger [76] also established the formal, mathematical identity of wave mechanics and matrix mechanics. It was subsequently demonstrated that if the wave function is regarded as a field in space and time, but treated as an operator subject to quantum conditions, the quantum formalism for particles and the formalism for waves are mathematically equivalent—this goes by the name of second quantization [77, 78]. In 1927, Schrödinger succeeded Max Planck at Friedrich-Wilhelms Universität (now Humboldt University) in Berlin, but, becoming disgusted with the Nazi anti-Semitism, left Germany in 1933 and went on to the Magdalen College in Oxford. Schrödinger shared the Nobel Prize in 1933 with Paul Dirac. Returning to Austria in 1936, he suffered harassment and dismissal from his position at the University of Graz on account of his opposition to Nazism. He then fled to Italy with his wife, and then went on to Oxford and Ghent, before settling in Dublin in 1940. He eventually returned to Vienna in 1956.

The time-dependent Schrödinger equation can be transformed into a pair of hydrodynamic equations: the continuity equation

$$\frac{\partial \rho}{\partial t} + \nabla \cdot \mathbf{j} = 0 \tag{2.32}$$

relating the temporal change of the scalar charge density ρ to the divergence of the vector current density $\mathbf{j}$:

$$\mathbf{j} = \frac{eh}{4\pi i m}(\psi^* \nabla \psi - \psi \nabla \psi^*), \tag{2.33}$$

and an Euler-type equation of motion. Using a wave function in the polar form

$$\psi = \sqrt{\rho}\mathrm{e}^{2\pi i S/h}, \tag{2.34}$$

where S is a phase factor, leads to a current density of the form

$$\mathbf{j} = \frac{e}{m}\rho \nabla S. \tag{2.35}$$

This leads to a fluid dynamical formulation of quantum mechanics [79–83], wherein the electron density is treated as a classical fluid, moving with velocity $\frac{e}{m}\nabla S$, under the influence of classical Coulomb forces augmented by a potential of quantum origin. The quantum potential was invoked by de Broglie [80] to explain interference and diffraction phenomena. The fluid dynamical formalism involves the solution of a set of nonlinear partial differential equations, instead of Equation 2.31.

2.8 THE STATISTICS OF ELECTRONS

In 1893, Pieter Zeeman at Leiden University set out to find out "*whether the light of a flame if submitted to the action of magnetism would perhaps undergo any change*." In August 1896, he observed "an immediate widening" (splitting) of the spectral lines, which persisted only as long as the magnetic field stayed on [84]. Stern and Gerlach [85] studied atomic beams of silver in a magnetic field, first at the University of Frankfurt and then at the Institute for Physical Chemistry in Hamburg. They found that the atomic beams were split into two components, with a conspicuous absence of any silver atoms in the center of the deflection pattern, that is, at the position of the undeflected beam. Similar splitting was subsequently observed with beams of other atoms. These experiments revealed the existence of an additional degree of freedom and an extra angular momentum component (which, for historical reasons, goes by the unfortunate name of "spin") for the electron in an atom. These were the problems addressed by Wolfgang Pauli while visiting Bohr in Copenhagen in the fall of 1922. After getting his PhD from the University of Munich under the supervision of Sommerfeld, Pauli had worked as an assistant to Max Born at Göttingen, and, in 1923, he accepted a Privatdozent position at Hamburg. He was already an expert on relativity, having written an article on the theory for the Encyklopdie der mathematischen Wissenschaften [86] at the age of 20 on Sommerfeld's request. Pauli now realized that the shell structure of atoms and the structure of the periodic table can be explained by labeling each electronic state with a set of four quantum numbers and **excluding** the possibility that more than one electron occupies any given level [87]. Pauli stated this exclusion principle as: "*There never exist two or more equivalent electrons in an atom which, in strong fields, agree in all quantum numbers*."

The Pauli exclusion principle, which was empirically obtained by Stoner [88], is a consequence of the antisymmetric wavefunction required by Fermi–Dirac statistics. An antisymmetric wavefunction for a system of N electrons can be written in the form of a determinant:

$$\psi(1, 2, \ldots, N) = (N!)^{-1/2} \begin{bmatrix} \psi_1(1) & \cdots & \psi_1(N) \\ \vdots & \ddots & \vdots \\ \psi_N(1) & \cdots & \psi_N(N) \end{bmatrix}, \tag{2.36}$$

known as a *Slater determinant*, first introduced by Dirac [57]. Here, the notation $\psi_i(j)$ denotes electron j represented by the *set* of quantum numbers i and $(N!)^{-1/2}$ is a normalization factor. That this determinantal form is really antisymmetric follows from the fact that a determinant is antisymmetric under an exchange of any two (rows or) columns, which is equivalent to interchanging the labels of two electrons. For such an antisymmetric wavefunction, there can be no stationary states with two or more electrons having the same set of quantum numbers, because the determinant vanishes when two (rows or) columns are identical. Thus, such an antisymmetric wavefunction naturally obeys the Pauli exclusion principle.

2.9 DOES GOD PLAY DICE?

For Max Born, reconciling the corpuscular aspects of electrons, as exhibited by the collision experiments in James Franck's laboratory at Göttingen, took primacy. Born introduced the statistical interpretation of the wave function, and it was primarily for this work that he was awarded the Nobel Prize in 1954. "*The statistical interpretation of de Broglie's waves was suggested to me by my knowledge of experiments on atomic collisions which I had learned from my experimental colleague James Franck*" [89]. In Born's conception, the squared wave amplitudes or intensities determine the probability of the presence of the corresponding quanta or their density. Thus, "*it was almost self-understood to regard* $|\psi|^2$ *as the probability density of particles*." **However, to reconcile the statistical interpretation with the existence of quantum interference phenomena, one needs to modify the classical laws for compounding probabilities**: If a wave function ψ_1 with probability density $P(\psi_1) = |\psi_1|^2$ is superposed with a wave function ψ_2 with probability density $P(\psi_2) = |\psi_2|^2$, the probability density of the superposed wave $\psi_1 + \psi_2$ is given by

$$P(\psi_1 + \psi_2) = |\psi_1 + \psi_2|^2 = |\psi_1|^2 + |\psi_2|^2 + {\psi_1}^*\psi_2 + {\psi_2}^*\psi_1, \quad (2.37)$$

NOT just the sum of the individual probabilities $|\psi_1|^2 + |\psi_2|^2$. The last two terms in Equation 2.37 are the interference terms. Seen in this way, **quantum mechanics is a modification of the basic laws of classical statistics.**

Consider the general solution

$$\psi(x,t) = \sum_n c_n \psi_n(x) \exp\left(\frac{2\pi i E_n t}{h}\right) \quad (2.38)$$

of the time-dependent Schrödinger equation (Eq. 2.31) and assume the $\psi_n(x)$ to be normalized. Further assuming that a system can be in only one stationary state at any given time (as postulated by Bohr), Born thus interpreted [90]

$$|c_n|^2 = |\textstyle\int \psi_n(x,t){\psi_n}^*(x,t)\,\mathrm{d}x|^2 \quad (2.39)$$

as the probability that the system is in state n at time t. For a system which is in the state n at time $t = 0$ and then evolves under the action of a time-dependent potential $V(x,t)$, integration of Equation 2.31 for any subsequent time t yields

$$\psi_n(x,t) = \sum_m b_{nm} \psi_m(x) \exp\left(\frac{2\pi i E_m t}{h}\right), \quad (2.40)$$

where $|b_{nm}|^2$ is the transition probability for the state, initially in state n at $t = 0$, to be found in state m at time t. In general, if the system is given by the solution (Eq. 2.38) at time $t = 0$, then the probability for it to be in state m at time t is $\sum_m |c_m b_{mn}|^2$. This is different from what would be expected on the basis of the classical theory of composition of probabilities of independent events, which would lead us to expect a transition probability $\sum_m |c_m|^2 |b_{mn}|^2$. Thus the transition probabilities between states cannot be regarded as independent events. The phases of the expansion coefficients now assume physical significance on account of the interference of probabilities. These results were also derived independently by Dirac [67].

Born reinforced the statistical nature of his theory by writing with regard to atomic collisions,

> One does not get an answer to the question, how probable is a given effect of the collision? ... From the standpoint of our quantum mechanics, there is no quantity which causally fixes the effect of a collision in an individual event. I myself am inclined to renounce determinism in the atomic world [90],

but acknowledged "*but that is a philosophical question for which physical arguments alone do not set standards*." Born's papers received a mixed reception from his peers. Among those who found it hard to abandon classical notions of causality and retained severe reservations about the statistical interpretation of the wave function were Schrödinger, de Broglie, and Einstein. Einstein expressed his philosophical distaste for Born's statistical interpretation of quantum mechanics by declaring that God does not play dice. Einstein's negative reaction was a severe blow for Born, as the two were close personal friends and maintained an active correspondence throughout their lives [29].

Owing to his strong pacifist views and Jewish ancestry, Born was stripped of his professorship by the Nazi regime and he emigrated from Germany in 1933. He took up a position at the University of Cambridge, and worked at the Indian Institute of Science in Bangalore in 1935 and 1936. Sir C. V. Raman persuaded him to stay on in India as Professor of Mathematical Physics, but the appointment fell through owing to political problems at the institute [91]. Max Born was then appointed Tait Professor of Natural Philosophy at the University of Edinburgh, where he stayed from 1936 until his retirement in 1953. He was elected a Fellow of the Royal Society of London in 1939. Born returned to Germany on retirement from Edinburgh at the age of 65. In 1954, nearly three decades after his seminal papers on quantum mechanics, two decades after his assistant Heisenberg was similarly recognized, and a quarter century after Einstein had first nominated them both, Max Born was awarded the Nobel Prize "for his fundamental research in quantum mechanics, especially for his statistical interpretation of the wavefunction."

Einstein's argument against the statistical interpretation of quantum mechanics was that it does not provide a "complete description" of the motion of an individual particle. Consider the problem of an electron diffracting through a slit and striking a detection screen. If we know that an electron arrives at a point A on the screen, we know instantly that it did not arrive at a different, distant point B $\neq$ A. This would imply an instantaneous action at a distance between A and B. But had not Einstein abolished just such an action at a distance from Newtonian mechanics with his theory of relativity? As we shall see, quantum mechanics denies locality and makes only statistical predictions concerning the probability of an electron arriving at a given point on the screen. Does this foreclose the possibility that a more detailed, deterministic description, one underlying the undoubtedly correct mathematical apparatus of quantum mechanics, might exist and might someday be found? Clearly, Einstein and others like David Bohm believed so. Born was inclined to doubt it, but considered it an open, philosophical question. Bohr, Heisenberg, and most other physicists maintained that the statistical description provided by quantum mechanics is complete and is the only one compatible with observational evidence. The famous debate between Bohr and Einstein was joined at the 1927 Solvay conference and would continue for years [29, 92]. It is widely believed today that Bohr won those debates. On the basis of the experimental evidence available at that time, however, there was no way to make a call. Max Born's position, that this was only a matter of interpretation and that the distinction was of no practical consequence, was off the mark too, but this would not be known until 1964 (with the proof of Bell's theorem [93, 94]).

By 1933, Einstein had left Germany just prior to Hitler coming to power. Einstein then immigrated with his family to the United States and settled at the Institute for Advanced Study in Princeton. In 1935, with Boris Podolsky and Nathan Rosen, he reformulated his objections in the clearest form yet [95]. Employing the noncommuting observables position and momentum for two particles that were initially interacting, Einstein, Podolsky, and Rosen showed that quantum mechanics predicted the existence of two-particle states with strong correlations between the positions of the two particles and strong correlations between their momenta. Measurements of position would always give values symmetric about the origin. So, a measurement on one particle also reveals with certainty the position of the other. The particle momenta are likewise correlated, owing to momentum conservation, and thus measurement of one particle's momentum is sufficient to know with certainty the momentum of the other. **These correlations survive even after the particles have separated by a large distance**. Schrödinger [96] coined the term *entanglement* to characterize this lack of factorability of the quantum wavefunction. Since position and momentum do not commute, they cannot be simultaneously determined with certainty. One has to choose between an accurate measurement of position and an accurate measurement of momentum. **But this choice can be made even after the particles have separated**. Such "delayed choice" experiments were first suggested by David Bohm and Yariv Aharonov [97]. Since a measurement on the first particle does not disturb the second distant particle, Einstein, Podolsky, and Rosen concluded that the individual particles must have had well-defined values of position and momentum even before the measurement, and thus that the quantum mechanical description (which disallows such simultaneously well-defined values of noncommuting variables) has to be incomplete. In other words, the proposition that the quantum mechanical description is complete is incompatible with the notion of "objective reality"—which they defined as follows:

> If without in any way disturbing a system we can predict with certainty (i.e. with a probability equal to unity) the value of a physical quantity, then there exists an element of physical reality corresponding to this physical quantity.

They emphasized that one

> would not arrive at our conclusion if one insisted that two ... physical quantities can be regarded as simultaneous elements of reality only when they can be simultaneously measured or predicted.

Bohr [98] responded to Einstein, Podolsky, and Rosen's argument by forbidding us to speak of properties of individual particles, even when they are distant from each other. Bohr emphasized the importance of a careful description of the details of the measuring apparatus, and "*the impossibility of any sharp distinction between the behavior of atomic objects and the interaction with the measuring instruments which serve to define the conditions under which the phenomena appear*" [92]. Others tried to relate the quantum measurement problem to the role of a conscious observer. In this view then, physics would not be about the world as it exists, but about our knowledge of the world. The paradox this leads to can be illustrated by the experiment suggested by Schrödinger and known nowadays as that of Schrödinger's cat [99], and the related one of Wigner's friend. Here, a cat is destined to be the hapless victim of a chemical weapon that can be set off by a radioactive trigger. When the radioactive atom decays, it triggers the release of a poison gas, which then kills the cat. The cat along with the entire apparatus is sealed in a box and hidden from view until someone opens the box to check on the status of the cat.

> A cat is penned up in a steel chamber, along with the following device (which must be secured against direct interference by the cat): in a Geiger counter, there is a tiny bit of radioactive substance, so small that perhaps in the course of one hour, one of the atoms decays, but also, with equal probability, perhaps none; if it happens, the countertube discharges, and through a relay releases a hammer that shatters a small flask of hydrocyanic acid. If one has left this entire system to itself for an hour, one would say that the cat still lives if meanwhile no atom has decayed. The psi-function of the entire system would express this by having in it the living and dead cat (pardon the expression) mixed or smeared out in equal parts [99].

The laws of quantum mechanics and radioactive decay only give us the probability of decay at any time. If we can describe the cat by a wave function, then we might say that the cat was initially in the "alive" state. After one half life of the radioactive material, the probability that the cat might still be alive is now reduced to 50% and its wavefunction can be described by the superposition $1/2$ "dead" + $1/2$ "alive." After we open the box and see what has happened, we know for certain whether the cat is alive or dead. The superposition wavefunction now collapses into one of the two states: either definitely "alive" or definitely "dead"—and as this does not happen until we open the box, any responsibility for killing the cat lies not with the cruel person who designed and set up the experiment, but with the curious observer who opened the box to check on the cat's well-being! Wigner devised a curious twist to this experiment, which brought out the role of the observer even more starkly. Suppose he were not to open the box himself, but to depute a friend to do so. Until and unless the friend reported the result to him, Wigner would still describe the state of the cat as $1/2$ "dead" + $1/2$ "alive." But Wigner's friend would describe the state of the same cat as either definitely alive or definitely dead. All this fueled considerable confusion and a fertile field for science fiction and metaphysics with questions such as "does a tree fall in a forest if there is nobody around to see or hear it?"

In his 1932 book on the *Mathematical Foundations of Quantum Mechanics* [100] John von Neumann provided a proof on the mathematical impossibility of a more detailed, deterministic description (known by the name of *hidden variables*) underlying quantum mechanics, and concluded: "*It is therefore not ... a question of reinterpretation of quantum mechanics—the present system of quantum mechanics would have to be objectively false in order that another description ... than the statistical one be possible*." The problem with von Neumann's proof was not immediately realized and most physicists assumed that the question of hidden variables was settled, until David Bohm [81, 101] explicitly constructed a model with hidden variables in 1952. Even so, in view of the remarkable successes of quantum mechanics, few physicists were willing to "waste" their time on conceptual or foundational problems in quantum theory, until John Bell came along, and it took another decade before his work was taken seriously enough to be put to experimental test. Bell started his career in accelerator design, but devoted most of his life to conceptual and theoretical questions. In 1964, he turned his attention to the correlations between measurements on a pair of entangled particles. Bell [93] investigated the possibility of allowing the results of measurements on each particle to depend on the hidden variables carried by that particle and on the setting of the apparatus employed for the measurement, with the restriction that the result of a measurement on the first particle should not depend on the setting of the second, distant measuring device. This locality restriction is a consequence of Einstein's relativity principle, whereby no physical signal can travel faster than light, in an experimental setup where the instrument settings are rapidly changed during the time of flight of the particles to the detectors, when the particles are no longer physically interacting with each other. Bell found that **any such hidden variable theory would predict correlations different from quantum mechanics for some settings of the measuring apparatus**. This remarkable result now goes by the name of Bell's theorem. Here, at last was the possibility of being able to experimentally resolve the Bohr–Einstein controversy by measuring the correlations between entangled quantum particles! It was only some 30 years after Einstein, Podolsky, and Rosen's paper that the first such experiments were actually performed [102–105]. The matter was finally settled only in the 1980s with the remarkable delayed choice experiments on correlated quantum particles performed by Alain Aspect and his coworkers [106–108]. All these experiments, and subsequent ones [109–114], have borne out the statistical correlations predicted by the quantum theory. The inescapable conclusion seems to be that **entangled quantum objects must be treated as a single entity, described by a correlated wavefunction that cannot be factorized into individual-particle components**. Aspect's experiments involved "super-luminal communication" between a pair of entangled quantum photons: the two polarization detectors were separated by a space-like separation, that is, one that no signal traveling at or below the velocity of light could connect, and the settings of the detectors were changed during the flight of the photons from the source to the detectors. Such "super-luminal communication" has since been demonstrated over distances more than 10 km [109–112] and forms the basis of what is now known as *quantum teleportation* [113, 114].

REFERENCES

1. Planck M. In: von Laue M, editor. *Physikalische Abhandlungen und Vorträge*. Braunschweig: Friedrich Vieweg; 1958; *A Survey of Physics*. New York: Dover; 1960.
2. Jammer M. Volume 12, *The Conceptual Development of Quantum Mechanics*, *The History of Modern Physics 1800–1950*. 2nd ed. New York: American Institute of Physics; 1989.
3. Planck M. Berl Ber 1899: 440–480; Ann Phys 1900;1:69–122; Verh Dtsch Phys Ges 1900;2:202–204.
4. Einstein A. Ann Phys 1905;17(6):132–148.
5. Pais A. *Subtle is the Lord: The Science and Life of Albert Einstein*. Oxford: Oxford University Press; 1982.
6. Einstein A. A new determination of molecular dimensions [PhD thesis]. University of Zurich; 1905.
7. Einstein A. Ann Phys 1905;17(8):549–560.
8. Einstein A. Ann Phys 1905;17(10):891–921.
9. Einstein A. Ann Phys 1905;18(13):639–641.
10. Hertz H. Wied Ann Phys 1887;31:982–1000.
11. Wiedemann E, Ebert H. Wied Ann Phys 1888;33:241–264.
12. Hallwachs W. Wied Ann Phys 1888;33:301–312.
13. Stotetow AG. Comptes Rendus 1888;106:1149–1152, 1593–1595; 1889;108: 1241–1243; J Phys 1889;9:468–473.
14. Lenard P. Wien Ber 1899;108:1649–1666; Ann Phys 1900;2:359–375; 1902;8: 149–198.
15. Elster J, Geitel H. Wied Ann Phys 1889;38:40–41; 1890;39:332–335; 1890;41: 161–165, 166–176; 1895;55:684–700.
16. Ladenburg E. Ann Phys 1903;12:558–578.
17. Hughes AL. Philos Trans R Soc Lond A 1912;212:575–594.
18. Millikan RA. Phys Rev 1914;4:73–75; 1915;6:55; 1916;7:355–388; Phys Z 1916; 17:217–221.
19. Klein MJ. *Paul Ehrenfest*. Amsterdam: North Holland Pub. Co.; 1970.

20. Ehrenfest P. Verh Dtsch Phys Ges 1913;15:451–457.
21. Bohr N. Philos Mag 1915;29:332–335.
22. Wilson EB. Philos Mag 1915;29:795–802.
23. Sommerfeld A. Ann Phys 1916;51:1–94,125–167.
24. Ehrenfest P. Ann Phys 1916;51:327–352.
25. Einstein A. Verh Dtsch Phys Ges 1916;18:318–323.
26. Rutherford E. Philos Mag 1900;49:1–14.
27. Einstein A. Phys Z 1917;18:121–128.
28. Mackay GW. *Mathematical Foundations of Quantum Mechanics*. New York, Amsterdam: W. A. Benjamin; 1963. p 61.
29. Born M. *The Born-Einstein Letters*. New York: Walker & Co.; 1971.
30. Eve AS. Philos Mag 1904;8:669–685.
31. Kleeman RD. Philos Mag 1908;15:638–663.
32. Sadler DA, Mesham P. Philos Mag 1912;24:138–149.
33. Madsen JPV. Philos Mag 1909;17:423–448.
34. Florance DCH. Philos Mag 1910;20:921–938.
35. Gray JA. Philos Mag 1913;26:611–623.
36. Compton AH. Phys Rev 1923;21:715.
37. Debye P. Phys Z 1923;24:161–166.
38. de Broglie L. J Phys 1922;3:422–428; Comptes Rendus 1922;175:811–813.
39. de Broglie L. Comptes Rendus 1923;177:507–510.
40. de Broglie L. Comptes Rendus 1923;177:548–550.
41. de Broglie L. Philos Mag 1924;47:446–458.
42. de Broglie L. *Recherches sur la théorie des quanta* [PhD thesis]. University of Paris; 1924; Ann Phys 1925;3:22–128.
43. Einstein A. Letter to H. A. Lorenz Dec. 16, 1924, quoted in Pais A. *Subtle is the Lord: The Science and Life of Albert Einstein*. Oxford: Oxford University Press; 1982.
44. Einstein A. Z Berl Ber 1925;3–14.
45. Davisson CJ, Kunssman CH. Science 1921;54:522–524; Phys Rev 1923;22: 242–258.
46. Ramsauer C. Ann Phys 1921;64:513–540; 1921;66:546–558; 1923;72:345–352.
47. Townsend JS, Bailey VA. Philos Mag 1922;43:593–600.
48. Chaudhuri RN. Philos Mag 1923;46:461–472.
49. Elsasser W. Naturwissenschaften 1925;13:711.
50. Davisson CJ, Germer LH. Phys Rev 1927;30:705–740.
51. Thomson GP, Reid A. Nature 1927;119:890.
52. Bose S. Z Phys 1924;26:178–181.
53. Einstein A. Z Berl Ber 1924;22:261–267.
54. London F. Nature 1938;141:643–644.
55. Anderson MH, Ensher JR, Matthews MR, Wieman CE, Cornell EA. Science 1995;269:198–201.

56. Davis KB, Mewes M-O, Andrews MR, van Druiten NJ, Durfee DS, Kurn DM, Ketterle W. Phys Rev Lett 1995;75:3969–3973.
57. Dirac PAM. Proc R Soc London, Ser A 1926;112:661–677.
58. Fermi E. Z Phys 1926;36:902.
59. Heisenberg W. Z Phys 1925;33:879–893.
60. Cayley AJ. Reine Angew Math 1855;50:272–317.
61. Born M, Jordan P. Z Phys 1925;34:858–888.
62. Born M. *Physics in My Generation*. London, New York: Pergamon Press; 1956. p 100.
63. Interview with Werner Heisenberg, 1963 Feb 15, in (Jammer, ref. 3, 1989), p. 281.
64. Born M, Heisenberg W, Jordan P. Z Phys 1926;35:557–615.
65. Pauli W. Z Phys 1926;36:336–363.
66. Dirac PAM. Proc R Soc London, Ser A 1925;109:642–653.
67. Dirac PAM. Proc R Soc London, Ser A 1926;110:561–579; 1926;111:281–305.
68. Dirac PAM. Proc Camb Philol Soc 1926;23:412–418.
69. Poisson SD. J Ecole Polytech 1809;8:266–344.
70. Einstein A. Letter to Paul Ehrenfest, Aug. 23, 1926.
71. Einstein A. *James Clerk Maxwell*. New York: Macmillan; 1931. p 66.
72. Schrödinger E. Ann Phys 1926;79:361–376.
73. Schrödinger E. Ann Phys 1926;79:489–527.
74. Schrödinger E. Ann Phys 1926;80:437–490.
75. Schrödinger E. Ann Phys 1926;81:109–139.
76. Schrödinger E. Ann Phys 1926;79:734–756.
77. Jordan P, Klein O. Z Phys 1927;45:751–765.
78. Jordan P, Wigner EP. Z Phys 1928;47:631–651.
79. Madelung E. Z Phys 1926;40:322–326.
80. de Broglie L. Comptes Rendus 1926;183:447–448.
81. Bohm D. *Quantum Theory*. Englewood Cliffs (NJ): Prentice-Hall; 1951.
82. (a) Ghosh SK, Deb BM. Phys Rep 1982;92:1–44; (b) Deb BM, Ghosh SK. J Chem Phys 1982;77:342–348.
83. Deb BM, Ghosh SK. In: March NH, Deb BM, editors. *The Single-Particle Density in Physics and Chemistry*. London: Academic Press; 1987. p220–284.
84. Zeeman P. Philos Mag 1897;43:226–239.
85. Stern O, Gerlach W. Z Phys 1922;8:110–111; 1922;9:349–355.
86. Pauli W. *Encyklopdie der Mathematischen Wissenschaften*. Leipzig: Tuebner; 1921.
87. Pauli W. Z Phys 1925;316:373–385.
88. Stoner EC. Philos Mag 1924;48:719–726.
89. Born M. *Experiment and Theory in Physics*. Cambridge: Cambridge University Press; 1943.
90. Born M. Z Phys 1926;38:803–827.
91. Venkataraman G. *Journey Into Light: Life and Science of C. V. Raman*. Bangalore: Indian Academy of Sciences; 1988.

92. Schilp PA, editor. *Albert Einstein: Philosopher Scientist*. Evanston (IL): Library of Living Philosophers; 1949.
93. Bell JS. Physics 1964;1:195–200.
94. Bell JS. *Speakable and Unspeakable in Quantum Mechanics*. United States: Cambridge University Press; 1987.
95. Einstein A, Podolsky B, Rosen N. Phys Rev 1935;47:777.
96. Schrödinger E. Proc Camb Philol Soc 1935;31:555–563.
97. Bohm D, Aharonov Y. Phys Rev 1957;108:1070–1076.
98. Bohr N. Phys Rev 1935;48:696–702.
99. Schrödinger E. Naturwissenschaften 1935;23:807–812,823–828,844–849.
100. von Neumann J. *Mathematische Grundlagen der Quanten-mechanik*. Berlin: Springer; 1932.
101. Bohm D. Phys Rev 1952;85:166–179.
102. Clauser JF, Horne MA, Shimony A, Holt RA. Phys Rev Lett 1969;23:880–884.
103. Freedman SJ, Clauser JF. Phys Rev Lett 1972;28:938–941.
104. Clauser JF, Horne MA. Phys Rev D 1974;10:526–535.
105. Clauser JF, Shimony A. Rep Prog Phys 1978;41:1881–1927.
106. Aspect A, Grangier P, Roger G. Phys Rev Lett 1981;47:460–463.
107. Aspect A, Grangier P, Roger G. Phys Rev Lett 1982;49:91–94.
108. Aspect A, Dalibard J, Roger G. Phys Rev Lett 1982;49:1804–1807.
109. Tittel W, Brendel J, Gisin B, Herzog T, Zbinden H, Gisin N. Phys Rev A 1998;57:3229–3232.
110. Weihs G, Jennewein T, Simon C, Weinfurter H, Zeilinger A Phys Rev Lett 1998;81:5039–5043.
111. Rowe MA, Kielpinski D, Meyer V, Sackett CA, Itano WM, Monroe C, Wineland DJ. Nature 2001;409(6822):791–794.
112. Salart D, Baas A, van Houwelingen JAW, Gisin N, Zbinden H. Phys Rev Lett 2008;100:220404.
113. Bouwmeester D, Pan J-W, Mattle K, Eibl M, Weinfurter H, Zeilinger A. Nature 1997;390:575–579.
114. Bennett CH. Phys Rev Lett 1993;70:1895–1899.

3

THE ELECTRON DENSITY

N. Sukumar and Sunanda Sukumar

3.1 MOLECULAR STRUCTURE

The concept of molecular structure is fundamental to the practice and understanding of chemistry. At the heart of this concept lies the Born–Oppenheimer separation of electronic and nuclear motions [1–3]. This separation introduces a great computational and practical simplification in the study of chemistry, but is neither essential to the conceptual formulation of molecular structure nor universally valid. The notion that a molecule has structure is fundamental to much of chemistry as practised today. In the words of Woolley [4], "The idea that molecules are microscopic, material bodies with more or less well-defined shapes has been fundamental to the development of our understanding of the physicochemical properties of matter, and it is now so familiar and deeply ingrained in our thinking that it is usually taken for granted—it is the central dogma of chemistry." Indeed, Woolley [4–9] has argued that the notion that a single molecule has a shape—a nearly fixed (relative) arrangement of nuclei in space—is a classical idea imposed on the quantum mechanical picture of matter. Practising chemists customarily envision molecular structure in terms of two-dimensional graphs and ball-and-stick-type molecular models. Such models are simple to visualize and have immense intuitive appeal. We shall define structure here from a statistical perspective, as that which distinguishes an object from an arbitrary collection of its parts, in this case, a molecule from the set of its constituent atoms. This definition generalizes the concept of molecular structure to situations where the relative

A Matter of Density: Exploring the Electron Density Concept in the Chemical, Biological, and Materials Sciences, First Edition. Edited by N. Sukumar.

spatial locations of the constituent atoms may not be fully known, or even well defined.

In 1927, Max Born and J. Robert Oppenheimer [1] effected an approximate separation of electronic and nuclear motions in molecules by means of a perturbation expansion in the ratio of the electronic to nuclear masses, and showed that molecular states can be approximately represented as products of electronic and nuclear functions.

$$\Psi(\mathbf{r}, \mathbf{R}) \approx \psi(\mathbf{r}; \mathbf{R})\chi(\mathbf{R}), \tag{3.1}$$

where $\Psi(\mathbf{r},\mathbf{R})$ is an eigenfunction of the molecular Hamiltonian $\mathcal{H}$:

$$\mathcal{H}\Psi(\mathbf{r}, \mathbf{R}) = \mathcal{E}\Psi(\mathbf{r}, \mathbf{R}), \tag{3.2}$$

$$\mathcal{H} = T_{\mathrm{n}} + T_{\mathrm{e}} + V_{\mathrm{ne}} + V_{\mathrm{ee}} + V_{\mathrm{nn}}. \tag{3.3}$$

Here, we use $\mathbf{r}$ to represent the electronic coordinates and $\mathbf{R}$ to represent the nuclear coordinates. T_{n} is the nuclear kinetic energy, T_{e} the interelectronic kinetic energy, while V_{ne}, V_{ee}, and V_{nn} include the electron-nuclear attraction, interelectronic repulsion, and internuclear repulsion terms, respectively. Equation 3.1 is the Born–Oppenheimer approximation. The electronic function $\psi(\mathbf{r};\mathbf{R})$ is obtained by solution of the electronic Schrödinger equation:

$$H\psi(\mathbf{r};\mathbf{R}) = E\psi(\mathbf{r};\mathbf{R}), \tag{3.4}$$

The electronic structure problem then involves solving for the eigenfunctions of an electronic Hamiltonian,

$$H = T_{\mathrm{e}} + V_{\mathrm{ne}} + V_{\mathrm{ee}}. \tag{3.5}$$

The semicolon denotes the parametric dependence of ψ on the nuclear coordinates. $\mathcal{E}$ is the total energy of the molecule, while E is termed the electronic energy. The nuclear function $\chi(\mathbf{R})$ satisfies an equation of motion:

$$[T_{\mathrm{n}} + E + V_{\mathrm{nn}}]\chi(\mathbf{R}) \approx \mathcal{E}\chi(\mathbf{R}). \tag{3.6}$$

The eigenvalues of the electronic Hamiltonian H (together with the internuclear repulsion terms V_{nn} in $\mathcal{H}$) thus form an effective potential energy surface on which the nuclei are envisioned to move. As the nuclei are several thousand times heavier than the electrons, Born and Oppenheimer argued that the electrons move much faster than the nuclei; consequently, at any given instant, the electrons feel the instantaneous effect of the nuclei, whereas the nuclei feel an average, smeared-out effect of the moving electrons. The physical picture that emerges from the Born–Oppenheimer approximation is that of this smeared-out electron cloud instantaneously adjusting to the motion of the slowly moving nuclei. Thus, although they move in accordance with the quantum Schrödinger equation (Eq. 3.6), the nuclei are visualized in a corpuscular framework, whereas the electrons are represented by an electron cloud or charge density. This picture is central to most chemical concepts today, but it is important to recognize the limitations on its validity.

In 1951, Max Born [2, 3] reexamined the problem without invoking a perturbation with respect to the ratio of electronic to nuclear masses, and instead recast it as an exact sum-over-states expansion:

$$\Psi(\mathbf{r}, \mathbf{R}) = \sum_i \psi_i(\mathbf{r}, \mathbf{R})\chi_i(\mathbf{R}); \tag{3.7}$$

where $\psi_i(\mathbf{r};\mathbf{R})$ is the ith electronic state wavefunction with eigenvalue E_i:

$$H\psi_i(\mathbf{r};\mathbf{R}) = E_i\psi_i(\mathbf{r};\mathbf{R}). \tag{3.8}$$

The set of electronic functions $\psi_i(\mathbf{r};\mathbf{R})$ for all i now form a basis for the expansion of the total wavefunction $\Psi(\mathbf{r},\mathbf{R})$. In coordinate representation,

$$\left[-\frac{1}{2}\nabla_e^2 + V_{ne} + V_{ee}\right]\psi_i(\mathbf{r};\,\mathbf{R}) = E_i\psi_i(\mathbf{r};\,\mathbf{R}). \tag{3.9}$$

Using the expansion (Eq. 3.7) in the Schrödinger equation (Eq. 3.2) and making use of the eigenvalue equation (Eq. 3.8),

$$\sum_j \sum_\alpha \frac{\mathbf{p}_\alpha{}^2}{2M_\alpha}\psi_j(\mathbf{r}, \mathbf{R})\chi_j(\mathbf{R}) + [E_i + V_{nn}]\psi_i(\mathbf{r};\,\mathbf{R})\chi_i(\mathbf{R}) = E_i\psi_i(\mathbf{r};\,\mathbf{R})\chi_i(\mathbf{R}), \tag{3.10}$$

where we have used

$$T_n = \sum_\alpha \frac{\mathbf{p}_\alpha{}^2}{2M_\alpha}, \tag{3.11}$$

$\mathbf{p}_\alpha = i\hbar\boldsymbol{\nabla}_\alpha$ are nuclear momentum operators and M_α are nuclear masses, nuclear variables being represented by Greek subscripts; $\hbar$ is an abbreviation for $h/2\pi$. In addition to the terms appearing in the Born–Oppenheimer equation (Eq. 3.4), we now have both diagonal and off-diagonal ($j \neq i$) terms in Equation 3.10, resulting from the action of the nuclear kinetic energy operator on the electronic basis functions. These terms take the form

$$\mathbf{A}_{ij} = \frac{i\hbar}{M_\alpha}\psi_j{}^*(\mathbf{r}, \mathbf{R})\boldsymbol{\nabla}_\alpha\psi_i(\mathbf{r}, \mathbf{R}), \tag{3.12}$$

$$\mathrm{B}_{ij}^\alpha = \frac{1}{2M_\alpha}\psi_j{}^*(\mathbf{r}, \mathbf{R})\boldsymbol{\nabla}_\alpha^2\psi_i(\mathbf{r}, \mathbf{R}). \tag{3.13}$$

The vector diagonal term $\mathbf{A}_{ii}^{\alpha}$ is a most interesting quantity, whose importance was not realized until recently. It is of relevance to the geometric phase and plays a crucial role in the density functional polarization theory of insulators. This quantity is discussed in detail in Chapter 5. The scalar diagonal term B_{ii}^{α} represents an adiabatic correction to the Born–Oppenheimer energy, while the off-diagonal Born coupling terms $\mathbf{A}_{ij}^{\alpha}$ and B_{ij}^{α} for $i \neq j$ serve to couple different electronic states [2, 10]. The vector term $\mathbf{A}_{ij}^{\alpha}$ (for $j \neq i$) appears in the equation of motion as a dot product with the nuclear momentum operator $\mathbf{p}_{\alpha}$:

$$\left[\sum_{\alpha} \frac{\mathbf{p}_{\alpha}{}^{2}}{2M_{\alpha}} + E_i + V_{\mathrm{nn}}\right] \chi_i(\mathbf{R}) + \sum_{j}\sum_{\alpha} [\mathbf{A}_{ij}^{\alpha}\, \mathbf{p}_{\alpha} + \mathrm{B}_{ij}^{\alpha}] \chi_j(\mathbf{R}) = \mathcal{E}_{\mathbf{i}} \chi_{\mathbf{i}}(\mathbf{R}). \tag{3.14}$$

The Born–Oppenheimer approximation is thus tantamount to neglecting the terms involving the Born coupling terms. As V_{ne} is the only term in the electronic Hamiltonian H that depends on nuclear coordinates, the expression (Eq. 3.12) may be rewritten using the eigenvalue equation (Eq. 3.8) as

$$\mathbf{A}_{\alpha}^{ij} = \frac{\mathrm{i}\hbar}{E_i - E_j} \psi_j{}^{*}(\mathbf{r}, \mathbf{R}) [\nabla_{\alpha} V_{\mathrm{ne}}] \psi_i(\mathbf{r}, \mathbf{R}). \tag{3.15}$$

The appearance of the energy difference denominator in Expression 3.15 for the Born coupling term $\mathbf{A}_{ij}^{\alpha}$ shows that a necessary, but not sufficient, condition for the validity of the Born–Oppenheimer approximation is that the electronic wavefunction must be nondegenerate at all points in the region of interest in nuclear configuration space. This can be seen through the terms B_{ij}^{α} and $\mathbf{A}_{ij}^{\alpha}$ that are related through the identity:

$$B_{ij}^{\alpha} = -\frac{1}{2}\hbar \left[\nabla_{\alpha} \cdot \mathbf{A}_{ij}^{\alpha} + \sum_{k} \mathbf{A}_{ik}^{\alpha} \cdot \mathbf{A}_{kj}^{\alpha}\right], \tag{3.16}$$

which is easily verified by differentiation of Expression 3.12 with respect to nuclear coordinates. This enables the molecular Hamiltonian to be written in the gauge-invariant matrix form [10–12]:

$$\mathcal{H} = \sum_{\alpha} \frac{1}{2M_{\alpha}} [\mathbf{p}_{\alpha} + \mathbf{A}_{\alpha}] \cdot [\mathbf{p}_{\alpha} + \mathbf{A}_{\alpha}] + [E + V_{\mathrm{nn}}]. \tag{3.17}$$

In addition, the nature of the intramolecular forces is also of importance in effecting the Born–Oppenheimer separation. Essén [13, 14] presented an alternative analysis of the Born–Oppenheimer separation, as instead a separation between collective and internal motions in a molecule, which, on account of the small ratio of electronic to nuclear masses, can be approximately identified with relative nuclear motions and electronic motions relative to fixed nuclei, respectively.

The two-dimensional graphical structure of a molecule generally forms the input to an electronic structure computation within the Born–Oppenheimer approximation. This has provoked some discussion [4–9, 15] as to whether molecular structure is really derivable from first principles quantum mechanics. There have been several recent efforts directed at first principles computation of molecular structure beyond the Born–Oppenheimer approximation. These nowadays go by the name of non-Born–Oppenheimer methods or NOMO (nuclear orbital plus molecular orbital) theory [16–18]. The earliest such studies were a pioneering series of computations at the Hartree–Fock (see Section 3.2) level of theory, performed by Thomas [19–24], for a series of second-row hydrides, putting electrons into electronic orbitals and protons into protonic orbitals, separately antisymmetrizing the electronic and protonic wavefunctions and iterating the Fock equations to self-consistency. While the solution of the electronic Schrödinger equation within the Born–Oppenheimer approximation takes the coordinates of the nuclei as input parameters, no such input is required for computing the eigenfunctions of the total molecular Hamiltonian (Eq. 3.5), which only takes the numbers of each kind of particle as input. Equilibrium bond lengths and bond angles emerge in this picture as peaks in the protonic radial distribution function. These computations yielded a remarkable structure for ammonia, very different from the expected pyramidal structure: the molecular structure of ammonia in this picture was a disk centered on the nitrogen atom! While the pyramidal structure is indeed the equilibrium ground state structure of ammonia obtained from the solution of the electronic Schrödinger equation within the Born–Oppenheimer approximation, it is not an eigenfunction of the total Hamiltonian of the isolated molecule. The ground state solution of the total molecular Schrödinger equation beyond the Born–Oppenheimer approximation is a symmetric linear combination of two pyramidal ammonia wavefunctions (ψ_1 and ψ_2) that are related by a reflection operation (Fig. 3.1). The wavefunction for the first excited state ψ_- of ammonia is the corresponding antisymmetric linear combination:

$$\psi_+ = \sqrt{\frac{1}{2}}[\psi_1 + \psi_2],$$
$$\psi_- = \sqrt{\frac{1}{2}}[\psi_1 - \psi_2]. \tag{3.18}$$

Indeed, excitation of a proton from ψ_+ to ψ_- gives rise to a line in the protonic spectrum that corresponds to the observed microwave line in ammonia; this is conventionally explained as due to umbrella-flipping transformation between ψ_1 and ψ_2 (Fig. 3.1). Woolley [4–9] has argued that when experimental structure determinations reveal broken symmetry in molecules, this must be due to weak intermolecular or other environmental perturbations. Such formulations provide attractive alternative means for deriving and visualizing molecular structure, treating electrons and protons on a similar footing, and without invoking the Born–Oppenheimer approximation.

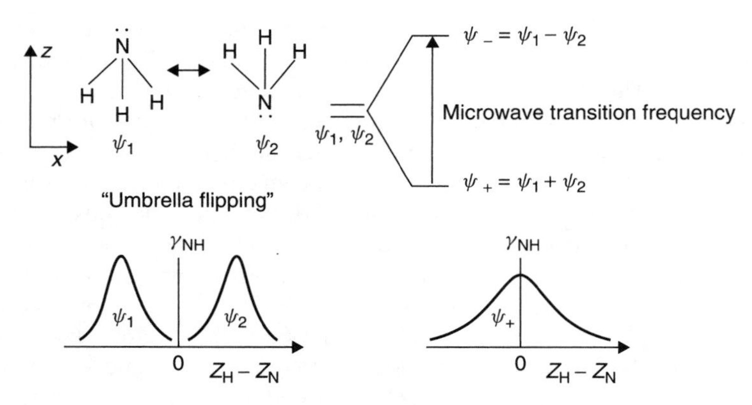

Figure 3.1 The structure and protonic spectrum of ammonia. The lower part of the figure shows the radial distribution functions for the protons as functions of the N–H distance [15]. (Reproduced with permission from Sukumar N. Found Chem 2009;11(1):7–20, Copyright 2009 Springer.)

3.2 SELF-CONSISTENT TREATMENT OF MANY-ELECTRON SYSTEMS

As early as 1922, Bohr [25] suggested that the electrostatic field through which an electron moves in an atom can be represented by a static distribution of the other electrons, or in other words by the averaged electron density of the remaining electrons. Thus, even before the discovery of the Schrödinger equation, many experimental results from atomic spectroscopy could be qualitatively understood by identifying the electronic energy levels with those of a single electron moving in the central field of the nucleus and other electrons. Hartree [26] then used this suggestion to develop a method for self-consistently including interelectronic repulsion in the Schrödinger equation. Writing the nonrelativistic electronic Hamiltonian for a one-electron atom of atomic number Z as

$$H_i(0) = -\frac{1}{2}\nabla_i^2 + \frac{Z}{r_i}, \tag{3.19}$$

where the subscript i labels the electron, the nonrelativistic electronic Hamiltonian for an N-electron atom is then

$$H = \sum_{i=1}^{N} H_i^{(0)} + \sum_{i<j} \frac{1}{r_{ij}}. \tag{3.20}$$

Here and in the remainder of this chapter, we use atomic units wherein the charge and mass of the electron are set to unity and further $\hbar = 1$. Assuming an approximate N-electron wavefunction, given as a simple product of one-electron functions (called *orbitals*),

$$\psi(1, 2, \ldots, N) = \prod_{i=1}^{N} \varphi_i(i), \tag{3.21}$$

Hartree demonstrated how to obtain these orbitals systematically. The electron i moves in an effective one-electron potential:

$$V_i^{\text{eff}} = \sum_{j \neq i}^{N} \int \frac{\varphi_j{}^*(j)\varphi_j(j)}{r_{ij}}\, \mathrm{d}r_j = \sum_{j \neq i} \left\langle \varphi_j(j) \left| \frac{1}{r_{ij}} \right| \varphi_j(j) \right\rangle, \tag{3.22}$$

where Dirac's bra and ket notation [27] of dual vector spaces has been used in the second identity. V_i^{eff} is the electrostatic interaction of the ith electron with the $N - 1$ other electrons. The ith electron is thus assumed to move in the potential field owing to the average charge distributions of all the other electrons in the atom. We now replace the interelectronic repulsion operator $\sum_{i<j} \frac{1}{r_{ij}}$ (second term in Equation 3.20) with the effective potential $\sum_{i=1}^{N} V_i^{\text{eff}}$. This enables the Schrödinger equation to be separated into contributions from individual electrons. The problem then reduces to the solution of N independent equations of the form

$$[H_i^{(0)} + V_i^{\text{eff}}]\varphi_i(i) = \varepsilon_i \varphi_i(i). \tag{3.23}$$

Defining the Coulomb operator

$$J_j(i) = \left\langle \varphi_j(j) \left| \frac{1}{r_{ij}} \right| \varphi_j(j) \right\rangle, \tag{3.24}$$

the effective potential becomes

$$V_i^{\text{eff}} = \sum_{j \neq i}^{N} J_j(i). \tag{3.25}$$

The Hartree equations then take the integrodifferential form

$$\left[H_i(0) + \sum_{j \neq i}^{N} J_j(i) \right] \varphi_i(i) = \varepsilon_i \varphi_i(i). \tag{3.26}$$

The effective potential V_i^{eff} in Equation 3.22 depends on the set of unknown orbitals $\{\varphi\}$, which are themselves obtained by solving the Hartee equations (Eq. 3.23). Hartree obtained both the orbitals and the effective potential by an iterative procedure, starting with a set of approximate orbitals $\{\varphi_i{}^{(0)}(i)\}$. This gives a first approximation to the Coulomb operator:

$$J_j^{(0)}(i) = \left\langle \varphi_j^{(0)}(j) \left| \frac{1}{r_{ij}} \right| \varphi_j^{(0)}(j) \right\rangle \tag{3.27}$$

and a first approximation to the effective Hamiltonian:

$$H_i{}^{(1)} = H_i{}^{(0)} + \sum_{j \neq i}^{N} J_j^{(0)}(i). \tag{3.28}$$

Solving the set of N equations

$$H_i{}^{(1)} \varphi_i(i) = \varepsilon_i{}^{(1)} \varphi_i{}^{(1)}(i) \tag{3.29}$$

gives the next level of approximation to the orbitals $\{\phi_i{}^{(1)}(i)\}$. These are then used to define the next level of approximation to the Coulomb operator:

$$J_j^{(1)}(i) = \left\langle \varphi_j^{(1)}(j) \left| \frac{1}{r_{ij}} \right| \varphi_j^{(1)}(j) \right\rangle, \tag{3.30}$$

which, in turn, leads to an effective Hamiltonian

$$H_i{}^{(2)} = H_i{}^{(0)} + \sum_{j \neq i}^{N} J_j^{(1)}(i), \tag{3.31}$$

and the next set of Hartree equations:

$$H_i{}^{(2)} \varphi_i(i) = \varepsilon_i{}^{(2)} \varphi_i{}^{(2)}(i), \tag{3.32}$$

and thus an improved set of orbitals $\{\varphi_i{}^{(2)}(i)\}$. The process is repeated until a set of orbitals yields an effective potential that, when used in the Hartree equations for the next level of approximation, yields back the same set of orbitals, a condition known as *self-consistency*. The electrons are now said to move in a self-consistent field (SCF) and the converged set of orbitals are known as *SCF orbitals*. The electronic energy of the N-electron atom in the Hartree SCF approximation is given by

$$E_{\mathrm{Hartree}} = \sum_{i=1}^{N} \langle \psi | H_i{}^{(0)} | \psi \rangle + \sum_{i<j} \left\langle \psi \left| \frac{1}{r_{ij}} \right| \psi \right\rangle. \tag{3.33}$$

If the SCF orbitals are normalized, the electronic energy in the Hartree SCF method is given by

$$E = \sum_{i=1}^{N} \varepsilon_i^{(0)} + \sum_{i<j} J_{ij}, \tag{3.34}$$

where

$$\varepsilon_i^{(0)} = \langle \varphi_i(i) | H_i^{(0)} | \varphi_i(i) \rangle \tag{3.35}$$

and J_{ij} is the Coulomb integral, defined as

$$J_{ij} = \langle \varphi_i(i) | J_j(i) | \varphi_i(i) \rangle = \langle \varphi_j(j) | J_i(j) | \varphi_j(j) \rangle = \left\langle \varphi_i(i)\varphi_j(j) \left| \frac{1}{r_{ij}} \right| \varphi_i(i)\varphi_j(j) \right\rangle, \tag{3.36}$$

J_{ij} represents the electrostatic repulsion between two charge clouds $|\varphi_i(i)|^2$ and $|\varphi_j(j)|^2$.

The electronic wavefunction in the Hartree method does not have the correct Fermi–Dirac statistics for a many-electron system, because the Hartree product (Eq. 3.21) is not antisymmetric with respect to permutation of electrons. Thus the Pauli principle is not satisfied. This was rectified by Fock [28] and Slater [29], who used an antisymmetrized product instead. Writing the wavefunction in the Slater determinant form (Eq. 2.36), we now have, in addition to the Coulomb integral (Eq. 3.36), an additional term arising from permutation of electrons in the same orbital, but with opposite spin. As these electrons differ in the spin quantum number, the Slater determinant does not vanish under such permutations. The additional term in the Hamiltonian is the exchange integral:

$$K_{ij} = \left\langle \varphi_i(1)\varphi_j(2) \left| \frac{1}{r_{12}} \right| \varphi_j(1)\varphi_i(2) \right\rangle, \tag{3.37}$$

which has no classical analog, and arises solely from the antisymmetry principle. The electronic energy in the Hartree–Fock approximation can be written as a sum over the $N/2$ spatial orbitals:

$$\begin{aligned} E_{\mathrm{HF}} &= 2\sum_{i=1}^{N/2} \varepsilon_i^{(0)} + \sum_{i<j}^{N/2} (4J_{ij} - 2K_{ij}) + \sum_{i=1}^{N/2} J_{ii} \\ &= 2\sum_{i=1}^{N/2} \varepsilon_i^{(0)} + \sum_{i,j}^{N/2} (2J_{ij} - K_{ij}) \end{aligned} \tag{3.38}$$

The second identity is obtained by noting that

$$J_{ij} = J_{ji} \qquad K_{ij} = K_{ji} \qquad J_{ii} = K_{ii}. \tag{3.39}$$

The Hartree–Fock equations can be obtained by considering arbitrary variations in the orbitals φ and employing the calculus of variations to give [30, 31]

$$[H_i^{(0)} + \sum_j (2J_j - K_j)]\varphi_i = \varphi_i \lambda_{ij}, \tag{3.40}$$

where the Coulomb and exchange integrals are now defined through the expressions

$$J_i(1)\varphi_j(1) = \left\langle \varphi_i(2) \left| \frac{1}{r_{12}} \right| \varphi_i(2) \right\rangle \varphi_j(1), \tag{3.41}$$

$$K_i(1)\varphi_j(1) = \left\langle \varphi_i(2) \left| \frac{1}{r_{12}} \right| \varphi_j(2) \right\rangle \phi_i(1), \tag{3.42}$$

and the Lagrange multipliers λ_{ij} are elements of a Hermitian matrix $\boldsymbol{\lambda}$, that is, $\lambda_{ij} = \lambda_{ji}{}^*$. The Hartree–Fock equations (Eq. 3.40) can be written in the convenient matrix form

$$\mathbf{F}\boldsymbol{\varphi} = \boldsymbol{\varphi}\boldsymbol{\lambda}, \tag{3.43}$$

where the Fock operator $\mathbf{F}$ is defined as

$$\mathbf{F} = H^{(0)} + \sum_j (2J_j - K_j) \tag{3.44}$$

and

$$\boldsymbol{\varphi} = [\varphi_i \; \varphi_2 \quad \cdots \quad \varphi_{N/2}]. \tag{3.45}$$

Diagonalization of the matrix $\boldsymbol{\lambda}$ yields the orbital energies ε

$$\varepsilon_i = \langle \varphi_i | F | \varphi_i \rangle = \varepsilon_i^{(0)} + \sum_j (2J_{ij} - K_{ij}). \tag{3.46}$$

3.3 DENSITY MATRICES AND ELECTRON CORRELATION

All terms in the nonrelativistic molecular Hamiltonian and the electronic Hamiltonian are either one-body- or two-body terms. Thus, knowledge of the full N-electron wavefunction is not necessary; one needs no more than the reduced two-electron density matrix for molecular electronic problems [32, 33]. We define the density operator using Dirac's bra and ket notation [27] as

$$\boldsymbol{\rho} = |\psi\rangle\langle\psi|, \tag{3.47}$$

and the full N-electron density matrix in coordinate representation as

$$\begin{aligned}\Gamma(\mathbf{r}_1, \mathbf{r}_2, \ldots, \mathbf{r}_N, \sigma_1, \sigma_2, \ldots, \sigma_N | \mathbf{r}'_1, \mathbf{r}'_2, \ldots, r'_N, \sigma'_1, \sigma'_2, \ldots, \sigma'_N) = \\ N!\psi^*(\mathbf{r}_1, \mathbf{r}_2, \ldots, \mathbf{r}_N, \sigma_1, \sigma_2, \ldots, \sigma_N)\psi(\mathbf{r}'_1, \mathbf{r}'_2, \ldots, \mathbf{r}'_N, \sigma'_1, \sigma_2, \ldots, \sigma'_N)\end{aligned} \tag{3.48}$$

Here, the spatial coordinates have been denoted by $\mathbf{r}$ and the spins by σ. The primes are used to indicate that operators act only on the primed variables; primed and unprimed variables are set equal to each thereafter. The density operator is Hermitian ($\boldsymbol{\rho} = \boldsymbol{\rho}^*$) and idempotent ($\boldsymbol{\rho}^2 = \boldsymbol{\rho}$). Integrating Γ over the coordinates of all but two electrons (and summing over their spins) yields the reduced two-electron density matrix $\Gamma^{(2)}$:

$$\Gamma^{(2)}(\mathbf{r}_1, \mathbf{r}_2, \sigma_1, \sigma_2|\mathbf{r}'_1, \mathbf{r}'_2, \sigma'_1, \sigma'_2) = N(N-1)\int d\mathbf{r}_3 \cdots \int d\mathbf{r}_N \sum_{\sigma_3} \cdots \sum_{\sigma_N} \psi^*(\mathbf{r}_1, \mathbf{r}_2, \ldots, \mathbf{r}_N, \sigma_1, \sigma_2, \ldots, \sigma_N)\psi(\mathbf{r}'_1, \mathbf{r}'_2, \ldots, \mathbf{r}'_N \sigma'_1, \sigma'_2, \ldots, \sigma'_N) \tag{3.49}$$

Similarly integrating Γ over the coordinates of all but one electron (and summation over their spins) yields the reduced one-electron density matrix $\Gamma^{(1)}$, a function of the coordinates and spin of a single electron:

$$\Gamma^{(1)}(\mathbf{r}_1\sigma_1|\mathbf{r}'_1\sigma'_1) = \gamma(\mathbf{r}|\mathbf{r}') = N \int d\mathbf{r}_2 \cdots \int d\mathbf{r}_\mathrm{N} \sum_{\sigma_2} \cdots \sum_{\sigma_N} \psi^*(\mathbf{r}_1, \mathbf{r}_2, \ldots, \mathbf{r}_N \sigma_1, \sigma_2, \ldots, \sigma_N)\psi(\mathbf{r}'_1, \mathbf{r}'_2, \ldots, \mathbf{r}'_N \sigma'_1, \sigma'_2, \ldots, \sigma'_N). \tag{3.50}$$

In an analogous manner, one can define the reduced density matrices in momentum space Π by considering the wavefunctions in the momentum representation and integrating over electronic momenta:

$$\prod^{(1)}(\mathbf{p}_1\sigma_1|\mathbf{p}'_1\sigma'_1) = N \int d\mathbf{p}_2 \cdots \int d\mathbf{p}_N \sum_{\sigma_2} \cdots \sum_{\sigma_N} \psi^*(\mathbf{p}_1, \mathbf{p}_2, \ldots, \mathbf{p}_N, \sigma_1, \sigma_2, \ldots, \sigma_N)\psi(\mathbf{p}'_1, \mathbf{p}'_2, \ldots, \mathbf{p}'_N, \sigma'_1, \sigma'_2, \ldots, \sigma'_N). \tag{3.51}$$

The diagonal elements of the first- and second-order density matrices are the electron density and the pair density, respectively

$$\rho(\mathbf{r}) = \gamma(\mathbf{r}|\mathbf{r}) = N \int d\mathbf{r}_2 \cdots \int d\mathbf{r}_N \sum_{\sigma} \cdots \sum_{\sigma_N} \psi^*(\mathbf{r}_1, \mathbf{r}_2, \ldots, \mathbf{r}_N, \sigma_1, \sigma_2, \ldots, \sigma_N)\psi(\mathbf{r}_1, \mathbf{r}_2, \ldots, \mathbf{r}_N, \sigma_1, \sigma_2, \ldots, \sigma_N), \tag{3.52}$$

$$\rho(\mathbf{r}_1, \mathbf{r}_2) = N(N-1) \int d\mathbf{r}_3 \cdots \int d\mathbf{r}_N \sum_{\sigma_1} \cdots \sum_{\sigma_N} \psi^*(\mathbf{r}_1, \mathbf{r}_2, \ldots, \mathbf{r}_N, \sigma_1, \sigma_2, \ldots, \sigma_N)\psi(\mathbf{r}_1, \mathbf{r}_2, \ldots, \mathbf{r}_N, \sigma_1, \sigma_2, \ldots, \sigma_N), \tag{3.53}$$

$$\pi(\mathbf{p}) = N \int d\mathbf{p}_2 \cdots \int d\mathbf{p}_N \sum_{\sigma} \cdots \sum_{\sigma_N} \psi^*(\mathbf{p}_1, \mathbf{p}_2, \ldots, \mathbf{p}_N, \sigma_1, \sigma_2, \ldots, \sigma_N)\psi(\mathbf{p}_1, \mathbf{p}_2, \ldots, \mathbf{p}_N, \sigma_1, \sigma_2, \ldots, \sigma_N). \tag{3.54}$$

Here, we have summed over all electron spins to give spin-free densities ρ, π and density matrices γ. Assuming the wavefunctions to be antisymmetric and normalized, the trace of the first-order density matrix gives the total number of electrons N, as does the integral over all space of the electron density:

$$\mathrm{Tr.}\gamma(\mathbf{r}|\mathbf{r}') = \int \rho(\mathbf{r})\, d\mathbf{r} = N, \tag{3.55}$$

$$\int d\mathbf{r}_1 \int d\mathbf{r}_2\, \rho(\mathbf{r}_1, \mathbf{r}_2) = N(N-1). \tag{3.56}$$

It is the electron densities $\rho(\mathbf{r})$ and $\pi(\mathbf{p})$ that will be the subject of the remainder of this book. $\rho(\mathbf{r})$ represents the probability of an electron being found at position $\mathbf{r}$; $\pi(\mathbf{p})$ the probability of an electron being found with momentum $\mathbf{p}$; $\rho(\mathbf{r}_1, \mathbf{r}_2)$ the probability of two electrons being simultaneously found, one at $\mathbf{r}_1$ and the other at $\mathbf{r}_2$. Unlike the wavefunction, the electron densities and density matrices, in position and momentum space, are amenable to direct experimental determination. In Section 3.4, we shall describe methods for experimental determination of $\rho(\mathbf{r})$.

As a consequence of the antisymmetry principle, the second-order density matrix (and all higher order matrices) is antisymmetric with respect to each set of electron indices:

$$\begin{aligned}\Gamma^{(2)}(\mathbf{r}_1, \mathbf{r}_2, \sigma_1, \sigma_2|\mathbf{r}_1', \mathbf{r}_2', \sigma_1', \sigma_2') &= -\Gamma^{(2)}(\mathbf{r}_2, \mathbf{r}_1, \sigma_2, \sigma_1|\mathbf{r}_1', \mathbf{r}_2', \sigma_1', \sigma_2') \\ &= -\Gamma^{(2)}(\mathbf{r}_1, \mathbf{r}_2, \sigma_1, \sigma_2|\mathbf{r}_2', \mathbf{r}_1', \sigma_2', \sigma_1'),\end{aligned} \tag{3.57}$$

whence

$$\begin{aligned}\Gamma^{(2)}(\mathbf{r}_1, \mathbf{r}_2, \sigma_1, \sigma_2|\mathbf{r}_1', \mathbf{r}_2', \sigma_1', \sigma_2') = 0 \quad &\text{for} \quad \{\mathbf{r}_1, \sigma_1\} = \{\mathbf{r}_2, \sigma_2\} \\ &\text{or} \quad \{\mathbf{r}_1', \sigma_1'\} = \{\mathbf{r}_2', \sigma_2'\}.\end{aligned} \tag{3.58}$$

This represents a Fermi hole for each electron at the position of any other electron of the same spin. Furthermore, because of Coulomb repulsion, $\Gamma^{(2)}$ should vanish even when two electrons of opposite spin $\sigma_1 \neq \sigma_2$ come to occupy the same spatial coordinates $\mathbf{r}_1 = \mathbf{r}_2$. This is known as the *Coulomb hole*.

It is apparent that all terms in the electronic Hamiltonian can be expressed in terms of the reduced second-order density matrix:

$$\langle T_e \rangle = -\frac{1}{2}\nabla_i^2 \gamma(\mathbf{r}|\mathbf{r}')|_{\mathbf{r}=\mathbf{r}'}, \tag{3.59}$$

$$\langle V_{ne} \rangle = \int d\mathbf{r} \sum_\alpha \frac{Z_\alpha \rho(\mathbf{r})}{|\mathbf{r} - R_\alpha|}, \tag{3.60}$$

$$\langle V_{ee} \rangle = \frac{1}{2} \int d\mathbf{r}_1 \int d\mathbf{r}_2 \frac{\rho(\mathbf{r}_1, \mathbf{r}_2)}{|\mathbf{r}_1 - \mathbf{r}_2|}. \tag{3.61}$$

The electronic kinetic energy depends only on the first-order density matrix, and the potential energy on the diagonal elements of the first- and second-order density matrices. In the Hartree–Fock approximation, the energy is given entirely in terms of the one-electron density matrix:

$$E_{\mathrm{HF}} = -\frac{1}{2}\nabla_i^2 \gamma(\mathbf{r}|\mathbf{r}')|_{\mathbf{r}=\mathbf{r}'} - \int \mathrm{d}\mathbf{r} \sum_{\alpha} \frac{Z_\alpha \rho(\mathbf{r})}{|\mathbf{r} - R_\alpha|} + \frac{1}{2}\int \mathrm{d}\mathbf{r}_1 \int \mathrm{d}\mathbf{r}_2 \frac{\rho(\mathbf{r}_1)\rho(\mathbf{r}_2) - \gamma(\mathbf{r}_1|\mathbf{r}_2)\gamma(\mathbf{r}_2|\mathbf{r}_1)}{|\mathbf{r}_1 - \mathbf{r}_2|}. \quad (3.62)$$

The Hartree–Fock approximation does not properly account for the tendency of electrons of opposite spin to avoid each other (the Coulomb hole); electrons of opposite spin behave, in this approximation, as quasi-independent particles, interacting only in the sense that each electron moves in the effective potential field owing to the other electrons. The tendency of electrons of the same spin to avoid each other (the Fermi hole) is accounted for through the antisymmetry of the Hartree–Fock wavefunction.

We now define [34] the correlation energy as "the difference between the exact eigenvalue of the Hamiltonian and its expectation value in the Hartree–Fock approximation." Thus for the nonrelativistic electronic Hamiltonian considered here, the correlation energy is the difference between the exact nonrelativistic energy (in general, an unknown quantity) and the Hartree–Fock energy. Correlation arises from the tendency of electrons to avoid each other beyond that represented by the Hartree–Fock approximation. While electrons of the same spin are kept apart by the antisymmetry principle, the average potential field in the Hartree–Fock approximation does not adequately separate electrons of opposite spin. The Hartree approximation, which neglects both exchange and correlation, does an equally poor job with paired as with unpaired electrons, but these errors (being of opposite sign) tend to cancel each other.

Popular methods of going beyond the Hartree–Fock approximation include configuration interaction (CI), perturbation theory, diagrammatic coupled cluster methods, and density functional theory (the subject of the next chapter). For a closed-shell system, the Hartree–Fock wavefunction takes the form of a Slater determinant, with each orbital occupied by either two electrons (occupied orbitals) or none (virtual orbitals [30]). Wavefunctions for excited states have more nodes than the ground state wavefunction. These nodes serve to keep electrons apart. Thus, an effective way of including correlation would be to mix in higher excited states. CI employs an expansion of the wavefunction for the electronic state of interest in terms of Slater determinantal configurations. Unfortunately, the convergence of the CI expansion is often disappointingly slow and a very large number of configurations are typically required to account for a significant fraction of the correlation energy. Constructing the density matrix from a CI wavefunction (or any other wavefunction beyond Hartree–Fock) and diagonalizing the first-order density matrix leads to a set of orbitals $\chi_i(\mathbf{r})$ (called *natural orbitals*) with nonintegral occupation numbers n_i:

$$\int \mathrm{d}\mathbf{r}\, \gamma(\mathbf{r}'|\mathbf{r})\chi_i(\mathbf{r}) = n_i \chi_i(\mathbf{r}). \quad (3.63)$$

Natural orbitals are thus eigenfunctions of the first-order density matrix; the occupation numbers n_i are the corresponding eigenvalues. Löwdin [32] showed that natural orbitals lead to the most rapid convergence of CI expansions.

3.4 EXPERIMENTAL DETERMINATION OF THE ELECTRON DENSITY

The most common experimental technique for the determination of electron densities is X-ray diffraction. X-rays were discovered by Wilhelm Conrad Röntgen in 1895, winning him the Nobel Prize in 1901, but it was Max von Laue who first discovered that electrons are capable of diffracting X-rays. von Laue was born in October 1879 at Pfaffendorf, near Koblenz, Germany, the son of Julius von Laue. After a semester at the University of Munich, he went to the University of Berlin in 1902 to work under Max Planck. Here, he attended lectures by Lummer on interference spectroscopy. After obtaining his doctorate in 1903, von Laue went to the University of Gottingen, where he was offered the post of assistant to Max Planck at the Institute for Theoretical Physics. He moved back to the University of Munich in 1909 and served as privatdozent there, before accepting appointment as professor of physics at the University of Zurich in 1912. The idea that crystals could be used as a diffraction grating arose during a conversation between Max von Laue and Paul Peter Ewald in 1912 in Munich. Using a crystal of zinc sulfide, von Laue in collaboration with Friedrich and Knipping found unequivocal evidence of regular diffraction of X-rays. For this work von Laue was awarded the Nobel Prize in 1914. von Laue served as professor of physics at the University of Berlin from 1919 to 1943, but was a vocal opponent of Nazi policies. When Nazi Germany invaded Denmark during World War II, the Hungarian chemist George de Hevesy dissolved the gold Nobel prizes of von Laue and James Franck in aqua regia to keep them hidden from the Nazis (export of gold was illegal at that time). The dissolved awards stayed undisturbed on a shelf in Hevesy's laboratory at the Neils Bohr Institute. After the war, Hevesy returned to Denmark and precipitated the gold out of the acid, whereupon the Nobel Foundation recast the Nobel prizes from the original gold.

The pattern of radiation scattered by any object, called the *diffraction pattern*, can be exploited to obtain the electron density and the crystal structure. X-rays are employed to produce the diffraction pattern because their wavelength λ is typically of the same order of magnitude (of the order of Ångstroms) as the spacing d between adjacent planes in the crystal. In crystals, nature has provided us with a diffraction grating ideally suited for the diffraction of X-rays. A beam of X-rays striking a crystal is scattered elastically (i.e., the scattered X-rays have the same wavelength as the incident X-rays) into many specific directions. von Laue observed that the reflected beams were only observed in certain directions. In studying this effect, William Lawrence Bragg realized that the origin of this selection of specific directions for the diffracted beam lies in the regularities of the crystal structure [35, 36]. While in most directions the scattered waves interfere

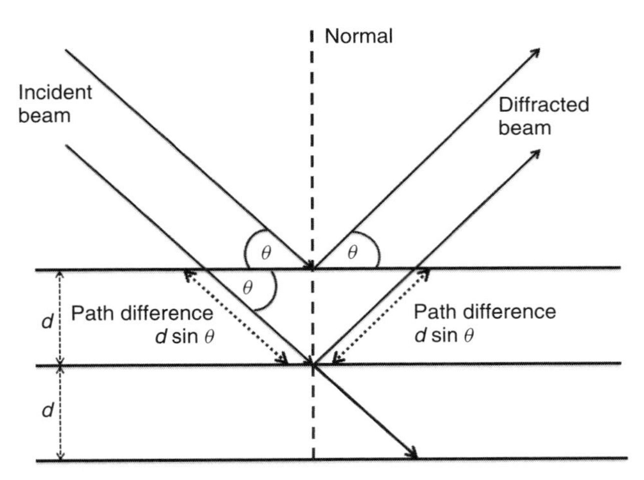

Figure 3.2 Illustration of the Bragg equation $n\lambda = 2d \sin \theta$. Since the path difference (causing phase differences) of waves scattered by two adjacent planes is $2d \sin \theta$, this must equal $n\lambda$ for reinforcement to occur to give a diffracted beam [44].

destructively with each other and hence cancel out, they add constructively in a few specific directions, as determined by Bragg's law:

$$n\lambda = 2d \sin \theta. \tag{3.64}$$

Here, d is the spacing between diffracting planes (Fig. 3.2), θ the incident angle, λ the wavelength of the beam, and n an integer. Waves scattered from adjacent lattice planes will be exactly in phase, that is the difference in the paths traveled by these waves will be an integral multiple $n\lambda$ of the wavelength, and will interfere constructively, only for those angles of scattering θ that satisfy Bragg's law. In analyzing the diffraction of X-rays by zinc sulfide, certain planes were found to reflect strongly, while other planes did not. Bragg explained this by assuming that the atoms in the crystal were arranged in a face-centered cubic lattice, rather than a simple cubic lattice. This was confirmed by Bragg and his father by analyzing several other crystal structures, both simple and complex [37–43]. The first atomic-resolution structure to be solved was that of sodium chloride in 1914 [37, 38], which revealed that the sodium and chlorine atoms are not associated as molecules, indicating that the atoms are ionized in the crystal structure and held together by electrostatic attraction of the oppositely charged ions. Bragg also solved the structure of diamond [39, 40], revealing the tetrahedral arrangement of its bonds. The Nobel Prize in physics was awarded in 1915 to the British father and son duo, Sir William Henry Bragg (the father) and William Lawrence Bragg (the son, also knighted in 1941, and at the age of 25, the youngest-ever Nobel laureate), "*for their service in the analysis of crystal structure by means of X-rays*".

Bragg's law does not interpret the relative intensities of the reflections. Thus to solve for the arrangement of atoms within the unit cell, a Fourier transform must be performed. In an X-ray diffraction measurement, a crystal is mounted on a goniometer and gradually rotated while being subjected to X-ray bombardment, which produces a diffraction pattern of regularly spaced dots (reflections). The two-dimensional diffraction patterns at different rotations are then converted, using Fourier transforms, into a three-dimensional model of the electron density $\rho(\mathbf{r})$ within the crystal:

$$\rho(\mathbf{r}) = \int \frac{\mathrm{d}\mathbf{q}}{(2\pi)^3} F(\mathbf{q})\, \mathrm{e}^{\mathrm{i}\mathbf{q}\cdot\mathbf{r}}. \tag{3.65}$$

The incident X-ray can be represented as a plane wave $A\mathrm{e}^{i\mathbf{k}\cdot\mathbf{r}}$, where $\mathbf{k}$ is the wave vector of the incident wave. At any position $\mathbf{r}$ within the sample, the density of scatterers $\rho(\mathbf{r})$ produces a scattered spherical wave with amplitude proportional to the local amplitude of the incident wave and to the number of scatterers in a small volume element $\mathrm{d}V$ around $\mathbf{r}$: that is, to $A\ \mathrm{e}^{i\mathbf{k}\cdot\mathbf{r}}\ \rho(\mathbf{r})\ \mathrm{d}V$. Consider a scattered wave with wave vector $\mathbf{k}'$ striking the detector at $\mathbf{r}'$. For elastic scattering $|\mathbf{k}| = |\mathbf{k}'|$. The change in the phase of the photon is thus $\mathrm{e}^{i\mathbf{k}'\cdot(\mathbf{r}'-\mathbf{r})}$. The net radiation arriving at $\mathbf{r}'$ is the sum of all scattered waves throughout the crystal: $A \int \mathrm{d}\mathbf{r}\rho(\mathbf{r})\mathrm{e}^{\mathrm{i}\mathbf{k}\cdot\mathbf{r}}\ \mathrm{e}^{\mathrm{i}\mathbf{k}'\cdot(\mathbf{r}'-\mathbf{r})} = A\ \mathrm{e}^{\mathrm{i}\mathbf{k}'\cdot\mathbf{r}'} \int \mathrm{d}\mathbf{r}\ \rho(\mathbf{r})\ \mathrm{e}^{\mathrm{i}(\mathbf{k}-\mathbf{k}')\cdot r} = A\ \mathrm{e}^{\mathrm{i}\mathbf{k}'\cdot\mathbf{r}'}\ F(\mathbf{q})$, where $\mathbf{q} = \mathbf{k}' - \mathbf{k}$. The measured intensity of reflected radiation is thus proportional to the square of the amplitude $|F(\mathbf{q})|^2$. The intensities of the reflections from an X-ray diffraction measurement yield the magnitudes $|F(\mathbf{q})|$, but not the corresponding phases. This is known as the *phase problem*. To obtain the phases, full sets of reflections are collected with known alterations to the scattering, using one of several methods (see below). Combining the magnitudes with the phases gives $F(\mathbf{q})$, the Fourier transform (Eq. 3.65) of which gives the electron density $\rho(\mathbf{r})$.

Nowadays, the structures of systems containing several hundred atoms involved in complicated structural permutations such as proteins can be solved using X-ray diffraction. X-ray crystallography of biological molecules was pioneered by Dorothy Crowfoot Hodgkin, who solved the structure of cholesterol in 1937, vitamin B_{12} in 1945, penicillin in 1954, and insulin in 1969 [45]; she was awarded the Nobel Prize in chemistry in 1964. Max Perutz and Sir James Cowdery Kendrew solved the first crystal structure of a protein, sperm whale myoglobin [46], for which they were awarded the Nobel Prize in chemistry in 1962. Since then, X-ray crystal structures of several tens of thousands of proteins and complexes of proteins with nucleic acids have been deposited in the protein data bank [47]. Indeed, there has hardly been a more prolific field of science, as the number of Nobel prizes in the field of X-rays (Table 3.1) indicates.

Before data collection can take place a suitable single crystal must be chosen. A suitable crystal must possess two attributes: uniform internal structure and proper size and shape. The first requirement is met if the crystal is pure at the

TABLE 3.1 Nobel Prizes in the Field of X-rays and/or Diffraction

1901	Wilhelm Conrad Röntgen, in physics, for the discovery of X-rays
1914	Max von Laue, in physics, for the discovery of X-rays by crystals
1915	William Henry Bragg and William Lawrence Bragg, in physics, for the determination of crystal structures using X-rays
1917	Charles Glover Barkla, in physics, for the discovery of the characteristic X-radiation of the elements
1924	Karl Manne Georg Siegbahn, in physics, for discoveries in the field of X-ray spectroscopy
1927	Arthur Holly Compton, in physics, for revealing the particle nature of X-rays in scattering experiments on electrons
1936	Peter Debye, in chemistry, for determining the molecular structures by X-ray diffraction in gases
1962	Max Ferdinand Perutz and John Cowdery Kendrew, in chemistry, for determining the structure of hemoglobin and myoglobin
1962	Francis Crick, James Watson, and Maurice Wilkins, in medicine, for their discoveries concerning the molecular structure of nucleic acids and their significance in information transfer in living material
1964	Dorothy Crowfoot Hodgkin, in chemistry, for the determination of the structure of penicillin and other important biochemical substances
1976	William N. Lipscomb, in chemistry, for the determination of the structures of boranes
1979	Allan M. Cormack and Godfrey N. Hounsfield, in medicine, for the development of computerized tomography
1981	Kai M. Siegbahn, in physics, for developing high resolution electron spectroscopy
1985	Herbert A. Hauptman and Jerome Karle, in chemistry, for the development of direct methods for X-ray crystallographic structure determination
1988	Johann Deisenhofer, Robert Huber, and Hartmut Michel, in chemistry, for the determination of protein structures crucial to photosynthesis
2009	Venkatraman Ramakrishnan, Thomas A. Steitz, and Ada E. Yonath, in chemistry, for studies of the structure and function of the ribosome
2011	Dan Shechtman, in chemistry, for the discovery of quasicrystals

molecular, atomic, or ionic level. The crystal must be a single crystal in that it should not be twinned or composed of microscopic subcrystals.

Crystals can be screened by examination with a polarizing microscope. If rotated about an axis normal to the polarizing material, the crystal should appear uniformly dark in all positions or be bright and extinguish, that is, appear uniformly dark, once every 90°. A suitable size of the crystal is generally 0.1–0.3 mm. As the intensities of the diffracted rays from a given crystal are proportional to the amount of material present in the specimen, there is an advantage in selecting as large a crystal as possible; however, because of absorption there is an optimum thickness to prevent a decrease in intensity and an increase in random errors. For single crystals, it is best to have the crystal mounted on the goniometer for proper alignment and centering; modern diffractometers do not require any orientation, only centering of the crystal.

The dimensions of the unit cell ($a, b, c, \alpha, \beta, \gamma$) may be found from the angles 2θ of deviations of given diffracted beams from the direction of the incident beam, as each value of 2θ at which a diffraction maximum is observed is a function only of the cell dimensions and of the radiation used.

3.4.1 Determination of the Unit Cell Constants and Their Use in Ascertaining the Contents of the Unit Cell

Cell dimensions may be determined, with radiation of a known wavelength, from values of 2θ for reflections of known indices, where 2θ is the deviation from the diffracted beam. The Bragg equation is then used.

Example: Monoclinic cell, $\alpha = \gamma = 90°$, $\beta = 100.12°$, $\sin\beta = 0.98445$, $\lambda = 1.5418$ Å.

h	K	L	$2\theta(°)$	$\theta(°)$	$\sin\theta$	$n\lambda/2\sin\theta$ (Å)	
20	0	0	85.68	42.84	0.67995	22.675	
22	0	0	96.82	48.41	0.74791	22.676	d_{100}
0	4	0	47.41	23.705	0.40203	7.670	d_{010}
0	0	10	104.14	52.07	0.78876	9.774	d_{001}

Unit cell b is perpendicular to the plane of the paper. d_{hkl} is the space between the crystal planes hkl.

At this time, it is wise to measure the density of the crystal. A technique that can be used to measure the density of the crystal is the flotation method. This consists of suspending the crystal in a mixture of liquids, one lighter and one heavier than the crystal and adjusting the proportion of the liquids dropwise until the crystal remains suspended in the medium.

Let W be the weight in grams of 1 gram-formula weight of the contents of the unit cell and V be the volume in cubic centimeters of this weight of the crystal.

$$\text{Cell volume} = 1726\ \text{Å}^3 = 1726 \times 10^{-24}\ \text{cm}^3$$

$$\text{Observed density (by flotation)} = 1.34\ \text{g/cm}^3$$

$$N_{\text{Avog}}\ \text{unit cells occupy}\ 1726 \times 10^{-24} \times 6.02 \times 10^{23}\ \text{cm}^3 = V = 1039\ \text{cm}^3$$

$$\text{Density} = W/V = W/(1726 \times 0.602)\ \text{g/cm}^3 = 1.34\ \text{g/cm}^3$$

Therefore,

$$W = 1.39 \times 10^3\ \text{g/cm}^3$$

but

$$W = (ZM + zm)$$

where Z is the number of molecules of the compound (molecular weight M) per unit cell, and z is the number of molecules of the solvent of crystallization (molecular weight m) per unit cell. In this example, M is known to be 340 and $m = 18$ (for water):

$$(Z \times 340) + (z \times 18) = 1.39 \times 10^3$$

The monoclinic symmetry of the unit cell suggests that Z is 4 or a multiple of 4, leading to the conclusion that $Z = 4$ and $z = 2$ ($W = 1396$) is the correct solution and that the solution $Z = 3$ and $z = 20$ ($W = 1380$), which is equally probable from the calculated weight alone, is much less likely.

Having obtained this preliminary information about the crystal, data collection can now commence. The result of the collection of X-ray diffraction data is a relative intensity I for each reflection with indices, hkl, together with the corresponding value of the scattering angle 2θ for that reflection. All the values of I are on the same relative scale. The angular positions at which the scattered radiation is observed (related to the scattering angle 2θ) depend only on the dimensions of the crystal lattice, while the intensities of the different diffracted beams depend only on the nature and arrangement of atoms within the unit cell. Each diffracted beam contains information on the entire atomic structure of the crystal and structure determination involves a matching of the observed intensity pattern to that calculated from a postulated model. If the atomic arrangement in a crystal is known, the intensities of reflections in the diffraction pattern can be calculated and relative phases of these reflections are computed at the same time. However, when the diffraction pattern is measured, phase information is not obtainable. When an X-ray diffraction pattern is intercepted by a photographic film or some other detecting devices, the phase relationships are lost; only the amplitudes of the diffracted beams are known. The task of the crystallographers is to recombine these waves mathematically with approximately correct phases to give an image of the structure that scattered them. To compute an electron density map and hence determine the crystal structure, phases must be calculated from a "trial structure" together with the measured intensities. In order to represent the diffracted waves, the exponential form $c_{\mathrm{r}}e^{i\alpha_{\mathrm{r}}}$) may be used to represent the total scattering, where the amplitude of the wave is c_{r} and the phase angle is α_{r}. This complex representation is merely a convenient way of representing two orthogonal vector components in one equation. As the electrons are the only components of the atom that scatter X-rays significantly and because they are distributed over atomic volumes with dimensions comparable to the wavelengths of X-rays used in structure analysis, X-rays scattered from one part of an atom interfere with those scattered from another at all angles of scattering greater than 0°. Only at $2\theta = 0$, all electrons in the atom scatter in phase, and the scattering power of the atom at this angle, expressed relative to the scattering power of a free electron, is equal to the number of electrons present. The amplitude of the scattering for an atom is known as the *atomic scattering factor* or *atomic form factor* and is symbolized as f. For most purposes in structure analysis, it is adequate to assume that atoms are spherically symmetrical, but with good data small departures from spherical symmetry attributable to covalent bonding are detectable. This means that the scattering by an assemblage of atoms—that is, by the structure—can be very closely approximated by summing the contributions to each scattered wave from each atom independently, taking appropriate account of the differences in phase. As the diffraction pattern is the sum of the scattering from all unit cells, and thus represents the average content of a single one of these unit cells, vibrations or disorder may be considered the equivalent of the smearing out of the electron density, so that there is a greater fall-off at a higher $\sin\theta/\lambda$. Neutrons are scattered by atomic nuclei, rather than by electrons around a nucleus, and, because the nucleus is so small (relative to the atom), the scattering for a nonvibrating nucleus is almost independent of the scattering angle.

3.4.1.1 Scattering by a Crystal The X-radiation scattered by one unit cell of a structure in any direction in which there is a diffraction maximum has a particular combination of amplitude and phase, known as the *structure factor* and symbolized by F or $F(hkl)$. It is measured relative to the scattering by a single electron and is the Fourier transform of the scattering density (electrons in the molecule) sampled at the reciprocal lattice point *hkl*. The intensity of the scattered radiation is proportional to the square of the amplitude $|F^2|$. The fall off in intensity with high scattering angle increases as the vibrations of atoms become greater, and these vibrations, in turn, increase with rising temperature. If the vibration amplitude is sufficiently high, essentially no diffracted intensity will be observed beyond some limiting value of the scattering angle; that is, the "slit" is effectively widened by the vibration and so the "envelope" is narrow.

After the diffraction pattern has been recorded and measured in some approximate manner, the next stage is solving the structure—that is, finding a suitable trial structure (approximate positions of most atoms in a unit cell of known dimensions and space group). This trial structure should be close enough to the true structure so that it can be smoothly refined to a good fit to the data set. This is done by the "direct methods"-analytical techniques for deriving an approximate set of phases, from which a first approximation to the electron density map can be calculated. Interpretation of this may then give a suitable trial structure. "Direct methods" make use of the fact that the intensities of reflections contain structural information and that the electron density of a real crystal cannot be negative. In practice, analytical methods of phase determination are carried out on "normalized structure factors"—that is, values of the structure factor $|F|$ modified to remove the falloff in the individual scattering factors f with increasing scattering angles. Once a table of electron density $|E|$ values has been prepared, it is usual to rank these E values in decreasing order of magnitude and work with the strongest 10% or so to calculate an E-map, which is an electron density map calculated with E values. If all has gone well the structure will be clear in this map. Sometimes, only part of the structure is revealed in an interpretable way and the rest may be found from successive electron density maps. Generally, these "direct methods" result in a structure that can be refined and so the structure may be considered to be determined.

3.4.1.2 Derivation of Trial Structures The intensity information that is obtained in measuring the diffraction pattern of a crystal can be analyzed in other ways than "direct methods". These methods are of particular use in the determination of certain structures with high symmetry within the asymmetric unit and in the determination of the structures of biological macromolecules such as proteins and polynucleotide molecules. Two other methods are the Patterson method and the isomorphous replacement method. It is recommended that the Patterson map of any structure with possible ambiguity from "direct methods" be determined to see if it is consistent with the proposed trial structure.

3.4.1.3 Patterson Map A powerful method of analysis of the intensity distribution in the diffraction pattern can be the study of the Patterson $|F|^2$ map. The technique is of great value in unraveling some complex structures, especially those of macromolecules and other molecules containing heavy atoms or into which heavy atoms can be readily substituted. The Patterson method consists of evaluating a Fourier series for which only the indices and the $|F|^2$ value of each diffracted beam are needed; these quantities are directly derivable from the primary experimental quantities—that is, the directions and intensities of the diffracted beams:

$$P(u, v, w) = \frac{1}{Vc} \sum_{\text{all}} \sum_{h,k,l} \sum |F|^2 \cos 2\pi(hu + kv + lw). \tag{3.66}$$

There is only one Patterson function $P(u, v, w)$ for a given crystal structure. The function is evaluated at each point u, v, and w of a three-dimensional grid that fills the space with the size and shape of a unit cell. No phase information is required for this map because $|F|^2$ is independent of phase. The Patterson function $P(u, v, w)$ at points u, v, and w is the sum of the appearances of the structure when one views it from each atom in turn. It may be considered to be obtained by multiplying the electron density at points x, y, and z with that at $x + u$, $y + v$, and $z + w$ and adding the resulting products for all values of x, y, and z. Thus the Patterson function at points u, v, and w may be thought of as a convolution of the electron density at all points (x, y, and z) in the unit cell with the electron density at points $x + u$, $y + v$, $z + w$. If any two atoms in the unit cell are separated by a vector (**u,v,w**), then there will be a peak in the Patterson map at u, v, and w. Thus the orientation and length of every interatomic vector in the structure is represented in the Patterson map.

The contributions of individual interatomic attractions to the heights of the peaks in this three-dimensional map are approximately proportional to the values of $Z_i Z_j$, where Z_i is the atomic number of the atom at one end of the vector and Z_j is that of the atom at the other end. In general, the value of P at every point u, v, and w corresponds to the sum of the situations at the ends of such a vector as it is laid down with its origin at every possible point in the structure. The usefulness of the map decreases markedly with complicated structures composed of many atoms of about equal atomic number. With crystals of very large molecules, such as proteins, the overlap becomes hopeless to resolve, except for the peaks arising from the interactions between atoms of very high atomic number. If a structure containing a complex molecule with a multitude of vectors contains a group for which the vectors are known (relative to one another) rather precisely—for example, a benzene ring in a phenyl derivative, then the vector map can be calculated and the arrangement of vectors can be compared with the arrangement of vectors in the original Patterson map. The fit of the calculated and observed Patterson maps can be optimized with a computer by making a rotational search to examine all possible orientations of one map with respect to the other.

3.4.1.4 The Heavy-Atom Method In the *heavy-atom method*, one or a few atoms in the structure have atomic number Z_i considerably greater than those of the other atoms present. The method is based on the premise that if one atom has a much greater atomic scattering factor than the others, then the phase angle for the whole structure will seldom be far from that of the single atom alone. A way of using this method, if the molecule of interest does not contain such an atom is by preparing a derivative containing, for example, bromine or iodine with the hope that the molecular structural features of interest will not be modified in the process. Heavy atoms can be usually located by analysis of a Patterson map, although this depends on how many are present and how heavy they are relative to the other atoms present. Some data relevant to an organic compound containing Co, a derivative of vitamin B_{12} with the formula $C_{45}H_{57}O_{14}N_5CoCl{\cdot}C_3H_6O{\cdot}3H_2O$, are given below.

The derivative crystallizes in the space group $P2_12_12_1$ and the atomic numbers of the atoms are Co 27, Cl 17, O 8, N 7, C 6, and H 1, respectively. The expected approximate relative heights of the typical peaks in the Patterson map are

$$\text{Co–Co} = 27 \times 27 = 729$$
$$\text{Co–Cl} = 27 \times 17 = 459$$
$$\text{Cl–Cl} = 17 \times 17 = 289$$
$$\text{Co–O} = 27 \times 8 = 216$$
$$\text{Co–C} = 27 \times 6 = 162$$
$$\text{O–O} = 8 \times 8 = 64$$
$$\text{H–H} = 1 \times 1 = 1$$

This map will then be dominated by the Co–Co and the Co–Cl vectors. The atomic positions are then known in the space group, and the interatomic vectors (**u,v,w**) between symmetry related atoms can then be formulated. Once the heavy atom has been located the assumption is made that it dominates the diffraction pattern, and the phase angle for each diffracted beam for the whole structure is approximated by that for the heavy atom. The atomic positions are then found from the Patterson map and a comparison between observed peak positions with the general expectation gives the atomic coordinates.

3.4.1.5 Isomorphous Replacement Method This is the best method for the experimental determination of phase angles and is a very practical method for solving very large structures such as proteins. If two crystals have the same space group and their unit cells and atomic arrangements are identical, they are isomorphous. If atoms are added or replaced in such a system, the added or replaced atoms may be found from Patterson maps and if the atoms are sufficiently heavy, differences in the intensities of the two isomorphs can be used to determine the approximate phase angle for each reflection. Thus, small differences in electron densities in isomorphous molecules can be used to determine the crystal structures. The existence of isomorphism between a protein and a heavy atom derivative may be demonstrated by the determination that their unit cell dimensions do not differ by more than about 0.5% and that there are differences in the diffraction intensity patterns. The structures of many proteins can thus be studied with great success. With noncentrosymmetric structures, the situation is greatly complicated by the fact that the phase angle may have any value from 0° to 360°. Here, insertion of a bulky heavy atom causes a displacement of the rest of the structure and thus the structure determined.

3.4.1.6 Anomalous Dispersion and Absolute Configuration Even when the chemical formula and the three-dimensional structure of a molecule such as tartaric acid are known there is ambiguity about the absolute configuration. Information about the absolute configuration is not contained in the diffraction pattern of the crystal as it is normally measured. The means of determining the absolute configuration of molecules can be provided by X-ray crystallographic studies. The absorption coefficient of an atom for X-rays shows discontinuities when plotted as a function of the incident X-radiation. These discontinuities, also termed *absorption edges*, are sufficient to excite an electron in a strongly absorbing atom to a higher quantum state or to eject the electron completely when the energy of the X-radiation is at or below but near the absorption edge. This has an effect on the phase change on scattering. The scattering factor for the atom becomes complex and the factor f is replaced by

$$f_i + \Delta f_i' + \mathrm{i}\Delta f_i'. \tag{3.67}$$

Thus, if an atom in the structure absorbs, at least moderately, the X-rays being used, then this absorption will result in a phase change for the X-rays scattered by that atom–which is equivalent in its effect to changing the path length through which the scattered radiation travels. As a result, there is an effect on the intensities. In 1930, Koster, Knol, and Prins were able to determine the absolute configuration of a zinc blende (ZnS) crystal. It was seen that the shiny (111) faces have layers of sulfur atoms on the surface and the dull (111) faces have layers of zinc atoms on the surface.

3.4.2 Methods of Charge Density Analysis

In solving their early crystal structures, the Braggs assumed that X-rays are diffracted from the center of each atom (as if the entire electron density is concentrated there) and, further, that the diffraction intensity is proportional to the atomic number. These are very good approximations because, for most heavy atoms, the electron densities are strongly peaked about the nuclear positions, and the angular variations around each nucleus are very small when compared to the total electron density. Nevertheless, high resolution X-ray diffraction measurements are capable of generating detailed maps of the electron density distributions and revealing subtle density variations as a consequence of differences in chemical environment. Topological analysis of the total charge density has been exploited to obtain net atomic moments, including charges, and to infer the nature of the chemical bonding directly from the electron density distributions. X-ray diffraction is now a unique tool for mapping the charge distribution and thereby elucidates the structures and chemistry of crystals.

Electron density deformation maps have provided a wealth of qualitative information on bonding. Such density deformation maps are defined as the difference between the experimental density and the promolecule density (the density corresponding to a superposition of spherical atoms [48]), both calculated by Fourier summation. Experimental electron density deformation maps are thermally smeared with the internal and external modes in the crystal. Thermal effects can be minimized by performing the experiments at the lowest possible temperatures, at which almost all modes are reduced to zero point motion. In the Hansen Coppens formalism [49], the experimental X-ray structure factors are typically fitted with core functions and an atom-centered expansion of multipolar (spherical harmonic) valence density functions [49]. The atoms as defined by the sum of the nucleus-centered multipoles are often referred to as *pseudoatoms*. Real spherical harmonics describe the anisotropy of the valence electron density through multipole expansion. The atomic core electron density is commonly fixed during the electron density fitting, assuming no perturbation of the core density due to chemical bonding.

3.5 CONCLUDING REMARKS

The electron density concept is central to modern chemistry. In subsequent chapters, we discuss how the study of electron density in position and momentum space, and properties derived from it, can be used to partition the molecular properties into those of atoms and functional groups and to estimate the molecular similarity. We also delve into density functional theory, its conceptual aspects, and the applications of these concepts to surfaces and interfaces and nanomaterials.

REFERENCES

1. Born M, Oppenheimer JR. Ann Phys 1927;84:457–484.
2. Born M. Nachr Akad Goettingen Math-Phys Kl 1951;6:1–3.
3. Born M, Huang K. *Dynamical Theory of Crystal Lattices*. Oxford: Clarendon Press; 1954.
4. Woolley RG. Isr J Chem 1980;19:30–46.
5. Woolley RG. Adv Phys 1976;25:27–52.
6. Woolley RG, Sutcliffe BT. Chem Phys Lett 1977;45(2):393–398.
7. Woolley RG. Chem Phys Lett 1978;55:443–446.
8. Woolley RG. J Am Chem Soc 1978;100:1073–1078.
9. Sutcliffe BT. J Mol Struct (Theochem) 1992;259:29–58.
10. Sukumar N. Born couplings in H_2^+, H_2 and H_3. [PhD dissertation]. Stony Brook: State University of New York; 1984.
11. Zygelman B. Phys Lett A 1987;125:476–481.
12. Neuheuser Th, Sukumar N, Peyerimhoff SD. Chem Phys 1995;194:45–64.
13. Essén H. Int J Quantum Chem 1977;12:721–735.
14. Essén H. In: Hinze J, editor. *Energy Storage and Redistribution in Molecules*. New York: Plenum; 1983. p 327–336.
15. Sukumar N. Found Chem 2009;11(1):7–20.
16. Nakai H. Int J Quantum Chem 2002;86:511–517.
17. Nakai H, Hoshino M, Hyodo S. J Chem Phys 2005;122:164101.
18. Hoshino M, Nakai H. J Chem Phys 2006;124:194110.
19. Thomas IL. Phys Rev 1969;185:90–94.
20. Thomas IL. Phys Rev A 1970;2:72–76.
21. Thomas IL. Phys Rev A 1970;2:1675–1680.
22. Thomas IL. Phys Rev A 1971;3:565.
23. Thomas IL. Phys Rev A 1972;5:1104–1110.
24. Thomas IL, Joy HW. Phys Rev A 1970;2:1200.
25. Bohr N. Philos Mag 1922;9:1.
26. Hartree DR. Proc Camb Philol Soc 1928;24:89.
27. Dirac PAM. *The Principles of Quantum Mechanics*. Oxford: Oxford University Press; 1930.
28. Fock V. Z Phys 1930;61:126–148.
29. Slater JC. Phys Rev 1930;35:210–211.
30. Roothaan CCJ. Rev Mod Phys 1951;23:69–89.
31. Roothaan CCJ, Hall GG. Proc R Soc London, Ser A 1951;205:541–542.
32. Löwdin PO. Phys Rev 1955;97:1474–1489.
33. McWeeny R. Rev Mod Phys 1960;32:335–369.
34. Löwdin PO. Adv Chem Phys 1959;2:207–322.
35. Bragg WL. Nature 1942;90(2250):410.
36. Bragg WL. Proc Camb Philol Soc 1913;17:43–57.
37. Bragg WL. Proc R Soc London, Ser A 1913;89(610):248–277.

38. Bragg WL, James RW, Bosanquet CH. Philos Mag 1921;41(243):309–337; 1921;42(247): 1; 1922;44(261):433–449.
39. Bragg WH, Bragg WL. Nature 1913;91(2283): 557.
40. Bragg WH, Bragg WL. Proc R Soc London, Ser A 1913;89(610):277.
41. Bragg WL. Philos Mag 1914;28(165):355–360.
42. Bragg WL. Proc R Soc London, Ser A 1914;A89(613):468–489.
43. Bragg WH. Philos Mag 1915;30(176):305–315.
44. Glusker JP, Trueblood KN. *Crystal Structure Analysis: A Primer*. 2nd ed.. Oxford: Oxford University Press; 1985. p 16.
45. Crowfoot Hodgkin D. Nature 1935;135(3415):591.
46. Kendrew JC, Bodo G, Dintzis HM, Parrish RG, Wyckoff H Phillips DC. Nature 1958;181(4610):662–666.
47. Available at http://wwpdb.org. Accessed 2012.
48. Hirshfeld FL. Theor Chim Acta 1977;44:129–138.
49. Coppens P, Iversen B, Larsen FK. Coord Chem Rev 2005;249:179–195.

4

ATOMS IN MOLECULES

N. Sukumar

The concept of an atom or a functional group in a molecule is central to chemical theory, predating our understanding of quantum mechanics, atomic structure, and the nature of chemical bonding. Understanding the properties of atoms and functional groups of atoms, as molecules undergo transformations and combinations, is central to the science of chemistry. We recognize that an atom does not lose every shred of its identity when it combines with another atom or group of atoms to form a molecular or condensed phase assembly. Were it otherwise, chemistry would not be a science, but merely an encyclopedic catalog of the properties of disparate molecules. While perfect transferability of an atom between different molecules is an unattainable limit [1, 2], it is the quasi-invariant subset of properties retained by an atom in different chemical environments [2, 3] that we seek when we refer to the characteristics of an "atom in a molecule" (AIM).

The term *atom* is a loaded one in chemistry and is used in multiple contexts. Chemists often talk of carbon and hydrogen in organic molecules, or of a halogen displacing hydrogen in a reaction. While philosophers of chemistry have criticized this loose usage of the term atom [4–8], it does not generally cause confusion among chemists; it is understood that what is meant in this context is not the bare atom or the elemental atom, but the "atom in the molecule." Nevertheless, while we may intuitively understand the concept of an AIM, its rigorous definition is not without controversy. Various prescriptions have been proposed and the merits and shortcomings of each have been debated spiritedly in the literature. In this chapter, several schemes for partitioning a molecule into atoms are explored with a view to deriving chemical insight from the AIM concept. The first requirement of an AIM might be that it contains the nucleus of the free atom. Another would

A Matter of Density: Exploring the Electron Density Concept in the Chemical, Biological, and Materials Sciences, First Edition. Edited by N. Sukumar.

be that the electron density of the AIM would reduce to that of the free atom as the other atoms are stripped away. For a many-electron atom, the core electron density would be more or less unperturbed compared to that of the atom in its free state; one would thus expect the core electron density to be carried largely intact into the atom in its molecular environment. It is in the valence regions of atoms that we should expect to see variations between atoms in different molecular environments.

In general, if $\rho(\mathbf{r})$ is the electron density at position $\mathbf{r}$ in the molecule and we associate a density $\rho_{\mathrm{A}}(\mathbf{r})$ with every atom A in the molecule, we can define a weight function $w_{\mathrm{A}}(\mathbf{r})$ such that

$$\rho_{\mathrm{A}}(\mathbf{r}) = w_{\mathrm{A}}(\mathbf{r})\rho(\mathbf{r}). \tag{4.1}$$

We require, of course, that the AIM densities add up to the total molecular density $\rho(\mathbf{r})$ at every point $\mathbf{r}$:

$$\sum\nolimits_{\mathrm{A}} \rho_{\mathrm{A}}(\mathbf{r}) = \rho(\mathbf{r}), \tag{4.2}$$

where the summation is over all atoms in the molecule. From Equations 4.1 and 4.2, we get

$$\sum\nolimits_{\mathrm{A}} w_{\mathrm{A}}(\mathbf{r}) = 1. \tag{4.3}$$

One can also define an atomic electron population N_{A} for each AIM:

$$N_{\mathrm{A}} = \int \rho_{\mathrm{A}}(\mathbf{r})\mathrm{d}\mathbf{r}, \tag{4.4}$$

and a corresponding atomic charge $Z_{\mathrm{A}} - \mathrm{N}_{\mathrm{A}}$, where Z_{A} is the nuclear charge of atom A. As the total density in the molecule integrates to the total number of electrons N

$$N = \int \rho(\mathbf{r})\ \mathrm{d}\mathbf{r}, \tag{4.5}$$

we have

$$\sum\nolimits_{\mathrm{A}} N_{\mathrm{A}}(\mathbf{r}) = N. \tag{4.6}$$

4.1 CRITICAL POINTS OF THE ELECTRON DENSITY

Concentration of electron density around the nuclei is the most obvious feature of X-ray diffraction patterns of crystals. There is usually a maximum or a cusp in the electron density at the position of each nucleus (the electron density has a cusp at the position of the nucleus in the point nucleus approximation, with the slope of the density at the cusp proportional to the atomic number $\nabla\rho(\mathbf{r})|_{\mathbf{r}=0}\ \alpha - Z$; if the finite size of the nucleus is taken into account, the electron density has a maximum with zero slope). This is the feature that makes it possible to deduce the molecular structure from visual inspection of X-ray diffraction patterns. It also forms the basis for the quantitative determination of crystal structure from diffraction experiments: the crystal structure is determined by least squares fitting of the amplitudes of the calculated structure factors to the experimentally observed ones (Chapter 3). Owing to their low atomic number, hydrogen atoms often have shallow maxima, which makes them hard to be identified in X-ray diffraction patterns (this is especially so in low resolution X-ray diffraction spectra of large molecules such as proteins), and sometimes an electron density maximum may be missing altogether in the case of a hydrogen atom bonded to a highly electronegative atom. But in all other cases, nuclei are associated with pronounced maxima/cusps in the electron density (Fig. 4.1a). The electron density in an isolated atom is a decreasing function as one moves away from the nucleus in any direction, and this feature is generally retained when atoms combine to form molecules (except for some hydrogen atoms bonded to highly electronegative atoms, as mentioned above). The nuclei in a molecule thus function as attractors of the electron density gradient vector field $\nabla\rho(\mathbf{r})$.

There are some special situations where the electron density may come to a three-dimensional maximum at points that are not atomic nuclei. Molecules with such nonnuclear maxima have been extensively studied (see below), but, in general, the electron density comes to a three-dimensional maximum (i.e., a maximum along any direction in $\mathcal{R}^3$ space) only at the nuclear positions. The steepest descent paths of the electron density form the gradient vector field $-\nabla\rho(\mathbf{r})$, which often suffices to reveal the pattern of covalent bonding in a molecule. Most paths of the gradient vector field originate at a nucleus and terminate at infinity (where the density decays to zero), but there is a path of $\nabla\rho(\mathbf{r})$ connecting each pair of bonded atoms (Fig. 4.1b). Along this path (called the *bond path* [9]), the density decreases away from either nucleus and comes to a minimum along the bond path $\nabla\rho(\mathbf{r})|_{\mathbf{r}_c} = 0$ at a point known as the *bond critical point* ($\mathbf{r}_c$). The bond critical point is a saddle point of the electron density: here the electron density is a minimum along the bond path and a maximum in the plane perpendicular to it.

Two other types of critical points, each with vanishing gradient of the electron density $\nabla\rho(\mathbf{r})|_{\mathbf{r}_c} = 0$, can be identified from the topology of the electron density field: at a ring critical point, the electron density is a minimum in the plane of the ring and a maximum perpendicular to the ring plane (Fig. 4.1a and b); at a cage critical point, the electron density is a minimum along any direction. Critical points of the electron density may thus be classified by the rank (number of eigenvalues) and signature (sum of the signs of the eigenvalues) of the Hessian of the electron density ($\nabla\nabla\rho$):

- Nuclear [3, −3] critical points: three negative eigenvalues of $\nabla\nabla\rho$ at $\mathbf{r}_c$ and $\nabla^2\rho(\mathbf{r})|_{\mathbf{r}_c} < 0$. While [3, −3] critical points are most often found at the

Figure 4.1 (a) Electron density contours of benzene. Nuclei are associated with pronounced maxima/cusps in the electron density. At a ring critical point, the electron density is a minimum in the plane of the ring and a maximum perpendicular to the ring plane. (b) Paths of the gradient vector field $\nabla\rho(\mathbf{r})$, showing also the special paths connecting each pair of bonded atoms. (c) Contours of the Laplacian $L(\mathbf{r})$ of the electron density of benzene. (d) Contours of the electron density and its (e) Laplacian for azulene.

positions of atomic nuclei, they are also occasionally found at other points in space, referred to as *nonnuclear attractors (NNA)*.

- Bond [3, −1] critical points: two negative and one positive eigenvalues of $\nabla\nabla\rho$ at $\mathbf{r}_c$.
- Ring [3, +1] critical points: one negative and two positive eigenvalues of $\nabla\nabla\rho$ at $\mathbf{r}_c$.
- Cage [3, +3] critical points: three positive eigenvalues of $\nabla\nabla\rho$ at $\mathbf{r}_c$ and $\nabla^2\rho(\mathbf{r})|_{\mathbf{r}_c} > 0$.

The numbers and types of each kind of critical points that can coexist in a molecule are governed by the Poincare–Hopf relationship:

$$n - b + r - c = 1, \tag{4.7}$$

where n is the number of [3, −3] critical points (nuclei and NNA when they exist), b is the number of [3, −1] critical points (bonds), r is the number of [3, +1] critical points (rings), and c is the number of [3, +3] critical points (cages). These topological features of the gradient vector field of the electron density are the fundamental quantities employed by Bader [9–12] in the construction of AIMs (QTAIM).

4.2 VIRIAL PARTITIONING OF THE ELECTRON DENSITY

We have thus far made no specification of the choice of the partition functions $w_A(\mathbf{r})$ that divide molecules into AIMs, or even whether they may take continuous real values or only discrete integer values. An attractive way to explore the properties of atoms in different chemical environments is to investigate the topology of the total molecular electron density. This is the approach pioneered by Richard Bader and his group [9–14] who, over the course of several decades, went on to develop a systematic quantum theory of atoms in molecules on the basis of the topology of the electron density distribution. Richard was born in 1931 to a family of modest means in Canada, and from his young age delighted in observing the natural world in his garden. One of the most important lessons his father taught him was never to quit. As a child, he was given his first tricycle. But tricycles are just toys and Richard wanted to play with a truck. So he punctured a metal gas can with a nail, and tied it to the back of his tricycle, thus creating a theoretical truck! [15] Discovering the joys of chemistry, he assembled a home laboratory in his basement and performed experiments, often involving foul smelling gases or small explosions. One of these went out of control one day, flooding Bader's house with hydrogen sulfide from a Kipp's apparatus, until his father dealt with the emergency by tossing the offending apparatus out of the house. Eventually, Richard won a scholarship to attend McMaster University, becoming the first in his immediate family to attend university. Here, he got a Master's degree in physical organic chemistry under the tutelage of A. N. Bourns, followed by a PhD from the Massachusetts Institute of Technology in 1958, studying reaction mechanisms under the guidance of C. G. Swain. He then went on to study theoretical chemistry at Cambridge University with H. C. Longuet-Higgins. Here, he was much influenced by the work of Berlin and A. C. Hurley using electrostatic force concepts, as determined through the Hellmann–Feynman theorem [16, 17], to explain chemical binding in terms of the electron density in real space, rather than in terms of orbitals or abstract Hilbert spaces.

Returning to McMaster, Richard decided early in his scientific career that the means to understanding chemistry at a fundamental level was through the electron density. His first single-author in paper *Molecular Physics* in 1960 on vibrational interaction constants employed orbital symmetry arguments years before they became popular in chemistry through the Woodward–Hoffmann rules. This was followed in the following decade by a series of papers on molecular charge distributions and chemical binding. Collaborating with P. E. Cade at the Laboratory for Molecular Structure and Spectra at the University of Chicago, Bader studied molecular electron density distributions computed from Hartree–Fock wavefunctions for hundreds of diatomic molecules. These studies led him to the realization that the topology of the electron density $\rho(\mathbf{r})$ provides a natural partitioning of the space of a molecule or crystal into mononuclear regions (AIMs) that are by-and-large transferable between similar molecules. Furthermore, this approximate transferability of $\rho(\mathbf{r})$ is paralleled by the transferability of several "atomic" properties between molecules.

Bader's topological atoms result from partitioning space exhaustively into disjoint regions Ω_A, Ω_B ... each associated with a particular AIM. This is equivalent to the specification

$$w_A(\mathbf{r}) = 1 \text{ for } \mathbf{r} \epsilon \Omega_A \qquad (4.8)$$
$$= 0 \text{ otherwise.}$$

The boundaries of the region Ω_A are specified in Bader's prescription by requiring that at every point on the surface S_A bounding the region Ω_A the "zero-flux condition":

$$\nabla\rho(\mathbf{r}) \cdot \mathbf{n}(\mathbf{r}) = 0 \qquad (4.9)$$

is satisfied, where $\mathbf{n}(\mathbf{r})$ is the normal to the surface S_A bounding the atomic region Ω_A. The surface integral of this quantity

$$\oint \mathrm{dS}_A \nabla\rho(\mathbf{r}) \cdot \mathbf{n}(\mathbf{r})\mathbf{dr} = 0 \qquad (4.10)$$

is a measure of the flux of the gradient vectors of the charge density ($\nabla\rho$) through the surface S_A and vanishes over every atomic region Ω_A. The interatomic bounding surface between any pair of atoms is formed by all the steepest descent paths of the electron density $-\nabla\rho(\mathbf{r})$ originating at the bond critical point and terminating at infinity.

Using the partitioning (Eq. 4.8) in Equation 4.1 gives

$$\begin{aligned}\rho_A(\mathbf{r}) &= \rho(\mathbf{r}) \text{ for } \mathbf{r}\epsilon\Omega_A \\ &= 0 \text{ otherwise.}\end{aligned} \tag{4.11}$$

Bader's definition of an AIM thus results in atoms defined as nonoverlapping objects in real space. Using Equation 4.11 in Equation 4.4 then yields for the atomic electron population N_A

$$N_A = \int_{\Omega A} \rho(\mathbf{r})\, d\mathbf{r}, \tag{4.12}$$

where the domain of integration is restricted to the region Ω_A, called the basin of atom A. As described in Chapter 3, the nuclei in a molecule function as attractors of the electron density field. Bader's topological atoms may thus be formally defined as the union of an attractor of the gradient vector field of the electron density with its corresponding basin. Applying Gauss's theorem to Equation 4.10 leads to the result

$$\int_{\Omega A} \nabla^2\rho(\mathbf{r})d\mathbf{r} = 0, \tag{4.13}$$

that is, the Laplacian of the electron density integrates to zero over every atomic domain defined through the "zero-flux condition" [9].

Equation 4.13 can also be derived by applying Schwinger's principle of stationary action [9, 18] to the open quantum system represented by the AIM

$$\delta W = 0, \tag{4.14}$$

where the action W is given by the time integral of the Lagrangian

$$W = \int \mathcal{L} \mathrm{dt}. \tag{4.15}$$

The Lagrangian is the difference between the kinetic (T) and potential (V) energies. For a quantum system of many particles interacting through a potential V, $\mathcal{L}$ can be written as a function of the wavefunction ψ and its spatial and temporal derivatives

$$\mathcal{L}[\psi,\ \Delta\psi,\ \dot{\psi},\ t] = \int \left\{ \left(\frac{i}{2}\right)(\psi^*\dot{\psi} - \dot{\psi}^*\psi) - \tfrac{1}{2}\nabla\psi^* \cdot \nabla\psi - V\psi^*\psi \right\} d\mathbf{r}, \tag{4.16}$$

where atomic units have been used, and a possible explicit dependence on time has been indicated. The variation of the corresponding action integral with respect to ψ and ψ^* yields Schrödinger's equations

$$\begin{aligned} i\dot{\psi} &= H\psi, \\ -i\dot{\psi}^* &= H\psi^*, \end{aligned} \tag{4.17}$$

where

$$H = -\tfrac{1}{2}\nabla^2 + V. \tag{4.18}$$

Schwinger's principle of stationary action [18] may also be used to derive Newton's and Hamilton's equations for a classical system. When the stationary condition (Eq. 4.14) is satisfied, Equation 4.13 follows from the form (Eq. 4.16). Partitioning a molecule according to Bader's prescription thus leads to atoms satisfying Schwinger's quantum mechanics of open systems [9].

The kinetic energy density may be written in either of two alternate forms: the Schrödinger form

$$K(\mathbf{r}) = -\tfrac{1}{4}(\nabla^2 + \nabla'^2)\Gamma^{(1)}(\mathbf{r};\mathbf{r}')|_{r=r'} = -\tfrac{1}{4}(\psi^*\nabla^2\psi + \psi\nabla^2\psi^*), \tag{4.19}$$

or the gradient form

$$G(\mathbf{r}) = -\tfrac{1}{2}(\nabla\cdot\nabla')\Gamma^{(1)}(\mathbf{r};\mathbf{r}')|_{r=r'} = -\tfrac{1}{2}\nabla\psi^*\cdot\nabla\psi, \tag{4.20}$$

where $\Gamma^{(1)}(\mathbf{r},\mathbf{r}')|$ is the reduced one-electron density matrix and the kinetic energy densities are evaluated at $\mathbf{r} = \mathbf{r}'$. These two forms (Eqs. 4.19 and 4.20) differ by a term proportional to the Laplacian of the electron density

$$K(\mathbf{r}) - G(\mathbf{r}) = L(\mathbf{r}) = -\tfrac{1}{4}\nabla^2\rho(\mathbf{r}). \tag{4.21}$$

Since the Laplacian integrates to zero (Eq. 4.13) over an atomic domain satisfying the zero-flux condition, the kinetic energy is uniquely defined through either G or K for an AIM

$$<T>_{\Omega} = \int_{\Omega} K(\mathbf{r})\mathrm{d}\mathbf{r} = \int_{\Omega} G(\mathbf{r})\mathrm{d}\mathbf{r}. \tag{4.22}$$

Of course, the Laplacian also integrates to zero globally, so that the two forms are globally equivalent and the total kinetic energy of a molecule is likewise uniquely defined. Anderson et al. (2010) have argued that the local kinetic energy can be defined through a variety of different forms, besides $K(\mathbf{r})$ and $G(\mathbf{r})$, yielding an infinite variety of AIMs, most of which are not useful. Zadeh and Shahbazian [19] have demonstrated that one can construct a variety of different quantum subsystems–some of which have rather weird topologies–all obeying

Equation 4.13, and thus Equation 4.22, but without satisfying the local zero-flux condition (Eq. 4.9). It is thus the local zero-flux condition (Eq. 4.9), rather than the net zero-flux condition (Eq. 4.13) or the equivalence of kinetic energy forms (Eq. 4.22), that should be taken as the definition of Bader's topological AIMs.

Any topological AIM satisfying Equations 4.13 and 4.21 also satisfies the virial theorem

$$2 < T > = < \mathbf{r} \cdot \nabla V >, \tag{4.23}$$

where the quantity

$$\mathcal{V} = - < \mathbf{r} \cdot \nabla V > = < \mathbf{r} \cdot \mathbf{F} > \tag{4.24}$$

is known as the virial of the forces (**F**), due to the nuclei and other electrons, acting on the electron at **r**. Hence, partitioning a molecule into atomic subsystems satisfying Equation 4.9 is also called *virial partitioning*. For electrostatic interactions, the potential energy takes the form $V \alpha \ \mathrm{r}^{-1}$, so that the virial theorem (Eq. 4.23) reduces to the form

$$2 \ < T > = - < V > . \tag{4.25}$$

The fundamental role of the electron density in understanding chemical bonding was pointed out by Fritz London in 1928 [20]. London was also the first to define a bond path as a "bridge of density" between atomic nuclei. In 1933, John Slater published a derivation of the virial theorem [21], extending it to describe a molecule displaced from its equilibrium geometry. The virial theorem is the Heisenberg equation of motion for the Dirac observable **r.p**, which has the dimensions of action, where $\mathcal{V}$ is the molecular virial of all the external forces exerted on the electron density. The Ehrenfest force theorem is the Heisenberg equation of motion for the Dirac observable $\mathbf{p} = -i\hbar\nabla$, where the force is the rate of change of momentum, as in classical mechanics. Richard Feynman's electrostatic theorem [17] is the Heisenberg equation of motion for the Dirac observable $-i\hbar\nabla_{\alpha}$, where ∇_{α} is the gradient with respect to the coordinates of the nuclei α. This theorem explains chemical bonding as due to the accumulation of electron density between the nuclei exerting an attractive force on the nuclei sufficient to overcome the force of electrostatic repulsion between them. These concepts of the Ehrenfest force acting on the electron density and the Feynman forces acting on the nuclei were employed by Richard Bader in developing his theory of atoms in molecules.

Defining the quantum stress tensor $\sigma(\mathbf{r})$ at any point in space as

$$\begin{aligned}\sigma(\mathbf{r}) &= {}^{1}\!/\!_{4}\{(\nabla\nabla + \nabla'\nabla') - (\nabla\nabla' + \nabla'\nabla)\}\Gamma^{(1)}(\mathbf{r}, \mathbf{r}')|_{\mathrm{r}=\mathrm{r}'} \\ &= {}^{1}\!/\!_{4}(\psi^{*}\nabla\nabla\psi + \psi\nabla\nabla\psi^{*}) - \nabla\psi^{*}\nabla\psi - \nabla\psi\nabla\psi^{*},\end{aligned} \tag{4.26}$$

and using the definitions (Eqs. 4.19 and 4.20) for the kinetic energy operators, one obtains for the trace of the stress tensor

$$Tr\,\sigma(\mathbf{r}) = -K(\mathbf{r}) - G(\mathbf{r}) = -2G(\mathbf{r}) - L(\mathbf{r}) \tag{4.27}$$

using Equation 4.21 for the Laplacian. Thus

$$\tfrac{1}{4}\nabla^2\rho(\mathbf{r}) = 2G(\mathbf{r}) + \mathcal{V}(\mathbf{r}), \tag{4.28}$$

where use has been made of the identity

$$\nabla \cdot (\mathbf{r} \cdot \sigma) = Tr\sigma(\mathbf{r}) + \mathbf{r} \cdot \nabla \cdot \sigma. \tag{4.29}$$

Equation 4.28 is the local form of the virial theorem and relates the Laplacian of the density to the balance between the kinetic and potential energies.

4.3 THE BOND PATH AND THE MOLECULAR GRAPH

A bond is a connection between (usually a pair of) atoms. Chemists have employed various conventions to define a bond between atoms, such as energetic criteria, or based on a Cartesian distance cutoff between a pair of atoms, by counting valencies, or by some other algorithm. Bader's theory of atoms in molecules [9] exploits the topology of the scalar electron density $\rho(\mathbf{r})$ field and its associated gradient vector field $\nabla\rho(\mathbf{r})$ to define the presence or absence of a bond path between any pair of atoms. A bond path is defined as *a trajectory of the gradient vector field* $\nabla\rho(\mathbf{r})$ connecting two atomic nuclei. In Bader's conception, a pair of bonded atoms is associated with a bond path and a bond critical point between them. The network of bond paths then defines a molecular graph, the nodes or vertices of which are atomic nuclei and the connections between them represent chemical bonds between the atoms.

The correspondence between the existence of a bond path connecting two nuclei with the presence of a chemical bond between them in Bader's theory has been criticized on the grounds that the existence of a bond path is "merely" a topological property. However, Bader [22] has clarified that bond paths are not the same as chemical bonds. A different definition of a bond could result in a different connectivity and thus a different molecular graph. For instance, if a rare gas atom is trapped at the center of a C_{60} cage, topology requires the existence of 60 bond paths between the rare gas atom and the 60 carbon atoms! However, most chemists would be reluctant to characterize the resultant complex as one with 60 bonds between the rare gas atom and the carbons, on the basis of energetic or other criteria. The mere existence of a bond path between a pair of atoms does not say anything about the strength of the chemical bond; this topological information needs to be combined with information about the magnitudes of the electron density and the Laplacian at the bond critical point before meaningful chemical conclusions can be drawn. Thus even hydrogen bonds [23–28] and intermolecular van der Waals complexes have recognizable bond paths and bond critical points associated with them (Fig. 4.2). Of course, obtaining reliable computational data on the critical points for some of the weaker interactions requires the inclusion of electron correlation and correction for basis set superposition error [29]. It should

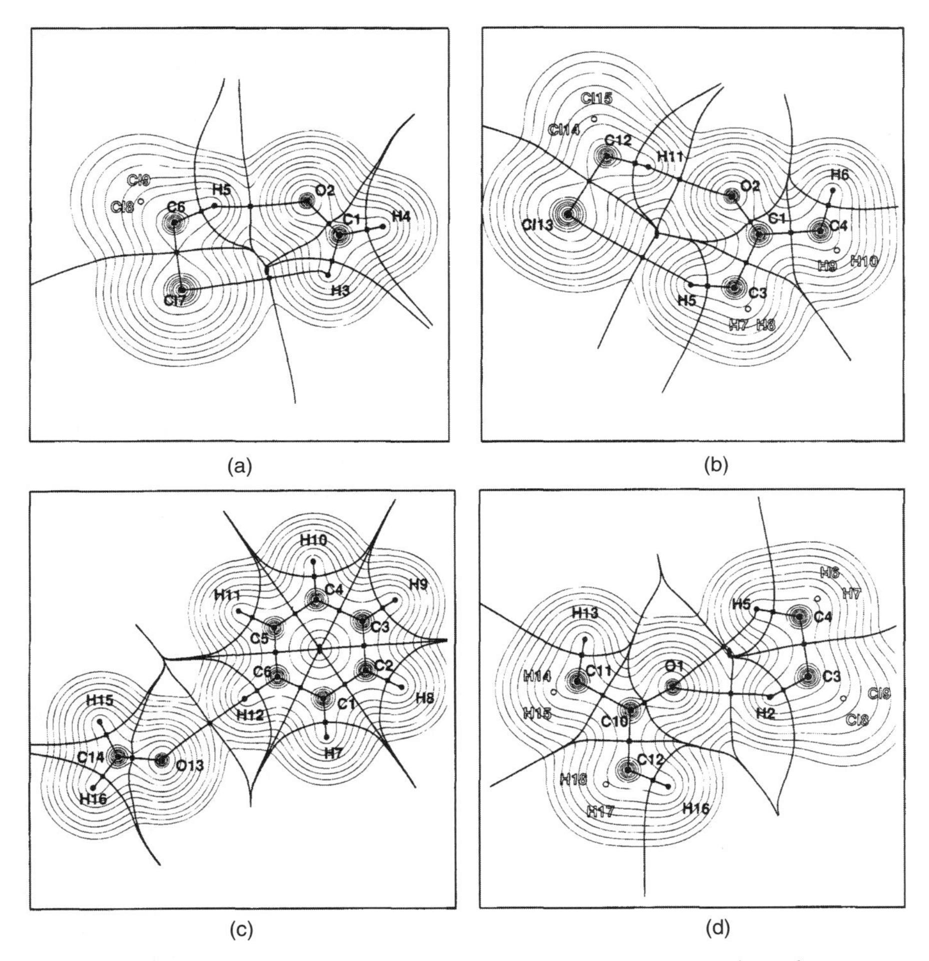

Figure 4.2 Hydrogen bonds and intermolecular van der Waals complexes have recognizable bond paths and bond critical points associated with them. Superposition of the contour lines (thin) of the charge density with the molecular graphs (bold) and interatomic surfaces (bold) of the van der Waals complexes: (a) formaldehyde–chloroform, (b) acetone–chloroform, (c) benzene–formaldehyde, and (d) 1,l-dichloroethaneacetone. Bond critical points are denoted by squares and the ring critical points by ellipses [25]. (Reproduced with permission from Koch U, Popelier PLA. J Phys Chem 1995;99:9747–9754, Copyright 1995 Springer.)

be realized that it is often merely for convenience that chemists exclude these kinds of weak interactions from the usual catalog of chemical bonds. Thus for instance, it is largely a matter of semantic convenience rather than accuracy to identify a system such as liquid water, which is connected through an extensive network of hydrogen bonds at room temperature, with a molecule of H_2O [5].

Despite these criticisms, the topology of the electron density and its network of critical points generally provides a fairly reliable view of molecular structure that corresponds with chemical concepts. The broad agreement between the molecular graphs generated using different criteria for defining bonds, for the vast majority of molecules, demonstrates the robustness of the concepts of the chemical bond and the molecular graph derived from it. Thus the molecular graphs for propellanes (Fig. 4.3a) display a bond critical point between the bridgehead atoms, while those for bicyclic molecules do not. Examination of the molecular graphs for strained rings and electron deficient molecules displays the feature of bent bond paths that chemists have come to expect from such systems (Fig. 4.3).

As mentioned earlier, nonnuclear attractors (NNA) of the electron density have been observed in several metal clusters (Fig. 4.4), as well as in systems containing a solvated electron (Fig. 4.5) [30–35]. Such attractors have their own basins of attraction and can be considered as topological pseudoatoms. The electron densities at nonnuclear maxima are not very different from those at the [3, −1] critical points connecting them to other maxima, and the integrated electron density in the basin of a pseudoatom is typically a small fraction of an electron charge. NNA are characterized by very low kinetic energy per electron and smooth electron density distributions.

4.4 CATASTROPHE POINTS IN THE CHANGE OF MOLECULAR STRUCTURE

The topology of the electron density during the process of bond formation and/or bond breaking reveals interesting insights into the mechanisms of chemical reactions. At this point, it is useful to distinguish between nuclear configurations and molecular structure. The network of bond paths in a molecule characterizes its molecular graph. Molecular structure is formed by an equivalence class of molecular graphs that share the same topology. Equivalence of molecular graphs is established with respect to the respective gradient vector fields of the electron density. Two vector fields are equivalent if each trajectory of one field can be mapped onto a corresponding trajectory of the other (and vice versa). Equivalence of the gradient vector fields also maps the critical points of one field onto the corresponding critical points of the other. A molecular structure is thus a region of nuclear configuration space, all points of which have the same molecular graph. In Bader's topological theory of molecular transformations [9, 36], nuclear configuration space is exhaustively partitioned into disjoint regions corresponding

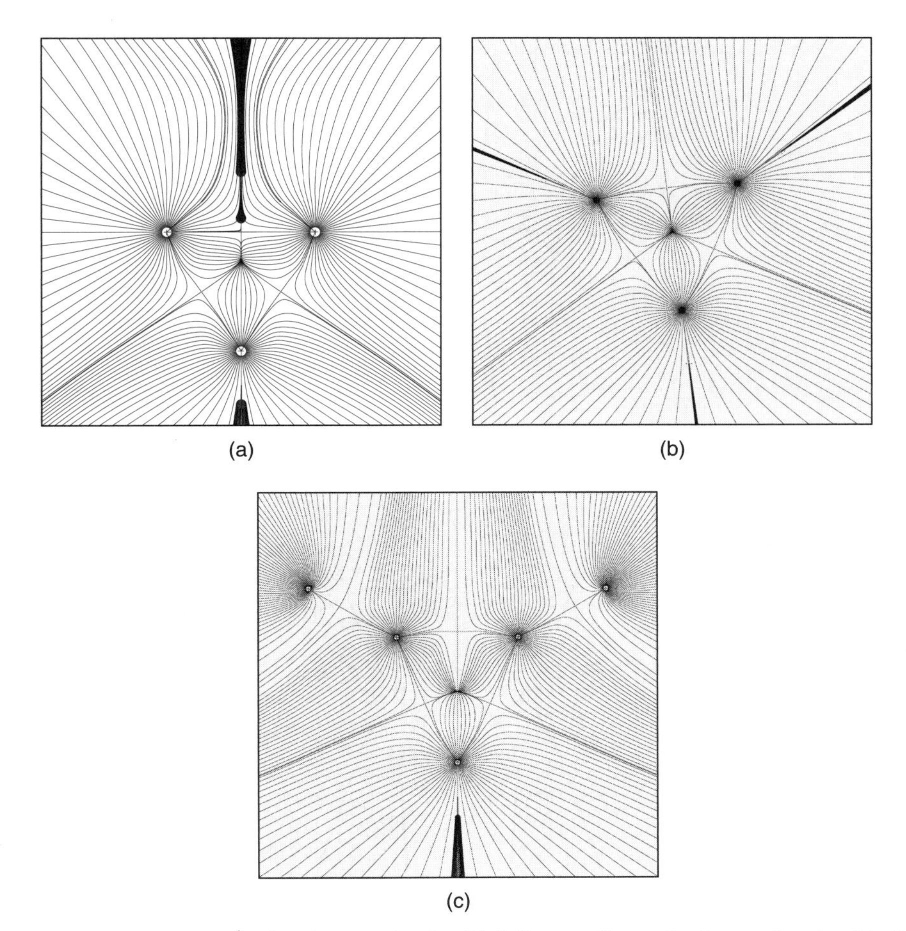

Figure 4.3 (a) The molecular graphs for [1.1.1] propellane displays a bond critical point between the bridgehead atoms. The molecular graphs for strained rings and electron deficient molecules such as (b) cyclopropane and (c) cyclopropane display the features of bent bond paths.

to distinct molecular graphs. Change of molecular structure is a discontinuous process of transformation of molecular graphs (through bond breaking and/or bond formation), but this discontinuous change is driven by continuous transformations of the coordinates in nuclear configuration space. Such topological transformations are described by Thom's catastrophe theory [37], which classifies the topological singularities (catastrophes) into various classes. Chemical transformations described by catastrophe theory include not only interconversion of isomers that correspond to the same stoichiometric formula,but also dissociation, association, substitution, and elimination reactions.

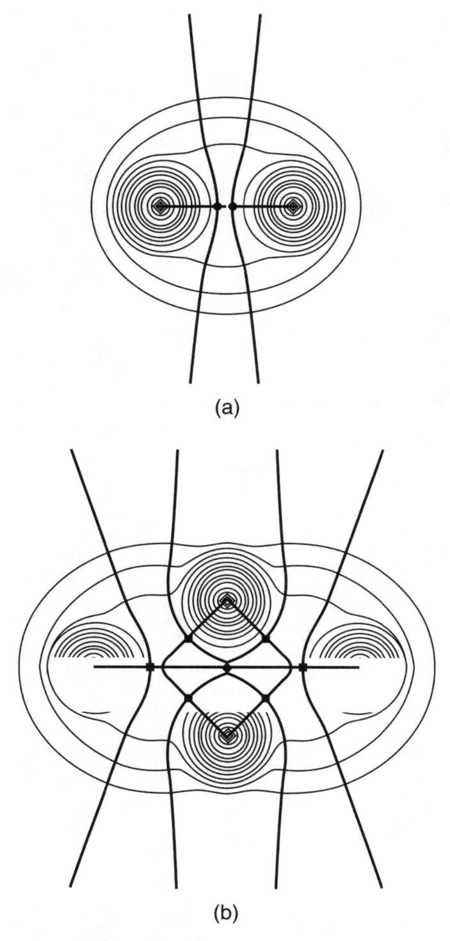

Figure 4.4 Nonnuclear attractors (NNA) of the electron density in the metal clusters (a) Na_2 and (b) Na_4 [30]. (Reproduced with permission from Cao WL, Gatti C, MacDougall PJ, Bader RFW. Chem Phys Lett 1987; 141: 380–385, Copyright 1987 Elsevier.) Contour maps of the charge densities overlaid with bond paths and lines denoting the intersections of the interatomic surfaces with the plane of the diagrams. There are two bonds in Na_2 and seven bonds in Na_4. The positions of the bond critical points are denoted by dots.

4.5 TOPOLOGY OF THE LAPLACIAN DISTRIBUTION

The Laplacian distribution $L(\mathbf{r})$ of a scalar function such as the electron density $\rho(\mathbf{r})$ distinguishes regions where this scalar function is locally concentrated ($\nabla^2\rho(\mathbf{r}) < 0$) or depleted ($\nabla^2\rho(\mathbf{r}) > 0$) (Fig. 4.1c). The Laplacian of the density details the balance between the kinetic and potential energies, as seen from the local form of the virial theorem (Eq. 4.28). The integral of the Laplacian, over all

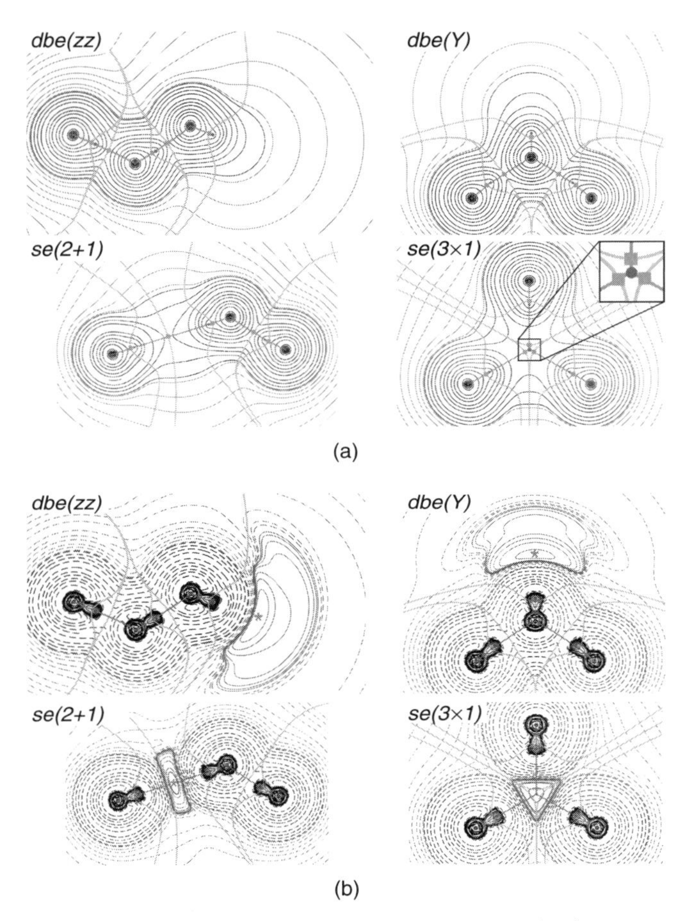

Figure 4.5 Nonnuclear attractors (NNA) of the electron density in systems containing a solvated electron [35]. (Reproduced with permission from Timerghazin QK, Peslherbe GH. J Chem Phys 2007;127:064108, Copyright 2007 American Institute of Physics.) (a) AIM plots for $(HF)_3^-$ isomers showing the electron density contours, overlaid with bond paths and interatomic surfaces (solid gray lines). The attractors of electron density are shown as circles and the bond critical points as boxes. (b) AIM plots for $(HF)_3^-$ isomers showing the Laplacian of the electron density contours, overlaid with bond paths and interatomic surfaces (solid gray lines.)

space or over a region bounded by the zero-flux surface (Eq. 4.9) defining the basin of a topological AIM, goes to zero (see the discussion leading to Eq. 4.13 above). The electron density along the bond path and the Laplacian at the bond critical point reveal clues as to the nature of bonding between the atoms in Bader's theory. In a covalent bond, the electron density accumulates in the region between the nuclei and along the bond path (Fig. 4.1c), with a shallow curvature along the bond path and a steep curvature perpendicular to it. Thus, $\nabla^2\rho(\mathbf{r}_c) < 0$ (two large negative and one small positive eigenvalues). Such bonds are characterized by large magnitudes of the potential energy of attraction between electrons and nuclei in the internuclear region. It is this Coulomb attraction of the nuclei for the accumulated electron density in the internuclear region that is responsible for the classical part of the binding stabilization (exchange provides the rest). In an ionic bond, on the other hand, there is no significant accumulation of electron density along the bond path; binding is primarily due to the electrostatic attraction of the net negative charge on the anion to the net positive charge on the cation (Fig. 4.6). The electron density is characterized by a steep curvature along the bond path and a shallow curvature perpendicular to it and thus $\nabla^2\rho(\mathbf{r}_c) > 0$. Ionic bonds are characterized by large kinetic energy in the internuclear region.

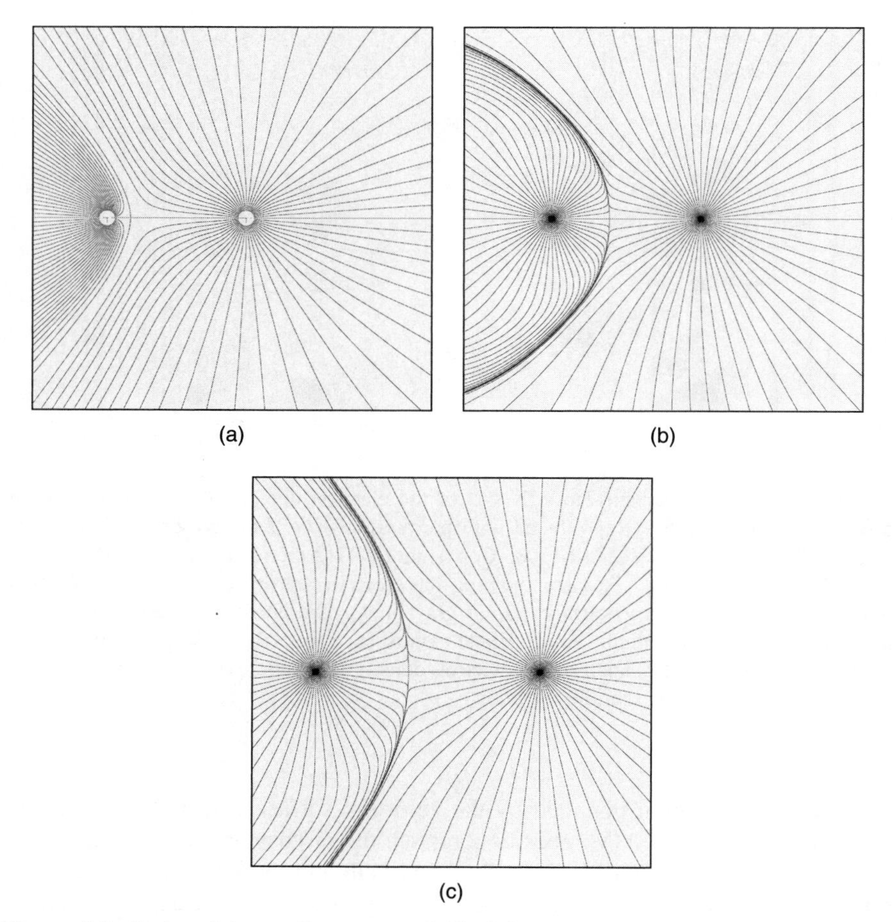

Figure 4.6 Paths of the gradient vector field of the electron density $\nabla\rho(\mathbf{r})$ for (a) HF, (b) LiF, and (c) NaF.

The Laplacian distribution $L(\mathbf{r})$ of the electron density is itself a scalar function and its topology can be studied in terms of the rank and signature of its own critical points, that is, where $\nabla L(\mathbf{r}) = 0$. Local minima or (3, −3) critical points of $L(\mathbf{r})$ are regions of bonded or nonbonded charge concentration, linked to each other by the unique pair of trajectories originating at an intervening (3, −1) critical point. The resultant graphs generally form polyhedra bounding each atom, with (3, +1) ring critical points at the faces and (3, +3) cage critical points at the nuclei. The topology of $L(\mathbf{r})$ reveals the characteristic shell structure of atoms, as well as regions of charge concentration in the valence shells of molecules (Fig. 4.1c), which forms the basis of the VSEPR model [38].

4.6 THE FERMI HOLE AND ELECTRON DELOCALIZATION

Ever since Gilbert Newton Lewis introduced the concept of the electron pair bond, much of the discussion of structure and bonding in chemistry deals with electron pairs [39, 40]. Electron pairing is a consequence of the Pauli exclusion principle, the tendency of electrons with the same spin to avoid each other, over and beyond the Coulomb repulsion between all electrons. The Fermi hole (Chapter 3) is the physical manifestation of the Pauli exclusion principle; it describes how the density of a reference electron at any point in space spreads out into the space of another electron of the same spin, thereby excluding an identical amount of same-spin density. The density of an uncorrelated α–β pair is

$$P^{\alpha\beta}(\mathbf{r}_1, \mathbf{r}_2) = \rho^{\alpha}(\mathbf{r}_1)\rho^{\beta}(\mathbf{r}_2). \tag{4.30}$$

The pair density for electrons of same spin, however, is less than the simple product of the one-particle densities. The difference is the density of the exchange–correlation hole ρ_{xc} introduced in Chapter 5, which, in the absence of Coulomb correlation, reduces to the Fermi hole

$$P^{\alpha\alpha}(\mathbf{r}_1, \mathbf{r}_2) = \rho^{\alpha}(\mathbf{r}_1)\{\rho^{\alpha}(\mathbf{r}_2) + h^{\alpha}(\mathbf{r}_1, \mathbf{r}_2)\} = \rho^{\alpha}(\mathbf{r}_1)\rho^{\alpha}(\mathbf{r}_2) + \rho^{\alpha}_{\mathrm{xc}}(\mathbf{r}_1, \mathbf{r}_2). \tag{4.31}$$

The density of the Fermi hole $h^{\alpha}(\mathbf{r}_1,\mathbf{r}_2)$, a negative quantity, decreases the amount of same-spin density at any point $\mathbf{r}_2$ by an amount determined by the delocalization of the Fermi hole away from the reference point $\mathbf{r}_1$. As Bader describes it [9], the Fermi hole is the electron's *doppelgänger*; it shadows the electron's motion and goes wherever the electron goes. If the Fermi hole is localized, so is the electron; and when the electron is delocalized, so is its Fermi hole. When integrated over all space for a fixed position of the reference electron, the Fermi hole corresponds to the removal of one electronic charge of same spin, that is, the integral of the Fermi hole density over all space equals −1

$$\int h^{\alpha}(\mathbf{r}_1, \mathbf{r}_2)\, \mathrm{d}\mathbf{r}_2 = -1. \tag{4.32}$$

The density of the Fermi hole at the position of the reference electron equals the negative of the total density of same-spin electrons, that is

$$h^{\alpha}(\mathbf{r}_1, \mathbf{r}_1) = -\rho^{\alpha}(\mathbf{r}_1), \tag{4.33}$$

ensuring that there is zero probability of finding another electron of same spin at the position of the reference electron. In Hartree–Fock theory, the density of the Fermi hole is given by

$$h^{\alpha}(\mathbf{r}_1, \mathbf{r}_2) = -\sum_i^{\alpha}\sum_j^{\alpha}\{\varphi_i{}^*(\mathbf{r}_1)\varphi_j(\mathbf{r}_1)\varphi_j{}^*(\mathbf{r}_2)\varphi_i(\mathbf{r}_2)\}/\rho^{\alpha}(\mathbf{r}_1), \tag{4.34}$$

where the summations run over α spinorbitals, and with an analogous expression for the Fermi hole due to β electrons. It is easily verified that the form (Eq. 4.34) satisfies the conditions (Eq. 4.32) and (Eq. 4.33). The quantity within curly brackets $\{\rho^{\alpha}(\mathbf{r}_2) + h^{\alpha}(\mathbf{r}_1, \mathbf{r}_2)\}$ in Equation 4.31 is the conditional same-spin density, the weighted probability of an α electron being at $\mathbf{r}_2$ when another is at $\mathbf{r}_1$ (weighted by the total number of α electrons). Equation 4.31 is thus a restatement of Bayes theorem (Chapter 1). From Equations 4.32 and 4.33, we find that the conditional probability of finding a second α spin electron at the reference point $\mathbf{r}_1$ vanishes and that the conditional probability integrates to $N^{\alpha} - 1$ over all space, where N^{α} is the number of electrons of α spin, that is, if an α spin electron is definitely at $\mathbf{r}_1$, then the total probability of finding another α spin electron elsewhere in the system is $N^{\alpha} - 1$.

With Hartree–Fock wavefunctions, ρ_{xc} reduces to the Hartree–Fock exchange density

$$\rho_{\mathrm{x}}{}^{\alpha}(\mathbf{r}_1, \mathbf{r}_2) = \rho^{\alpha}(\mathbf{r}_1)h^{\alpha}(\mathbf{r}_1, \mathbf{r}_2) = -\sum_i^{\alpha}\sum_j^{\alpha}\{\varphi_i{}^*(\mathbf{r}_1)\varphi_j(\mathbf{r}_1)\varphi_j^*(\mathbf{r}_2)\varphi_i(\mathbf{r}_2)\}. \tag{4.35}$$

Double integration of this exchange density over the coordinates of both electrons gives the total Fermi correlation for electrons of α spin, which equals the negative of the number of electrons of α spin

$$\int\int \rho_{\mathbf{x}}{}^{\alpha}(\mathbf{r}_1, \mathbf{r}_2)\,\mathrm{d}\mathbf{r}_1\mathrm{d}\mathbf{r}_2 = -N^{\alpha}. \tag{4.36}$$

Restricting the double integrations to the basin of atom A yields the total Fermi correlation within A

$$F^{\alpha}(\mathrm{A,A}) = \int_{\Omega A}\int_{\Omega A} \rho_{\mathbf{x}}{}^{\alpha}(\mathbf{r}_1, \mathbf{r}_2)\mathrm{d}\mathbf{r}_1\mathrm{d}\mathbf{r}_2. \tag{4.37}$$

The limiting value of F^{α}(A,A) is $-N^{\alpha}$(A), the negative of the number of electrons with α spin in atom A; this corresponds to all α spin electrons of atom A being localized within the atomic basin of A and all remaining α spin density excluded from A. The Fermi hole thus acts to exclude α electron density from the space surrounding a reference α electron. Since the Pauli principle does not exclude electrons of β spin from this space, electrons will tend to form spatially localized α–β pairs and exclude all other electrons, of α and β spin, from the space surrounding the reference pair. We thus define the localization index $\lambda(\mathrm{A})$ as

$$\lambda(\mathrm{A}) = |F^{\alpha}(\mathrm{A,A}) + F^{\beta}(\mathrm{A,A})| = \int_{\Omega\mathrm{A}}\int_{\Omega\mathrm{A}} \rho_{\mathrm{x}}(\mathbf{r}_1, \mathbf{r}_2)\,\mathrm{d}\mathbf{r}_1\mathrm{d}\mathbf{r}_2 \tag{4.38}$$

$$= 2\sum\nolimits_{i,j} S_{ij}(\mathrm{A})^2 \qquad \text{for a restricted Hartree–Fock wavefunction,}$$

$$= \sum\nolimits_{i,j} n_i^{1/2} n_j^{1/2} S_{ij}(\mathrm{A})^2 \qquad \text{in terms of natural orbitals,}$$

where S_{ij} is the overlap integral between spinorbitals i and j in the basin of atom A, n_i and n_j are the natural orbital occupation numbers, and where the summations now run over all occupied spinorbitals. While perfect localization is possible only for an isolated system, one finds near-complete localization for core electrons and for some simple ionic hydrides.

Conversely, the exchange of electrons between the basins of atoms A and B is given by

$$F^{\alpha}(\mathrm{A,B}) = \int_{\Omega\mathrm{A}}\int_{\Omega\mathrm{B}} \rho_{\mathrm{x}}{}^{\alpha}(\mathbf{r}_1, \mathbf{r}_2)\,\mathrm{d}\mathbf{r}_1\mathrm{d}\mathbf{r}_2 \tag{4.39}$$

$$= -\sum\nolimits_i^{\alpha}\sum\nolimits_j^{\alpha}\int_{\Omega\mathrm{A}}\int_{\Omega\mathrm{B}} \{\varphi_i{}^*(\mathbf{r}_1)\varphi_j(\mathbf{r}_1)\varphi_j{}^*(\mathbf{r}_2)\varphi_i(\mathbf{r}_2)\}\,\mathrm{d}\mathbf{r}_1\mathrm{d}\mathbf{r}_2$$

$$= -\sum\nolimits_i^{\alpha}\sum\nolimits_j^{\alpha} S_{ij}(\mathrm{A})S_{ji}(\mathrm{B}).$$

$F^{\alpha}(\mathrm{A,B})$ is then used to define the delocalization index $\delta(\mathrm{A,B})$, also known as the *shared electron density index*

$$\delta(\mathrm{A,B}) = 2|F^{\alpha}(\mathrm{A,B}) + F^{\beta}(\mathrm{A,B})| = 2\int_{\Omega\mathrm{A}}\int_{\Omega\mathrm{B}} \rho_{\mathrm{x}}(\mathbf{r}_1, \mathbf{r}_2)\,\mathrm{d}\mathbf{r}_1\mathrm{d}\mathbf{r}_2 \tag{4.40}$$

$$= 4\sum\nolimits_i\sum\nolimits_j S_{ij}(\mathrm{A})S_{ji}(\mathrm{B}) \text{ for a restricted Hartree–Fock wavefunction,}$$

$$= 2\sum\nolimits_i\sum\nolimits_j n_i^{1/2} n_j^{1/2} S_{ij}(\mathrm{A})S_{ji}(\mathrm{B}) \text{ in terms of natural orbitals,}$$

where the summations again run over all occupied spinorbitals. The localization and delocalization indices for an atom add up to the atomic population

$$N(\mathrm{A}) = \lambda(\mathrm{A}) + {}^1\!/\!_2\sum\nolimits_{\mathrm{B}\neq\mathrm{A}} \delta(\mathrm{A,B}). \tag{4.41}$$

The valence of atom A can be defined, for closed-shell wavefunctions, as the sum of all the delocalization indices from A to all other atoms

$$V(\mathrm{A}) = \sum\nolimits_{\mathrm{B}\neq\mathrm{A}} \delta(\mathrm{A,B}) = 2|N(\mathrm{A}) - \lambda(\mathrm{A})|. \tag{4.42}$$

As a quantitative measure of electron sharing between atomic basins, the delocalization index between atoms connected by a bond path has been found to correlate well with other measures of bond order for organic [41–43] and non-polar inorganic molecules [44].

The delocalization index has also been employed to construct aromaticity measures [45–48]. While aromaticity is a multidimensional phenomenon, manifested in bond length equalization, resonance stabilization, and ring currents, it is a consequence of delocalization of electrons over a ring. Bader [45] observed that, in conjugated organic molecules, the delocalization of the Femi hole density between atoms not connected by a bond path decreases with the distance between them. However, in benzene and other aromatic systems, the π-electron density is considerably more delocalized between the basins of carbon atoms para to each other (the atoms that are spin paired in the Dewar resonance structures) than between those meta to each other, despite the fact that the meta atoms are much closer. This observation inspired a proposal to define an aromaticity measure in terms of the average of the delocalization indices of all atoms para to each other in a given ring. Thus, naphthalene has a lower para delocalization index than benzene; and the inner rings in straight-chain acenes (anthracene and naphthacene) have lower para delocalization indices than the outer rings. This is also the case for the inner six-member rings in fullerenes, while the situation is reversed in the staggered acenes (phenanthrene, chrysene, and triphenylene). However, the para delocalization index as a measure of aromaticity is useful only for six-member rings. A different aromaticity measure constructed from delocalization indices is the π-fluctuation aromatic index, defined as the divergence from the average of π-delocalization indices for all bonded pairs of atoms in a ring. Electron density-based aromaticity measures for several polycyclic hydrocarbons are displayed in Figure 4.7.

4.7 ELECTRON LOCALIZATION FUNCTION

A local measure of electron localization can be derived from the Fermi hole through Taylor expansion of the spherically averaged conditional pair probability, that is, the quantity $\{\rho^{\alpha}(\mathbf{r}_2) + h^{\alpha}(\mathbf{r}_1, \mathbf{r}_2)\}$ in Equation 4.31 [49–54], which contains the leading term

$$\Delta(\mathbf{r}) = G(\mathbf{r}) - T_{\mathrm{w}}(\mathbf{r}), \tag{4.43}$$

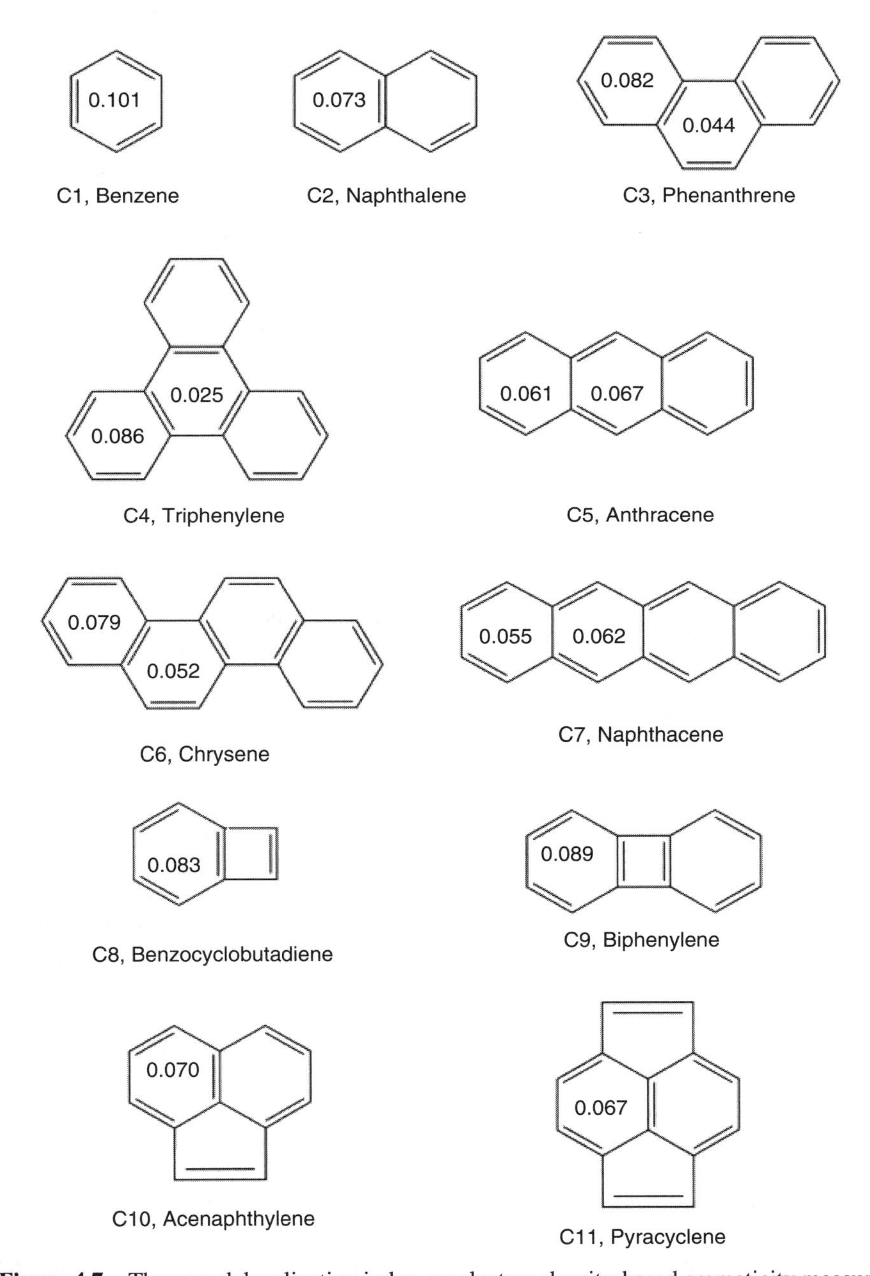

Figure 4.7 The para delocalization index, an electron density-based aromaticity measure, for several polycyclic hydrocarbons [47]. (Reproduced with permission from Poater J, Fradera X, Duran M, Sola M. Chem—Eur J 2003; 9 (2): 400–406, Copyright 2003 Wiley.)

where $T_w(\mathbf{r})$ is the Weiszäcker kinetic energy density

$$T_w(\mathbf{r}) = \frac{\nabla\rho(\mathbf{r}) \cdot \nabla\rho(\mathbf{r})}{8\rho(\mathbf{r})}. \tag{4.44}$$

$\Delta(\mathbf{r})$ is a measure of the local excess kinetic energy due to Pauli repulsion. It vanishes for one-electron systems and for regions within multielectron systems dominated by a single, localized spinorbital. This prompted Becke and Edgecombe [51, 52] to interpret this quantity as a measure of electron localization. In order to obtain a function that increases with increasing electron localization and is bounded from above, Becke and Edgecombe defined the electron localization function [51]

$$\text{ELF}(\mathbf{r}) = \{1 + [\Delta(\mathbf{r})/\Delta_0(\mathbf{r})]^2\}^{-1}, \tag{4.45}$$

where $\Delta_0(\mathbf{r})$ is the kinetic energy of a uniform electron gas with density $\rho(\mathbf{r})$

$$\Delta_0(\mathbf{r}) = (3/5)(6\pi^2)^{2/3}\ \rho(\mathbf{r})^{5/3}. \tag{4.46}$$

$\Delta(\mathbf{r})/\Delta_0(\mathbf{r})$ is thus a localization index defined with respect to the uniform electron gas as reference. ELF($\mathbf{r}$) as defined above is restricted to the range [0,1]. ELF $= 1/2$ corresponds to the uniform electron gas, while the upper limit ELF $= 1$ corresponds to perfect localization.

Since ELF($\mathbf{r}$) is a scalar function, its gradient field enables partitioning of the molecular space into adjacent nonoverlapping basins of attraction. An atom with more than two electrons (i.e., beyond helium in the periodic table) contains one or more inner core basins, surrounding the nucleus and localized on the atom, and one or more outer valence basins. ELF($\mathbf{r}$) distributions for atoms reveal the expected characteristic atomic shell structure (Fig. 4.8). A valence basin in a molecule may be localized on an atom or shared between two or more atoms. Valence basins are thus characterized by the synaptic order, the number of atomic valence shells in which they participate. The valence shell of a molecule is the union of its valence basins. The topology of ELF($\mathbf{r}$) has been employed extensively in studies of chemical bonding, aromaticity, chemical reactivity [47], and intermolecular interactions. Surfaces at successively higher values of ELF($\mathbf{r}$) can be used to generate bifurcation diagrams that yield clues to the nature of bonding in molecules and hydrogen-bonded intermolecular complexes. At sufficiently low values of ELF($\mathbf{r}$), the entire system is contained within a single connected envelope or domain. At higher values of ELF($\mathbf{r}$), separate disconnected domains corresponding to the core, valence nonbonding (monosynaptic), and valence bonding (di- or polysynaptic) regions separate out. The order in which this happens is governed by the nature of the bonding in the system. For a weakly hydrogen-bonded complex, the localization domain first bifurcates into separate atomic domains, each of which subsequently bifurcates into valence and core domains. For stronger hydrogen-bonded systems, the core–valence bifurcation occurs first.

4.8 THE SOURCE FUNCTION

We have interpreted the Laplacian $\nabla^2\rho(\mathbf{r})$ as mapping out regions of concentration or depletion of electron density and also as representing the balance between the kinetic and potential energies at **r**. Another way to look on the Laplacian is as the generator of the electron density distribution, by virtue of Poisson's equation

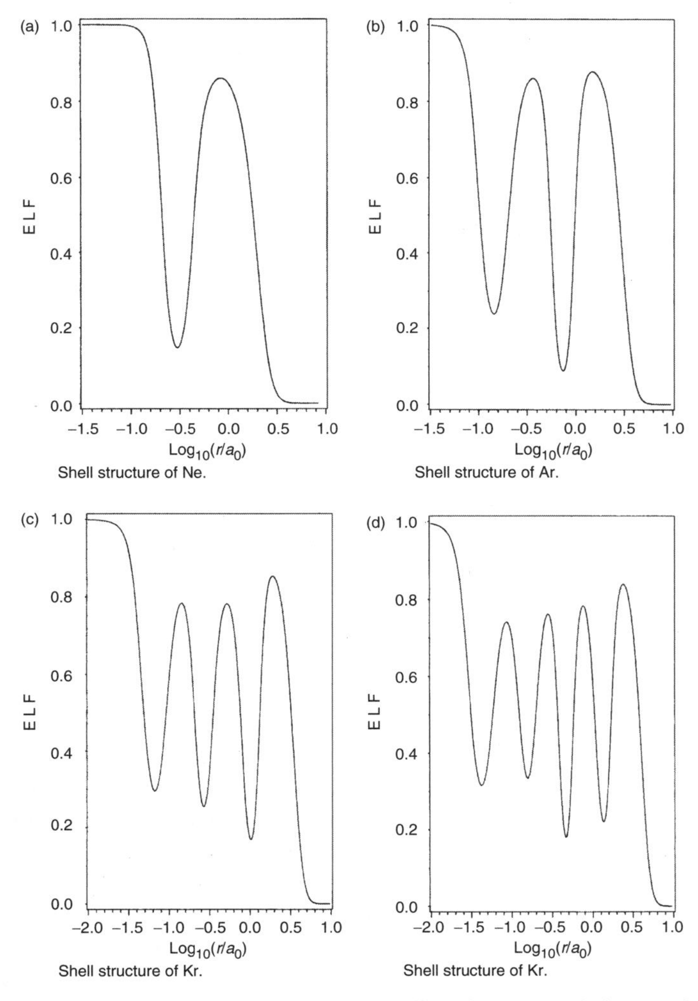

Figure 4.8 ELF(**r**) distributions for atoms, revealing the expected characteristic atomic shell structure, for (a) Ne, (b) Ar, (c) Kr, (d) Xe, (e) Rn, and (f) Zn [51]. (Reproduced with permission from Becke A, Edgecombe KE. J Chem Phys 1990;92(9): 5397–5403, Copyright 1990 American Institute of Physics.)

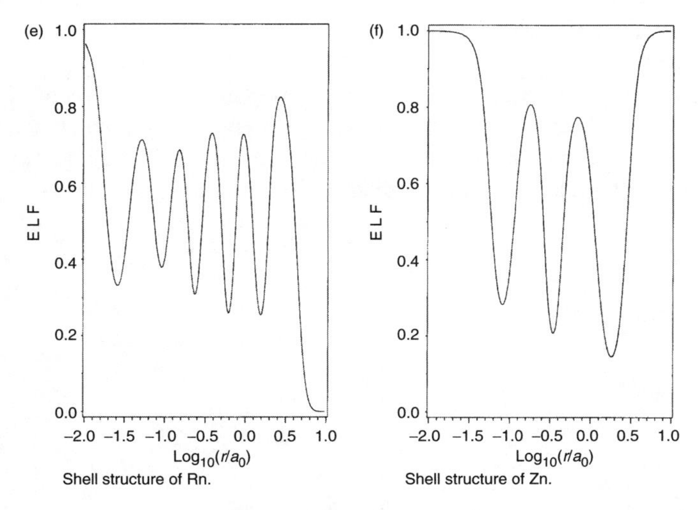

Figure 4.8 *(Continued)*

$$\rho(\mathbf{r}) = \frac{-1}{4\pi} \int d\mathbf{r}' \frac{\nabla^2 \rho(\mathbf{r}')}{|\mathbf{r} - \mathbf{r}'|}. \tag{4.47}$$

The integrand in Equation (4.47) is called the *local source function* [53, 54]

$$LS(\mathbf{r}, \mathbf{r}') = \frac{-1}{4\pi} \frac{\nabla^2 \rho(r')}{|\mathbf{r} - \mathbf{r}'|}, \tag{4.48}$$

representing the effectiveness of the concentration (or depletion) $\nabla^2 \rho(\mathbf{r}')$ at $\mathbf{r}'$ in functioning as a source (or sink) for the electron density at $\mathbf{r}$ modulated by the Green's function or influence function $(4\pi |\mathbf{r} - \mathbf{r}'|)^{-1}$. Using Equation 4.28, the local source may be rewritten as

$$LS(\mathbf{r}, \mathbf{r}') = -\left(\frac{1}{\pi}\right) \frac{2G(\mathbf{r}') + V(\mathbf{r}')}{|\mathbf{r} - \mathbf{r}'|}. \tag{4.49}$$

Any region where the electron density is locally concentrated ($\nabla^2 \rho(\mathbf{r}') < 0$) in a molecule, and where the potential energy dominates the kinetic energy, acts as a source for the electron density at other points in the molecule, while a region where the electron density is locally depleted ($\nabla^2 \rho(\mathbf{r})' > 0$) and where the kinetic energy dominates acts as a sink, removing electron density from $\mathbf{r}$.

The integral of the local source over the basin Ω of an atom or functional group is called the *source function* $S(\mathbf{r}, \Omega)$ contribution from that atom or functional group to $\rho(\mathbf{r})$

$$S(\mathbf{r}, \Omega) = \int_{\Omega} LS(\mathbf{r}, \mathbf{r}') d\mathbf{r}'. \tag{4.50}$$

Using Equation 4.50 in Equation 4.47 gives

$$\rho(\mathbf{r}) = \int_{\Omega} LS(\mathbf{r}, \mathbf{r}')\, d\mathbf{r}' + \sum\nolimits_{\Omega' \neq \Omega} \int_{\Omega'} LS(\mathbf{r}, \mathbf{r}')\, d\mathbf{r}'$$
$$= S(\mathbf{r}, \Omega) + \sum\nolimits_{\Omega' \neq \Omega} S(\mathbf{r}, \Omega'). \quad (4.51)$$

The electron density at any point in an AIM may thus be decomposed into a contribution arising from sources within the basin of the atom and a contribution arising from sources external to the atom. The source function is a measure of the relative contribution of an atom or group to the density at any point.

The source function is a very sensitive measure of the chemical transferability of atoms and functional groups between different molecules. It is, in fact, a more sensitive index of transferability than is the atomic energy or integrated electron population. Perfect transferability of atomic/group properties requires not just the transferability of the corresponding electron density but also that the sum of the contributions to this density from the remaining atoms or groups in the molecule (second term in Eq. 4.51) be constant. While one may study the source functions at any point $\mathbf{r}$ in a molecule, it is customary and instructive to compare the source function contributions of different atoms and groups to the electron density at a bond critical point. Table 4.1 shows the source function contributions of different groups to the density at the bond critical point of the terminal C–H bond (ρ_b) in a series of n-alkanes, where the terminal methyl group is known to have transferable properties (beyond ethane in the series). The source function contribution of the neighboring methylene group to ρ_b is constant after ethane. Furthermore, the ethyl group in propane contributes the same amount to ρ_b as the propyl group in butane or the butyl group in pentane. Thus the sum of the source contributions to ρ_b from groups external to the methyl group is constant.

For a series of diatomic hydrides, the source contribution from the hydrogen atom to the electron density at the bond critical point has been shown [55] to decrease with increasing electronegativity of the atom bonded to the hydrogen.

TABLE 4.1 Source Function Contributions of Different Groups to the Density at the Bond Critical Point of the Terminal C–H Bond in a Series of *n*-Alkanes[a]

Ethane	H-CH_2	CH_2	H		
	0.2704	0.0100	0.0026		
Propane	H-CH_2	CH_2	CH_3		
	0.2701	0.0091	0.0035		
Butane	H-CH_2	CH_2	CH_2	CH_3	
	0.2701	0.0091	0.0020	0.0016	
Pentane	H-CH_2	CH_2	CH_2	CH_2	CH_3
	0.2702	0.0090	0.0019	0.0008	0.0009

[a] Adapted from Reference 53.

Studies of lithium clusters [55] show that the source function also clearly discriminates between nuclear and nonnuclear maxima of $\rho(\mathbf{r})$: the internal source contribution to the density at the nuclear maxima in all clusters is more than 99.96%, while it is no more than 74% for NNAs (Table 4.2), the remaining 26% arising from external source contributions, mostly from the closest linked Li basins. The source function contributions to the electron density at the bond critical points linking pairs of NNAs arise 60–70% from the NNAs, with the remainder from more remote basins. This delocalized distribution of sources reflects the loosely bound character of the valence electrons in the negatively charged pseudoatoms. The source function has been employed [55] to classify hydrogen bonds on the basis of characteristic source contributions to the density at the hydrogen bond critical point. Study of the source function can also shed light on metal–metal bonding, such as in $Mn_2(CO)_{10}$ where the Mn atoms act as sinks for the electron density at the metal–metal bond critical point, with the carbonyl oxygens acting as sources, indicating the highly nonlocalized nature of the metal–metal bond in this complex.

4.9 STOCKHOLDER PARTITIONING OF THE ELECTRON DENSITY

Let us now turn to a different way of partitioning the electron densities between atoms in a molecule. The stockholder partitioning scheme was first proposed by Fred Hirshfeld in 1977 [56]. Here, we first form a promolecule from the superposition of the electron densities of the isolated atoms, with the nuclei at the positions of the corresponding nuclei in the molecule. The molecular electron density is then partitioned among the atoms, such that each AIM contributes to the molecular density $\rho(\mathbf{r})$ at each point in proportion to its contribution to the promolecule's density $\rho^0(\mathbf{r}) = \sum_A \rho_A^0(\mathbf{r})$ at that point

$$w_A(\mathbf{r}) = \frac{\rho_A^0(\mathbf{r})}{\rho^0(\mathbf{r})} = \frac{\rho_A(\mathbf{r})}{\rho(\mathbf{r})}, \tag{4.52}$$

where $\rho_A^0(\mathbf{r})$ are the reference atomic densities to which the AIM densities $\rho_A(\mathbf{r})$ are compared. The second equality in Equation 4.52 is merely a restatement of the general definition (4.1) for the weight function. This partitioning is analogous to the way a company's shares (and profits) are apportioned among its stockholders—in proportion to each stockholder's original contribution to the corpus of the company. Hirshfeld's partitioning scheme results in overlapping atoms, in contrast to Bader's partitioning. This allows for an understanding of chemical bonding in terms of the overlap electron density between atoms. Nalewajski and Parr [57] showed that Hirshfeld's partitioning scheme follows from requiring that the loss of information (entropy deficiency functional)

$$\Delta S[\rho/\rho^0] = \sum_A \int \rho_A(\mathbf{r}) \ln\left\{\frac{\rho_A(\mathbf{r})}{\rho_A^0(\mathbf{r})}\right\} d\mathbf{r} \tag{4.53}$$

TABLE 4.2 Percent Source Function Contributions of Different Atoms and Pseudoatoms (NNAs) to the Density at the Nuclear and Nonnuclear Maxima in Several Lithium Clusters[a]

Cluster		Critical point								
		S(Li1)	S(Li2)	S(Li3)	S(Li4)	S(Li5)	S(Li6)	S(NNA1)	S(NNA2)	S(NNA3)
Li_2	Li1	99.96	0.03	—	—	—	—	0.03	—	—
	NNA	12.43	12.43	—	—	—	—	74.28	—	—
	Li-NNA BCP	29.47	8.52	—	—	—	—	61.19	—	—
Li_4	Li1	99.96	0.00	0.00	0.00	—	—	0.02	0.02	—
	Li3	0.00	0.00	99.96	0.00	—	—	0.03	0.00	—
	NNA	2.65	2.65	12.18	1.48	—	—	73.29	5.41	—
	NNA-NNA BCP	8.52	8.52	5.16	5.16	—	—	34.47	34.47	—
Li_6	Li1	99.96	0.00	0.00	0.00	0.00	0.00	0.03	0.00	0.00
	Li4	0.00	0.00	0.00	99.95	0.00	0.00	0.02	0.02	0.00
	NNA	11.39	1.10	1.10	2.10	2.10	0.44	73.48	2.28	2.28
	NNA-NNA BCP	6.00	6.00	4.08	4.89	3.04	3.04	26.97	26.97	12.49

[a] Adapted from Ref. 55.

be minimized on formation of the molecule from neutral atoms. Minimization of the functional (Eq. 4.52) subject to a set of constraints

$$F_k[\rho] = F_{\mathrm{k}}^0 \tag{4.54}$$

is equivalent to solving

$$\delta\{\Delta S[\rho/\rho^0] + \sum_{\mathrm{k}} \lambda_{\mathrm{k}} F_{\mathrm{k}}[\rho]\} = 0, \tag{4.55}$$

where the λ_{k} are Lagrange multipliers. For two densities ρ and ρ^0, ΔS gives the information distance between ρ and ρ^0 or the information deficiency of ρ relative to ρ^0. Thus minimizing the functional ΔS (Eq. 4.53) subject to the constraints

$$\sum_{\mathrm{A}} \rho_{\mathrm{A}}(\mathbf{r}) = \rho(\mathbf{r}) \quad \text{and} \quad \sum_{\mathrm{A}} \rho_{\mathrm{A}}^0(\mathbf{r}) = \rho^0(\mathbf{r}) \tag{4.56}$$

leads to

$$\sum_{\mathrm{A}} \left[\ln \left\{ \frac{\rho_{\mathrm{A}}(\mathbf{r})}{\rho_{\mathrm{A}}^0(\mathbf{r})} \right\} + \lambda(\mathbf{r}) - 1 \right] \delta\rho_{\mathrm{A}}(\mathbf{r}) = 0. \tag{4.57}$$

Hence,

$$1 - \lambda(\mathbf{r}) = \frac{\rho_{\mathrm{A}}(\mathbf{r})}{\rho_{\mathrm{A}}^0(\mathbf{r})} = \frac{\rho(\mathbf{r})}{\rho^0(\mathbf{r})}, \tag{4.58}$$

from which the Hirshfeld condition (Eq. 4.52) immediately follows. However, it should be noted that other information-theoretic measures do not necessarily reduce to the Hirshfeld partitioning.

The Hirshfeld partitioning scheme and its information-theoretic interpretation have been criticized on several grounds and extended by Ayers [58] and by Bultinck and coworkers [59, 60]. The first serious problem is the arbitrariness involved in choosing neutral atoms as the reference densities [61] in defining the entropy loss functional. This choice has no strict theoretical basis and is adopted merely for computational convenience. For heteronuclear, strongly ionic molecules, in particular, it might seem more reasonable to use the ionic fragments as reference densities. For instance, constructing the promolecule LiF from a combination of neutral Li and F atomic densities results in atomic charges of ± 0.57. If instead, the Li^+ and F^- ions are employed to build the promolecule, the atomic charges of ± 0.98 are obtained. Using the opposite combination of Li^- and F^+ ions leads to atomic charges of ± 0.30 (with the positive charge still on Li). The Hirshfeld AIM electron populations thus depend sensitively on the choice of the promolecule density.

These ambiguities are further exacerbated when dealing with ionic molecules. One can still retain a neutral promolecule at the cost of having a promolecule density that does not integrate to the molecular density, but this makes the connection to information theory suspect. Furthermore, interpretation of the function (Eq. 4.53) as an information entropy measure requires that

$$N_{\mathrm{A}} = \int \rho_{\mathrm{A}}(\mathbf{r})d\mathbf{r} = \int \rho_{\mathrm{A}}^{0}(\mathbf{r})\, d\mathbf{r} = N_{\mathrm{A}}^{0} \tag{4.59}$$

for every atom, that is, the promolecule atom should have the same electronic population as the AIM. This requirement is not generally satisfied by the Hirshfeld procedure.

Defining a shape function for each atom as a density per electron

$$\sigma_{\mathrm{A}}(\mathbf{r}) = \frac{\rho_{\mathrm{A}}(\mathbf{r})}{N_{\mathrm{A}}}, \tag{4.60}$$

with an analogous expression for the promolecular atoms $\sigma_{\mathrm{A}}^{0}(\mathbf{r})$, the information loss function (Eq. 4.53) can be recast [58, 62] as

$$I = \sum_{\mathrm{A}} N_{\mathrm{A}} \int \sigma_{\mathrm{A}}(\mathbf{r}) \ln \left\{ \frac{\sigma_{\mathrm{A}}(\mathbf{r})}{\sigma_{\mathrm{A}}^{0}(\mathbf{r})} \right\} d\mathbf{r} + \sum_{\mathrm{A}} N_{\mathrm{A}} \ln \left(\frac{N_{\mathrm{A}}}{N_{\mathrm{A}}^{0}} \right). \tag{4.61}$$

The first term in Equation 4.61 represents the information loss because of the change in shape of the electron density (polarization) on molecule formation, while the second term is an entropy of mixing, reflecting the transfer of electronic charge between "atoms" on molecule formation. Note that the shape functions are normalized to unity by virtue of Equation 4.1

$$\int \sigma_{\mathrm{A}}(\mathbf{r})d\mathbf{r} = \int \sigma_{\mathrm{A}}^{0}(\mathbf{r})\, d\mathbf{r} = 1, \tag{4.62}$$

supporting the interpretation of I as an information entropy. The polarization contributions to the entropy are always nonnegative. The entropy of mixing term arises from the violation of the condition (Eq. 4.59). Each term also satisfies the condition of statistical independence (additivity of independent events), as would be expected of any measure of information.

This reformulation enabled Bultinck and coworkers [59, 60] to address some of the ambiguities and shortcomings of the original Hirshfeld procedure using an alternative, iterative scheme for defining the weight functions in Equation 4.1: Starting from some freely chosen promolecule with atomic densities $\rho^{0}{}_{\mathrm{A}}(\mathbf{r})$ and atomic populations $N_{\mathrm{A}}{}^{0}$, the Hirshfeld partitioning procedure is applied and the atomic densities $\rho^{1}{}_{\mathrm{A}}(\mathbf{r})$ and populations $N_{\mathrm{A}}{}^{1}$ are computed. The densities

$\rho^1{}_\mathrm{A}(\mathbf{r})$ are used to construct the promolecule in the next iteration. The weighting functions in each iteration are given by

$$W^i_\mathrm{A}(\mathbf{r}) = \frac{\rho^i{}_\mathrm{A}(\mathbf{r})}{\rho^{i-1}(\mathbf{r})}. \tag{4.63}$$

A self-consistent set of charges is obtained by forcing the promolecular atoms to have the same electron populations as in the AIM, that is, the process is continued until $N_\mathrm{A}{}^i = \mathrm{N}_\mathrm{A}{}^{i-1}$ for all atoms to within a self-consistency threshold. Once this point is reached, there is thus no further net charge transfer between the atoms and the entropy of mixing term no longer contributes. Since the requirement (Eq. 4.59) is now satisfied for every reference atom that contributes to the promolecule, the function (Eq. 4.53) or (Eq. 4.59) represents a mathematically proper measure of information entropy. It is only after the requirement (Eq. 4.59) is satisfied that one may identify the $\rho^i{}_\mathrm{A}(\mathbf{r})$ with AIM densities. This formulation eliminates of the arbitrariness in the choice of the promolecule, allowing iterated Hirshfeld atoms to be defined for charged as well as neutral molecules. One still has the arbitrariness as to which electronic states of the isolated atoms to use in constructing the promolecule—and alternative partitioning schemes have been proposed to address this [60]. The iterative Hirshfeld procedure has been shown to converge rapidly [59]. Furthermore, the self-consistent iterative Hirshfeld atomic charges are independent of the initial choice of promolecule, only weakly dependent on the basis set and correlate well [59, 60] with atomic charges obtained by fitting the electrostatic potential on a grid (CHELPG [63]).

4.10 ATOMS IN MOMENTUM SPACE

We now investigate AIM partitioning in momentum space. The fundamental quantity analogous to $\rho(\mathbf{r})$ in momentum space is the electron momentum density $\gamma(\mathbf{p})$

$$\gamma(\mathbf{p}) = N \int d\mathbf{p}_2 \ldots \int d\mathbf{p}_N \Phi^*(\mathbf{p}, \mathbf{p}_2, \ldots, \mathbf{p}_N)\Phi(\mathbf{p}, \mathbf{p}_2, \ldots, \mathbf{p}_N). \tag{4.64}$$

However, $\rho(\mathbf{r})$ and $\gamma(\mathbf{p})$ are related to each other only through Fourier-Dirac transformation of the full N-electron wavefunction

$$\Phi(\mathbf{p}, \mathbf{p}_2, \ldots, \mathbf{p}_N) = (2\pi)^{-3N/2} \int d\mathbf{r}_1 \int d\mathbf{r}_2 \ldots \int d\mathbf{r}_N \exp\left(-i \sum_{i=1}^{N} \mathbf{p}_i \cdot \mathbf{r}_i\right) \times \psi(\mathbf{r}_1, \mathbf{r}_2, \ldots, \mathbf{r}_N). \tag{4.65}$$

The attractiveness of the momentum space formalism is that the kinetic energy functional is known exactly. It is also helpful in interpreting the results of electron momentum spectroscopy and Compton scattering experiments. However, atomic momentum densities are not spherically symmetric. The absence of features centered on the nuclei means that partitioning the momentum density of a molecule with respect to zero-flux surfaces of $\nabla\gamma(\mathbf{p})$ does not result in basins that can be identified with atoms. Nevertheless, interesting information can be gleaned from study of the Laplacian distributions $\nabla^2\gamma(\mathbf{p})$ in momentum space (Chapter 10).

Hirshfeld partitioning of molecular momentum densities may be used to define AIMs in momentum space. Analogous to Equations 4.1 and 4.52, we may define

$$\gamma_{\mathrm{A}}(\mathbf{p}) = \frac{\gamma_{\mathrm{A}}^{0}(\mathbf{p})}{\sum_{\mathrm{A}} \gamma_{\mathrm{A}}^{0}(\mathbf{r})}\gamma(\mathbf{p}), \tag{4.66}$$

where $\gamma_{\mathbf{A}}(\mathbf{p})$ and $\gamma_{\mathrm{A}}^{0}(\mathbf{p})$ are the momentum densities of the AIM and the free atom A, respectively, centered at $\mathbf{p} = 0$. This scheme yields integrated atomic charges that agree well with atomic charges obtained from the corresponding Hirshfeld AIMs in coordinate space [64].

4.11 DENSITY MATRIX PARTITIONING

Although only the total energy of a system is an observable and fragment energies are not, chemical explanations are often formulated in terms of energies of interaction. This is due to the fact that the only meaningful energetic quantities in chemical reactions are energy differences. Schemes for the partitioning of energy in molecules are thus popular and have pedagogic value. For a molecular system of electrons and nuclei interacting through only Coulomb forces, the energy is a function of the first-order (nondiagonal) and second-order (diagonal) density matrices (Chapter 3)

$$\gamma^{(1)}(\mathbf{r};\mathbf{r}') = N\int d\mathbf{r}_2\ldots\int d\mathbf{r}_N\Psi^*(\mathbf{r},\mathbf{r}_2,\ldots,\mathbf{r}_N)\Psi(\mathbf{r}',\mathbf{r}_2,\ldots,\mathbf{r}_N); \tag{4.67}$$

$$\gamma^{(2)}(\mathbf{r}_1,\mathbf{r}_2) = N\int d\mathbf{r}_3\ldots\int d\mathbf{r}_N\Psi^*(\mathbf{r}_1,\mathbf{r}_2,\mathbf{r}_3,\ldots,\mathbf{r}_N)\Psi(\mathbf{r}_1,\mathbf{r}_2,\mathbf{r}_3,\ldots,\mathbf{r}_N). \tag{4.68}$$

An exact energy partitioning based on decomposition of these density matrices into intraatomic and interatomic components can be performed with respect to both Bader's topological and Hirshfeld's stockholder AIMs. Li and Parr [65] employed a partitioning similar to Equations 4.1 and 4.2 for the nondiagonal first-order density matrix

$$\gamma^{(1)}(\mathbf{r};\mathbf{r}') = \sum_{\mathrm{A}} w_{\mathrm{A}}(\mathbf{r})\gamma^{(1)}(\mathbf{r};\mathbf{r}'), \tag{4.69}$$

where the $w_A(\mathbf{r})$ satisfy Equation 4.3. Note that the diagonal part of $\gamma^{(1)}$ is simply the one-electron density: $\gamma^{(1)}(\mathbf{r};\mathbf{r}) = \rho(\mathbf{r})$. The second-order density matrix is obtained similarly through a double partitioning

$$\gamma^{(2)}(\mathbf{r}_1,\mathbf{r}_2) = \sum_A \sum_B w_A(\mathbf{r}) w_B(\mathbf{r}) \gamma^{(2)}(\mathbf{r}_1,\mathbf{r}_2). \tag{4.70}$$

The total molecular Born-Oppenheimer energy (Chapter 3), for a system of electrons and nuclei interacting through only Coulomb forces, is the sum of the electronic energy E and the internuclear repulsion energy $\mathcal{V}_{nn}$

$$\mathcal{E}_{BO} = \mathrm{E} + \mathcal{V}_{nn}, \tag{4.71}$$

$$E = \langle T \rangle + \langle V \rangle + \langle U \rangle, \tag{4.72}$$

$$\langle T \rangle = \int \frac{1}{2} \nabla \cdot \nabla' \gamma^{(1)}(\mathbf{r};\mathbf{r}')|_{r=r'} d\mathbf{r}, \tag{4.73}$$

$$\langle V \rangle = -\sum_A Z_A \int d\mathbf{r} \frac{\rho(r)}{|\mathbf{r} - \mathbf{R}_A|}, \tag{4.74}$$

$$\langle U \rangle = 1/2 \int d\mathbf{r}_1 \int d\mathbf{r}_2 \frac{\gamma^{(2)}(\mathbf{r}_1,\mathbf{r}_2)}{\mathbf{r}_{12}}, \tag{4.75}$$

$$\mathcal{V}_{nn} = 1/2 \sum_A \sum_{B\neq A} V_{nn}{}^{AB} = 1/2 \sum_A \sum_{B\neq A} \frac{Z_A Z_B}{R_{AB}}. \tag{4.76}$$

Using the density matrix partitions (Eqs. 4.69 and 4.70) in Equations 4.71–4.76 enables the partitioning of all the energy terms into intraatomic and interatomic components

$$\langle T \rangle = \sum_A T_A = \sum_A \int 1/2\, w_A(\mathbf{r})\, \nabla \cdot \nabla' \gamma^{(1)}(\mathbf{r};\mathbf{r}')|_{r=r'} d\mathbf{r}, \tag{4.77}$$

$$\langle V \rangle = \sum_A \sum_B V_{AB} = -\sum_A \sum_B Z_A \int d\mathbf{r} \frac{\rho_B(\mathbf{r})}{|\mathbf{r} - \mathbf{R}_A|}, \tag{4.78}$$

$$\langle U \rangle = \sum_A U_{AA} + 1/2 \sum_A \sum_{B\neq A} U_{AB}, \tag{4.79}$$

where

$$U_{AA} = 1/2 \int d\mathbf{r}_1 \int d\mathbf{r}_2\, w_A(\mathbf{r}_1)\, w_A(\mathbf{r}_2) \frac{\gamma^{(2)}(\mathbf{r}_1,\mathbf{r}_2)}{\mathbf{r}_{12}} \tag{4.80}$$

and

$$U_{AB} = \int d\mathbf{r}_1 \int d\mathbf{r}_2\, w_A(\mathbf{r}_1)\, w_B(\mathbf{r}_2) \frac{\gamma^{(2)}(\mathbf{r}_1,\mathbf{r}_2)}{\mathbf{r}_{12}}. \tag{4.81}$$

Note that $V_{AB} \neq V_{BA}$ in Equation 4.78 above. Using relations (Eqs. (4.76)–(4.81)) in Equation 4.71 yields the expression for the total Born-Oppenheimer energy

$$\mathcal{E}_{BO} = \sum_{A} E_A + \tfrac{1}{2} \sum_{A} \sum_{B \neq A} E_{AB}^{\text{int}}, \tag{4.82}$$

where E_A, the internal energy of AIM A, contains all the intraatomic energy contributions

$$E_A = T_A + V_{AA} + U_{AA}, \tag{4.83}$$

and E_{AB}^{int} represents the energy of interaction between AIMs A and B

$$E_{AB}^{\text{int}} = V_{AB} + V_{BA} + U_{AB} + V_{nn}^{AB}. \tag{4.84}$$

Note that the internal energy E_A of AIM A as defined above is different from the energy obtained by integrating the electronic energy functional $(T + V + U)$ over the basin of AIM A; the latter also includes a partitioning of some of the interaction energy $(V_{AB} + V_{BA} + U_{AB})$ included in Equation 4.84 between the atomic basins. As can be seen in Figure 4.9, the expressions (Eqs. 4.83 and 4.84) for the internal energy E_A and the interaction energy E_{AB}^{int} both involve large cancellations between the individual energy components [66].

The difference between the AIM energy E_A and the energy of the corresponding isolated atom A is the energy of deformation of the electron density

$$E_A^{\text{def}} = E_A - E_A^{0}, \tag{4.85}$$

while the difference between the total Born-Oppenheimer energy and the sum of the energies of the isolated neutral atoms is the binding energy of the molecule

$$E^{\text{bind}} = \mathcal{E}_{BO} - \sum_{A} E_A^{0} = \sum_{A} E_A^{\text{def}} + \tfrac{1}{2} \sum_{A} \sum_{B \neq A} E_{AB}^{\text{int}}. \tag{4.86}$$

It is also instructive to examine the classical Coulomb and exchange–correlation contributions to the interaction energy (Fig. 4.9c) or to the binding energy. Since the second-order density matrix can be partitioned into Coulomb (first term in Eq. 4.87) and exchange–correlation components (second term in Eq. 4.87)

$$\gamma^{(2)}(\mathbf{r}_1, \mathbf{r}_2) = \rho(\mathbf{r}_1)\rho(\mathbf{r}_2) + \gamma_{xc}^{(2)}(\mathbf{r}_1, \mathbf{r}_2), \tag{4.87}$$

U_{AA} and U_{AB} in Equations 4.80 and 4.81 can be likewise partitioned

$$U_{AA} = U_{AA}^{\text{Coul}} + U_{AA}^{\text{xc}}, \tag{4.88}$$

$$U_{AB} = U_{AB}^{\text{Coul}} + U_{AB}^{\text{xc}}, \tag{4.89}$$

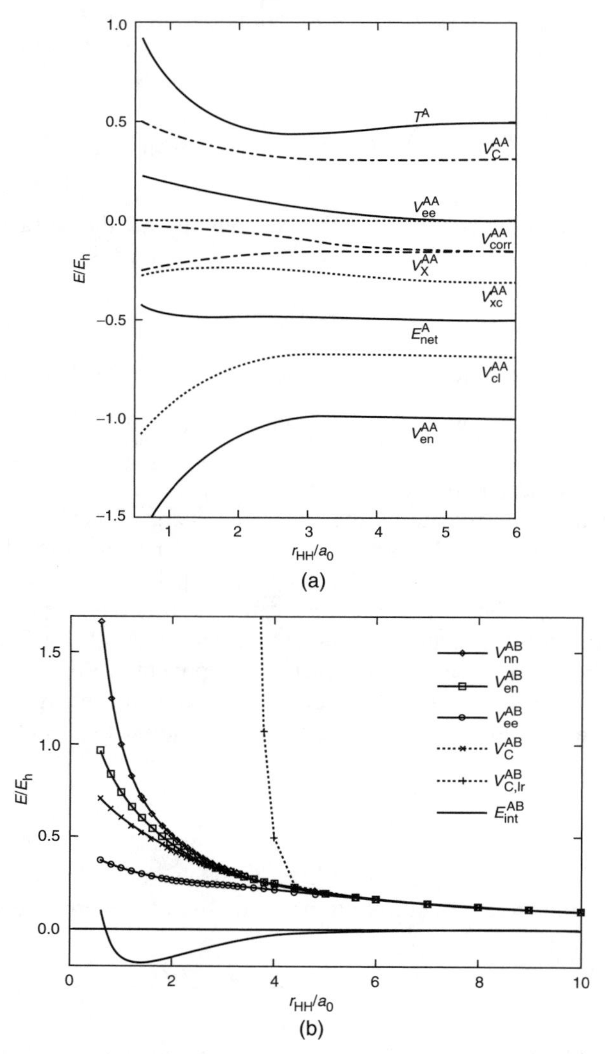

Figure 4.9 Plots of (a) the internal energy E_A and (b) the interaction energy $E_{AB}{}^{int}$, for H_2 as functions of internuclear distance, showing the large cancellations between the individual energy components. (c) Partition of the interaction energy ($E_{AB}{}^{int}$) into classical ($V_{AB}{}^{Class}$) and exchange–correlation ($U_{AB}{}^{xc}$) contributions for H_2 as functions of internuclear distance [66]. (Reproduced with permission from Blanco MA, Martin Pendas A, Francisco E. J Chem Theor Comput 2005; 1096–1109, Copyright 2005 American Chemical Society.)

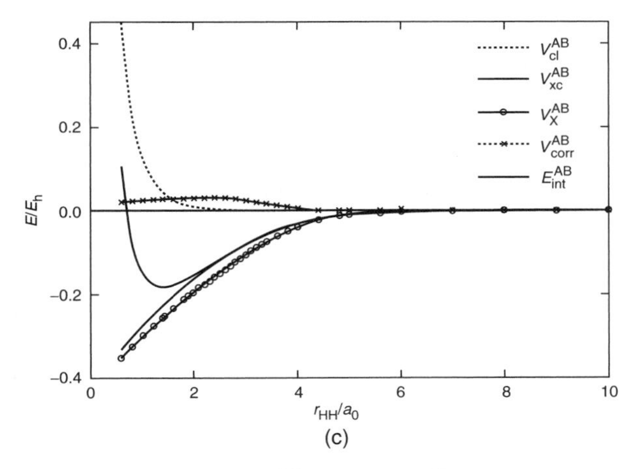

Figure 4.9 (*Continued*)

where

$$
\begin{aligned}
U_{\mathrm{AA}}{}^{\mathrm{Coul}} &= {}^{1}\!/\!_{2}\int d\mathbf{r}_1 \int d\mathbf{r}_2\; w_{\mathrm{A}}(\mathbf{r}_1) w_{\mathrm{A}}(\mathbf{r}_2) \frac{\rho(\mathbf{r}_1)\rho(\mathbf{r}_2)}{\mathbf{r}_{12}} \\
U_{\mathrm{AA}}{}^{\mathrm{xc}} &= {}^{1}\!/\!_{2}\int d\mathbf{r}_1 \int d\mathbf{r}_2 w_{\mathrm{A}}(\mathbf{r}_1) w_{\mathrm{A}}(\mathbf{r}_2) \frac{\gamma^{(2)}_{\mathrm{XC}}(\mathbf{r}_1,\mathbf{r}_2)}{\mathbf{r}_{12}} \\
U_{\mathrm{AB}}{}^{\mathrm{Coul}} &= \int d\mathbf{r}_1 \int d\mathbf{r}_2 w_{\mathrm{A}}(\mathbf{r}_1) w_{\mathrm{B}}(\mathbf{r}_2) \frac{\rho(\mathbf{r}_1)\rho(\mathbf{r}_2)}{\mathbf{r}_{12}} \\
U_{\mathrm{AB}}{}^{\mathrm{xc}} &= \int d\mathbf{r}_1 \int d\mathbf{r}_2 w_{\mathrm{A}}(\mathbf{r}_1) w_{\mathrm{B}}(\mathbf{r}_2) \frac{\gamma^{(2)}_{\mathrm{XC}}(\mathbf{r}_1,\mathbf{r}_2)}{\mathbf{r}_{12}},
\end{aligned}
\tag{4.90}
$$

so that

$$E_{\mathrm{AB}}{}^{\mathrm{int}} = V_{\mathrm{AB}}{}^{\mathrm{Class}} + U_{\mathrm{AB}}{}^{\mathrm{xc}}, \tag{4.91}$$

where

$$V_{\mathrm{AB}}{}^{\mathrm{Class}} = V_{\mathrm{AB}} + V_{\mathrm{BA}} + U_{\mathrm{AB}}{}^{\mathrm{Coul}} + V_{\mathrm{nn}}{}^{\mathrm{AB}} \tag{4.92}$$

includes all the classical electrostatic terms. For molecules that are essentially covalent, there are large cancellations in both the intraatomic and in the classical interatomic terms that constitute the binding energy. The picture that emerges with Bader's topological atoms satisfying the condition (Eq. 4.8) is that of essentially neutral atoms, largely unchanged from the isolated atoms, that is, with small deformation energies [66]. Binding in homopolar diatomics is almost exclusively due to the quantum effects of exchange–correlation interaction. Partitions based on overlapping atomic densities (as in the Hirshfeld scheme) result in larger intraatomic and interatomic energy components [67]. The binding energy is then governed by the detailed balance between intraatomic and interatomic energy contributions, each considerably larger than the binding energy itself.

When there is significant charge transfer, as in heteronuclear diatomics, the intraatomic deformation energy terms are large and positive for the atom losing charge, irrespective of the partitioning scheme employed [67]. For the negatively charged atom, the effects of electron population and the intrinsic deformation of the atomic density largely cancel out. The classical electrostatic interaction plays a stabilizing role for ionic bonds, more than compensating for the energy required to deform the atomic densities. Most of the classical electrostatic interaction is well reproduced by a point charge model, irrespective of the partitioning scheme, but in general, both classical and quantum effects are relevant in understanding the binding.

Vanfleteren et al. [68] employed a double-atom partitioning for the molecular first-order density matrix

$$\gamma^{(1)}(\mathbf{r};\mathbf{r}') = \sum_{\mathrm{A}}\sum_{\mathrm{B}}[w_{\mathrm{A}}(\mathbf{r})w_{\mathrm{B}}(\mathbf{r}') + w_{\mathrm{B}}(\mathbf{r})w_{\mathrm{A}}(\mathbf{r}')]\gamma^{(1)}(\mathbf{r};\mathbf{r}'). \tag{4.93}$$

Terms in (Eq. 4.93) diagonal in the atomic indices (A=B) correspond to atomic density matrices, with eigenvalues between 0 and 2. However, the atomic density matrices alone do not sum to the total number of electrons but correlate well with the atomic localization indices $\lambda(\mathrm{A})$. The remaining electrons are in the off-diagonal ($\mathrm{A} \neq \mathrm{B}$) terms of (93), which correspond to bond matrices. These matrices were shown to have good localization properties, even when used with fuzzy atomic boundaries, in contrast to single-index partitioning based on Equation 4.69 [69]. The total electron populations in the bond matrices correlate with the shared electron density indices $\delta(\mathrm{A,B})$ [68].

4.12 CONCLUDING REMARKS

While the very concept of the electron density is a consequence of the Born–Oppenheimer separation of electronic and nuclear motions in molecules (Chapter 3), there has recently been a preliminary study [70] of AIM partitioning on molecular wavefunctions computed without making the Born–Oppenheimer approximation. In these schemes, both electrons and nuclei are treated on a similar footing, using multicomponent molecular orbitals. Such computations include the effects of nuclear motion. Topological analysis of the resulting single-particle density is able to distinguish between atomic basins containing different isotopes.

A variety of schemes based on the electron density have been proposed for the partitioning of molecules into atoms, some using the topology of the electron density function in real space and justified on the basis of the virial theorem and Schwinger's action principle, others using interpenetrating atoms with fuzzy boundaries, justified on information-theoretic grounds. Some of the insight provided by different schemes is definitely complementary. The study of AIMs has proved its potential in providing conceptual understanding of many key concepts in chemistry and biochemistry, such as molecular graphs, transferability of atomic and functional group properties, the chemical bond, energetics of covalent and ionic bonds, chemical reactivity, the topology of molecular transformations, electron localization and delocalization, valence shell concentrations of electrons, bond orders, aromaticity, resonance, and the genetic code itself. AIM partitioning has demonstrated rich applications in organic chemistry, bonding in the solid state, metal–metal bonding, hydrogen bonding, molecular and bioisosteric similarity analysis, QSAR/QSPR (quantitative structure-activity/property relationships), and drug design. Some of these applications are reviewed in other chapters herein (Chapters 7 and 8) and elsewhere [13,71–73].

4.13 EPILOGUE

Richard Bader continued to teach chemistry at McMaster, his alma mater, for 30 years until 1996, when he officially retired. Thereafter, as an Emeritus Professor, he published 60 (out of a total of 223) refereed articles and book chapters on theoretical chemistry and physics, his last article published after his eightieth birthday [74]. Richard Bader passed away in Burlington, Ontario, on Sunday January 15, 2012, at age 80, after a difficult struggle with lung disease. He is survived by his wife Pam, three daughters, and grandson.

According to Professor Claude Lecomte, using the electron density (ρ), a Dirac quantum mechanical observable, to describe interatomic interactions was the missing bridge between quantum mechanics and crystallography that QTAIM has provided. Richard gave another dimension to the concept of electron density applied to chemical bonding. As a consequence of the QTAIM theory, Richard was the promoter of a very intense and constructive scientific dialog between crystallographers and quantum chemists (and thus between theoreticians and experimentalists) on the basis of quantum physics. This dialogue may have never been as fruitful and constructive (or even existing) had it not been for Richard Bader (C. Lecomte, private communications; 2012). Over and beyond Richard's scientific achievements and the outstanding theoretical developments flowing from his theory, this communication between theoreticians and experimentalists in the shared waters of the electron density distribution can be considered one of the most important consequences of his work, beneficial for both communities (E. Espinosa, private communications; 2012).

Among the seminal contributions that Richard made to science is passion for science and its communication. This was the additional spice that he brought to scientific debate (A. Pinkerton, private communications; 2012). His passion in defending the theory of Atoms in Molecules at conferences and the impossibility of session chairs to stop the discussions will be remembered by all who attended any conference with Bader. This characteristic passion was not reserved just for conferences, seminars, and distinguished lectures but also marked his undergraduate classes at McMaster. It was also not uncommon for Richard to lead a few graduate students into his office to continue their Socratic discussion after a class had ended. Soon everybody would have chalk dust on their hands. The passion with which he shared those ideas will continue to jump out and grab attention when future students watch his archived lectures (P. Macdougall, private communications; 2012).

Richard never had much respect for authority, as Professor Jack Dunitz recalls. It is said about Wolfgang Pauli that on his ascent to Heaven, he was given a unique privilege: God himself was to explain the fundamental mysteries of the Universe. When God came to the part where He started discussing the nature of light and its interactions with matter, Pauli shook his head and cried out "Falsch!" One likewise hopes that Richard will not get involved in any controversy up there, as he might win! (J. Dunitz, private communications; 2012.) Owing to the force of his personality and the popular Star Wars trilogy, Richard was often teased with Darth Bader jokes, to which he would respond "May the zero-flux be with you" [15].

REFERENCES

1. Riess I, Münch W. Theor Chim Acta 1981;58:295–300.
2. Bader RFW, Becker P. Chem Phys Lett 1988;148(5):452–458.
3. Lorenzo L, Mosquera RA. Chem Phys Lett 2002;356:305–312.
4. Paneth FA. Br J Philos Sci 1962;13(49):1–14; 1962;13(50):144–160.
5. Needham P. Found Chem 2005;7(1):103–118.
6. Earley JE. Found Chem 2005;7(1):85–102.
7. Hendry RF. Philos Sci 2006;73(5):864–875.
8. Earley JE. Found Chem 2009;11(2):65–77.
9. Bader RFW. *Atoms in Molecules: A Quantum Theory*. Oxford: Oxford Press; 1990.
10. Bader RFW, Beddall PM. J Chem Phys 1972;56:3320–3329.
11. Bader RFW, Beddall PM. J Am Chem Soc 1972;95(2):305–315.
12. Bader RFW. Int J Quantum Chem 1994;49:299–308.
13. Matta CF, Boyd RJ, editors. *The Quantum Theory of Atoms in Molecules: From Solid State to DNA and Drug Design*. Weinheim: Wiley-VCH; 2007.
14. Popelier PL. *Atoms in Molecules: An Introduction*. London: Prentice Hall; 2000.
15. Macdougall P. Stories in the News. 2012. Ketchikan, Alaska. Available at http://www.sitnews.us/. Accessed 2012.
16. Hellmann H. Z Phys 1933;44:455.
17. Feynman RP. Phys Rev 1939;56:340–343.
18. Schwinger J. Phys Rev 1951;82:914–927.
19. Zadeh FH, Shahbazian S. Theor Chem Acc 2011;128:175–181.
20. London F. Z Phys 1927;44:455–472.

21. Slater JC. J Chem Phys 1933;1:687–691.
22. Bader RFW. J Phys Chem A 2009;113:10391–10396.
23. Carroll MT, Bader RFW. Mol Phys 1988;65:695–722.
24. Breneman CM, Rhem M, Thompson TR, Dung MH. In: Smith D, editor. *Modeling the Hydrogen Bond, ACS Symposium Series*. Washington (DC); American Chemical Society; 1994. p 152–174.
25. Koch U, Popelier PLA. J Phys Chem 1995;99:9747–9754.
26. Krokidis X, Noury S, Silvi B. J Phys Chem A 1997;101:7277–7282.
27. Krokidis X, Vuilleumier R, Borgis D, Silvi B. Mol Phys 1999;96(2):265–273.
28. Fuster F, Silvi B. Theor Chem Acc 2000;104:13–21.
29. Boys SF, Bernardi F. Mol Phys 1970;19:553–566.
30. Cao WL, Gatti C, MacDougall PJ, Bader RFW. Chem Phys Lett 1987;141:380–385.
31. de Vries RY, Briels WJ, Feil D, te Velde G, Baerends EJ. Can J Chem 1996;74: 1054–1058.
32. Bader RFW, Platts JA. J Chem Phys 1997;107:8545–8553.
33. Pendas AM, Blanco MA, Costales A, Mori-Sanchez P, Luana V. Phys Rev Lett 1999;83:1930–1933.
34. Taylor A, Matta CF, Boyd RJ. J Chem Theor Comput 2007;3:1054–1063.
35. Timerghazin QK, Peslherbe GH. J Chem Phys 2007;127:064108.
36. Bader RFW, Nguyen-Dang TT, Tal Y. J Chem Phys 1979;70(9):4316–4329.
37. Thom R. *Structural Stability and Morphogenesis*. Reading (MA): Benjamin; 1975.
38. Bader RFW, Gillespie RJ, MacDougall PJ. J Am Chem Soc 1988;110:7329–7336.
39. Bader RFW. In: Maruani J, editor. *Molecules in Physics, Chemistry and Biology*. Dordrecht: Kluwer Academic; 1989.
40. Bader RFW, Johnson S, Tang T-H, Popelier PLA. J Phys Chem 1996;100: 15398–15415.
41. Matta CF, Hernandez-Trujillo J, Bader RFW. J Phys Chem A 2002;106:7369–7375.
42. Matta CF, Hernandez-Trujillo J. J Phys Chem A 2003;107:7496–7504.
43. Hernandez-Trujillo J, Matta CF. Stuct Chem 2007;18:849–857.
44. Firme CL, Antunes OAC, Esteves PM. Chem Phys Lett 2009;468:129–133.
45. Bader RFW, Johnson S, Wang T-H, Popelier PLA. J Chem Phys 1996;100:15398–15415.
46. Matta CF, Bader RFW. Proteins 2002;48(3):519–538.
47. Poater J, Fradera X, Duran M, Sola M. Chem—Eur J 2003;9(2):400–406.
48. Matito E, Poater J, Sola M, Duran M, Salvador P. J Phys Chem A 2005;109: 9904–9910.
49. Becke AD. Int J Quantum Chem 1983;23:1915.
50. Becke AD. Int J Quantum Chem 1985;27:585–594.
51. Becke A, Edgecombe KE. J Chem Phys 1990;92(9):5397–5403.
52. Tal Y, Bader RFW. Int J Quantum Chem 1978;12:153–168.
53. Bader RFW, Gatti C. Chem Phys Lett 1998;287:233–238.
54. Gatti C, Bertini L. Acta Crystallogr 2004;A60:438–449.
55. Gatti C, Cargnoni F, Bertini L. J Comput Chem 2003;24(4):422–436.

56. Hirshfeld FL. Theor Chim Acta 1977;44:129–138.
57. Nalewajski RF, Parr RG. Proc Natl Acad Sci U S A 2000;97:8879–8882.
58. Ayers PW. Theor Chem Acc 2006;115:370–378.
59. Bultinck P, Ayers PW, Fias S, Tiels K, Van Alsenoy C. Chem Phys Lett 2007;444: 205–208.
60. Bultinck P, Cooper DL, Van Neck D. Phys Chem Chem Phys 2009;11:3424–3429.
61. Davidson ER, Chakravorty S. Theor Chim Acta 1992;83:319–330.
62. Parr RG, Ayers PW, Nalewajski RF. J Phys Chem A 2005;109:3957–3959.
63. Breneman CM, Wiberg KB. J Comput Chem 1990;11(3):361–373.
64. Balanarayan P, Gadre SR. J Chem Phys 2006;124:204113.
65. Li L, Parr RG. J Chem Phys 1986;84:1704–1711.
66. Blanco MA, Martin Pendas A, Francisco E. J Chem Theor Comput 2005;1: 1096–1109.
67. Francisco E, Martin Pendas A, Blanco MA. J Chem Theor Comput 2006;2:90–102.
68. Vanfleteren D, Van Neck D, Bultinck P, Ayers PW, Waroquier M. J Chem Phys 2010;132:164111.
69. Mayer I, Salvador P. J Chem Phys 2009;130:234106.
70. Goli M, Shahbazian S. Theor Chem Acc 2011;129:235–245.
71. Matta CF. *From Atoms in Amino Acids to the Genetic Code and Protein Stability, and Backwards, in Quantum Biochemistry*. Weinheim, Germany: Wiley-VCH; 2010.
72. Matta CF, Arabi AA. Future Med Chem 2011;3(8):969–994.
73. Sukumar N, Breneman CM, Matta CF, Boyd RJ. *QTAIM in Drug Discovery and Protein Modeling, in The Quantum Theory of Atoms in Molecules: From Solid State to DNA and Drug Design*. Weinheim, Germany: Wiley-VCH; 2007. p 471–498.
74. Bader RFW. J Phys Chem A 2011;115:12667–12676.

5

DENSITY FUNCTIONAL APPROACH TO THE ELECTRONIC STRUCTURE OF MATTER

N. SUKUMAR

5.1 THE HOHENBERG–KOHN THEOREMS

We have seen in Chapter 3 that, in principle, we do not need the full many-electron wave function to determine the energy and other properties of quantum mechanical systems. As nonrelativistic quantum mechanics of molecular systems involves only two-body Coulomb forces, the two-particle density matrix contains all the information to determine these properties exactly. In fact, the one-electron density $\rho(\mathbf{r})$ is often sufficient for most purposes. This general result was shown by Walter Kohn and his student Pierre Hohenberg in 1964 [1]. Walter Kohn was born in Vienna in 1923 to Salomon and Gittel Kohn. Gittel's parents, Rappaport, were orthodox Jews. Salomon ran a business producing and selling quality art postcards. Gittel was highly educated, with a flair for languages. At the Akademische Gymnasium, Walter's favorite subject was Latin; mathematics earned him his only C. Following the union (Anschluss) of Austria with the German Third Reich in 1938, the Kohn family business was confiscated and Walter was expelled from school. His sister, Minna, managed to emigrate to England and Walter too in August 1939, where they were taken in by Charles and Eva Hauff of Sussex; but Salomon and Gittel were unable to leave Austria and were both murdered during the holocaust. In May 1940, Churchill ordered male "enemy aliens" to be interned, and thus Walter found himself, at the age of 17, interned in various camps, before being shipped off to Canada.

A Matter of Density: Exploring the Electron Density Concept in the Chemical, Biological, and Materials Sciences, First Edition. Edited by N. Sukumar.

At various camps in Quebec and New Brunswick, Walter managed to study physics, chemistry, and mathematics with the help of camp educational programs taught by other internees. By working as a lumberjack, he was able to save money to buy textbooks in mathematics and chemical physics and passed McGill University's junior matriculation examination. On his release from internment in January 1942, Walter was welcomed to the home of Professor Bruno Mendel and his wife Hertha in Toronto. Even so, attending the University of Toronto, he was not allowed to enroll in chemistry courses on account of his German nationality. After $2\frac{1}{2}$ years in the undergraduate program, while also serving in the Canadian Army, Walter received a bachelor's degree in applied mathematics and then a master's. Receiving a Lehman fellowship, he joined Harvard for a Ph.D. under Julian Schwinger. Here Kohn developed a variational principle for three-body scattering, receiving his Ph.D. in 1948. He then stayed on at Harvard, dividing his time between research and teaching one of the first broad courses on solid state physics in the United States. In 1951, Kohn was at Neils Bohr's Institute in Copenhagen on a National Research Council fellowship, before returning to the United States to teach at the Carnegie Institute of Technology (now Carnegie Mellon University). Regular visits and summer jobs at Bell Labs provided opportunity for interaction and/or collaboration with researchers on the cutting edge of solid state physics, such as John Bardeen, G. Wannier, Phillip Anderson, and Quin Luttinger.

Kohn moved to the University of California at San Diego in 1960 and, in the fall of 1963, spent a sabbatical semester at the École Normale Supérieure in Paris. Here, reading some metallurgical literature, he became interested in the concept of effective charge of an atom in an alloy, which describes the charge transfer between atomic cells locally in coordinate space (in contrast to the delocalized momentum space picture used in solid state physics). Wondering whether an alloy is characterized completely or only partially by its electronic density distribution, Kohn proved that the external potential $v(\mathbf{r})$ is determined, within a trivial additive constant, by the distribution of electron density $\rho(\mathbf{r})$. As $\rho(\mathbf{r})$ determines the number of electrons,

$$N = \int \rho(\mathbf{r})d\mathbf{r}, \tag{5.1}$$

it follows that $\rho(\mathbf{r})$ also uniquely determines the ground state wave function ψ, the ground state electronic energy,

$$E = E[\rho(\mathbf{r})], \tag{5.2}$$

the molecular structure and all other electronic properties of the molecule. In 1981, Riess and Münch [2] extended the Hohenberg and Kohn theorem to subdomains of a bounded quantum system, showing that the ground state density of an arbitrary subdomain uniquely determines the ground state properties of this or any other domain. In fact, any nonzero volume part of the nondegenerate ground state electron density contains all information about the molecule. Paul Mezey [3, 4] has termed this *the holographic electron density theorem*. Furthermore, all information about molecular properties not exhibited by a given molecular structure but exhibited by the same molecule in a different state or conformation (such as the response of a molecule to a specific interaction) is fully encoded in any nonzero volume of the nondegenerate ground state electron density. This principle provides theoretical justification for the study of structure–activity relationships (Chapters 7 and 8), as most biological activities deal not with properties of isolated molecules in their equilibrium geometries but with the response of molecules to complex intermolecular interactions.

Kohn established an existence theorem for the unique energy functional $E[\rho(\mathbf{r})]$, for the case of a nondegenerate ground state, by *reductio ad absurdum*: assume that there exists another potential $v'(\mathbf{r})$ with ground state ψ' that gives rise to the same density $\rho(\mathbf{r})$ as given by the potential $v(\mathbf{r})$ with ground state ψ. Except in the case where $v(\mathbf{r}) - v'(\mathbf{r}) = \text{constant}$, we must have $\psi' \neq \psi$, as they satisfy different Schrödinger equations:

$$E = <\psi|H|\psi> \quad \text{and} \quad E' = <\psi'|H'|\psi'>, \tag{5.3}$$

and the ground states were assumed nondegenerate. Here E and E' are the respective ground state energies of the electronic Hamiltonians H and H':

$$H = T + V + U \quad \text{and} \quad H' = T + V' + U, \tag{5.4}$$

where

$$< \psi|V|\psi > = \int v(\mathbf{r})\rho(\mathbf{r})\mathrm{d}\mathbf{r}, \tag{5.5}$$

and T and U are the kinetic energy and the energy of interelectron repulsion, respectively. From the variational theorem for the Hamiltonian H', we have

$$E' = <\psi'|H'|\psi'> \; < \; <\psi|H'|\psi > = <\psi|H + V' - V|\psi>, \tag{5.6}$$

so that

$$E' < E + \int [v'(\mathbf{r}) - v(\mathbf{r})]\rho(\mathbf{r})\mathrm{d}\mathbf{r}. \tag{5.7}$$

Similarly, from the variational theorem for the Hamiltonian H, we have

$$E = <\psi|H|\psi> \; < \; < \psi'|H|\psi'> = <\psi'|H' + V - V'|\psi'>, \tag{5.8}$$

or

$$E < E' + \int [v(\mathbf{r}) - v'(\mathbf{r})]\rho(\mathbf{r})\mathrm{d}\mathbf{r}. \tag{5.9}$$

Adding Equations 5.7 and 5.9, the terms under the integrals cancel out, leading to an inconsistency, unless the original presumption was in error. Thus $v(\mathbf{r})$ and $v'(\mathbf{r})$ can differ by at most a trivial constant. So $v(\mathbf{r})$, and thus H, is a unique functional of $\rho(\mathbf{r})$.

Kohn then set his new American student in Paris, Pierre Hohenberg, to prove a variational principle for the energy functional $E[\rho(\mathbf{r})]$: for a trial density $\rho'(\mathbf{r})$ satisfying

$$\rho'(\mathbf{r}) \geq 0 \tag{5.10}$$

and

$$N = \int \rho'(\mathbf{r})d\mathbf{r}, \tag{5.11}$$

$$E_0 \leq E[\rho'(\mathbf{r})], \tag{5.12}$$

where E_0 is the true ground state energy. This result was also published in the same 1964 paper.

The Hohenberg–Kohn theorems hinge on questions of N-representability and v-representability: the density $\rho(\mathbf{r})$ must be derivable from an antisymmetric wave function (N-representability), which must be the ground state solution of the Schrödinger equation corresponding to some potential $v(\mathbf{r})$ (v-representability). It was shown by Gilbert [5] in 1975 that any positive density function satisfying Equations 5.10 and 5.11 and the condition:

$$\int |\nabla\rho'(\mathbf{r})^{1/2}|^2\, d\mathbf{r} < \infty \tag{5.13}$$

can be derived from an antisymmetric wave function, thereby solving the N-representability problem, but v-representability is, in general, not guaranteed by an arbitrary trial density. Around the same time, Mel Levy devised what is now known as *the constrained search formalism* [6–8] by considering the functional $F[\rho(\mathbf{r})]$ of the density that minimizes the expected value of $T + U$ over all antisymmetric wave functions ψ, which yield that density $\rho(\mathbf{r})$:

$$F[\rho(\mathbf{r})] = \min_{\psi\rightarrow\rho} <\psi|T + U|\psi>. \tag{5.14}$$

$F[\rho(\mathbf{r})]$ can then be found by minimization of the expectation value $<\psi|\mathrm{T} + \mathrm{U} + \mathrm{V}|\psi>$ with respect to arbitrary variations of the wave function ψ. Here, the one-electron potential operator V of Equation 5.5 acts as the Lagrange multiplier for the constraint that the wave function ψ yields the correct density $\rho(\mathbf{r})$. To tackle the v-representability problem, Mel Levy similarly designed a constrained search formalism [6–8] for the first-order density matrix:

$$\gamma(r|r') = N \int d\mathbf{r}_2 \ldots \int d\mathbf{r}_N \psi^*(\mathbf{r}, \mathbf{r}_2 \ldots \mathbf{r}_N)\psi(\mathbf{r}', \mathbf{r}_2 \ldots \mathbf{r}_N) \tag{5.15}$$

in terms of a functional

$$W[\gamma(r|r')] = \min_{\psi \to \gamma} <\psi|U|\psi> \tag{5.16}$$

of the first-order density matrix that minimizes the expectation value of the interelectron repulsion U over all wave functions ψ that yield that density matrix $\gamma(r|r')$.

5.2 THE CHEMICAL POTENTIAL

We can thus write the Hohenberg–Kohn energy functional in terms of the Levy functional F[ρ(**r**)]:

$$E[\rho(\mathbf{r})] = F[\rho(\mathbf{r})] + \int v(\mathbf{r})\rho(\mathbf{r})d\mathbf{r}. \tag{5.17}$$

Then, from the second Hohenberg–Kohn theorem, Equation 5.12, we obtain for arbitrary variations in the density $\rho'(\mathbf{r})$, with constant $v(\mathbf{r})$:

$$\delta\{E[\rho'(\mathbf{r})] - \mu N[\rho'(\mathbf{r})]\} = 0, \tag{5.18}$$

where μ is the Lagrange multiplier for the normalization constraint on the density, Equation 5.11, and can be defined as the functional derivative of the energy functional with respect to the density:

$$\mu = \frac{\delta E[\rho(\mathbf{r})]}{\delta\rho(\mathbf{r})}. \tag{5.19}$$

In 1978, Robert Parr and coworkers [9, 10] identified this chemical potential as the negative of the electronegativity familiar to chemists:

$$\chi = -\mu = \left(\frac{\partial E}{\partial N}\right)_v \tag{5.20}$$

$$\approx {}^1\!/\!_2(I + A), \tag{5.21}$$

where I is the ionization energy and A the electron affinity. Parr's electronegativity formula thus reduces to Mulliken's electronegativity definition [11] in the finite difference approximation. Parr had earlier worked with DuPont chemist Rudolph Pariser on the molecular orbital theory of π-electron systems, originating the Pariser–Parr–Pople semiempirical method [12]. Now at the University of North Carolina, Parr was among the first theoretical chemists to work on density functional theory and is perhaps the person most responsible for the immense popularity that this theory has gained among chemists. Parr and coworkers also showed that the equilibrium condition requires equalization of chemical potentials in a molecule, a fact that had been recognized earlier by Sanderson [13] and termed the principle of equal orbital electronegativities.

We can pursue the thermodynamic analogy further to obtain a generalization of the Hellmann–Feynman theorem:

$$dE = \mu dN + \int \rho(\mathbf{r}) dv(\mathbf{r}) d\mathbf{r}. \tag{5.22}$$

Taking functional derivatives of the Levy functional, Equation 5.17 can be rewritten as,

$$E[\rho] = N\mu + F[\rho] - \int \left(\frac{\delta F[\rho]}{\delta \rho(\mathbf{r})}\right) \rho(\mathbf{r}) d\mathbf{r}. \tag{5.23}$$

Defining the sum of the last two terms as $-Q$, that is,

$$-Q = F[\rho] - \int \left(\frac{\delta F[\rho]}{\delta \rho}\right) \rho(\mathbf{r}) d\mathbf{r} \tag{5.24}$$

and differentiating yields:

$$dE = N d\mu + \mu dN - dQ, \tag{5.25}$$

or using Equation (5.22),

$$N d\mu = dQ + \int \rho(\mathbf{r}) dv(\mathbf{r}) d\mathbf{r}, \tag{5.26}$$

which is a generalization of the Gibbs–Duhem equation [14].

The functional derivative of the chemical potential with respect to the density is the chemical hardness (or band gap in solids):

$$\eta = \left.\frac{\delta \mu}{\delta \rho(\mathbf{r})}\right|_{v(\mathbf{r})} = \left.\frac{\delta^2 E[\rho(\mathbf{r})]}{\delta \rho(\mathbf{r})^2}\right|_{v(\mathbf{r})} = \left(\frac{\partial^2 E}{\partial N^2}\right)_v \approx \tfrac{1}{2}(I - A), \tag{5.27}$$

which appears in Pearson's principle of hard and soft acids and bases [15, 16]. The other functional derivatives have been identified as the Fukui function [17–20].

$$f = \left.\frac{\delta \mu}{\delta v(\mathbf{r})}\right|_N = \frac{\delta^2 E[\rho(\mathbf{r})]}{\delta \rho(\mathbf{r}) \delta v(\mathbf{r})} = \left.\frac{\partial \rho(\mathbf{r})}{\partial N}\right|_v, \tag{5.28}$$

which measures propensity for chemical reactivity [21], and the response function $\delta\rho(\mathbf{r})/\delta v(\mathbf{r})|_N = \delta^2 E\rho(\mathbf{r})]/\delta v(\mathbf{r})\delta v(\mathbf{r}')|_{\rho(\mathbf{r})}$. Further discussions on these chemical concepts derived from density functional theory is relegated to Chapter 7, and applications of these concepts to drug design can be found in Chapter 8.

5.3 THE EXCHANGE-CORRELATION HOLE

The interelectron repulsion U appearing in Equation 5.4 is a two-body operator with the expectation value:

$$\langle\psi|\mathrm{U}|\psi\rangle = \tfrac{1}{2}\int \mathbf{dr}\int \mathbf{dr}'\,\frac{\rho(\mathbf{r},\mathbf{r}')}{|\mathbf{r}-\mathbf{r}'|}, \tag{5.29}$$

where the pair density,

$$\rho(\mathbf{r},\mathbf{r}') = N(N-1)\int \mathbf{dr}_3\ldots\int \mathbf{dr}'_N|\psi(\mathbf{r},\mathbf{r}',\mathbf{r}_3\ldots\mathbf{r}_N)|^2, \tag{5.30}$$

gives the probability of simultaneously finding an electron at the point $\mathbf{r}$ within volume element $\mathbf{dr}$ and another electron at the point $\mathbf{r}'$ within volume element $\mathbf{dr}'$. Here, we have assumed the electron coordinates to include both space and spin, and the integrals over the electron coordinate to include summations over the spins as well. From Equations 5.1 and 5.30, the electron density $\rho(\mathbf{r})$ is then given by

$$\rho(\mathbf{r}) = \frac{1}{N-1}\int \mathbf{dr}'\rho(\mathbf{r},\mathbf{r}'), \tag{5.31}$$

while double integration of Equation 5.30 leads to the condition

$$\int \mathbf{dr}\int \mathbf{dr}'\rho(\mathbf{r},\mathbf{r}') = N(N-1). \tag{5.32}$$

In Chapter 2, we have seen that in a classical description, the motions of the electrons are not correlated, so that the classical simultaneous probability of finding an electron at $\mathbf{r}$ and another electron at $\mathbf{r}'$ is simply the product of the individual probabilities:

$$P^{\mathrm{class}}(\mathbf{r},\mathbf{r}') = \rho(\mathbf{r})\rho(\mathbf{r}'). \tag{5.33}$$

Substituting this classical definition into Equation 5.29 gives

$$\langle\psi|U^{\mathrm{class}}|\psi\rangle = \tfrac{1}{2}\int \mathbf{dr}\int \mathbf{dr}'\,\frac{\rho(\mathbf{r})\rho(\mathbf{r}')}{|\mathbf{r}-\mathbf{r}'|}, \tag{5.34}$$

which is the classical Coulomb repulsion or the Hartree energy, treated in Section 3.2. But electrons obey Fermi–Dirac statistics, and thus electrons with the same spin are kept apart from the Pauli exclusion principle. Furthermore, owing to Coulomb repulsion, even electrons of opposite spin tend to avoid each other more than suggested by the average description above. These two effects mean that each electron reduces the probability of finding another electron around it and thus creates a depletion or hole of electron density around itself. This is known as *the exchange-correlation hole*, with a density $\rho_{XC}(\mathbf{r}, \mathbf{r}')$.

The quantum mechanical pair density can now be written in terms of this exchange-correlation hole as

$$\rho(\mathbf{r}, \mathbf{r}') = \rho(\mathbf{r})\rho(\mathbf{r}') + \rho(\mathbf{r})\rho_{XC}(\mathbf{r}, \mathbf{r}'). \tag{5.35}$$

From normalization, we thus see that every electron is surrounded by a hole in the electron density of equal and opposite charge:

$$\int \mathrm{d}\mathbf{r}'\rho_{XC}(\mathbf{r}, \mathbf{r}') = -1. \tag{5.36}$$

This normalization condition is also known as *the sum rule*. It is easy to see that the exchange part of the hole (also known as *the Fermi hole*) has the charge of -1, while the correlation hole (also known as *the Coulomb hole*) has a net zero charge.

5.4 THE KOHN–SHAM EQUATION

Let us apply Levy's constrained search formalism to define a functional $T_0[\rho(\mathbf{r})]$ of the density that determines the smallest expectation value of the kinetic energy operator T:

$$T_0[\rho(\mathbf{r})] = \min_{\psi\to\rho} <\psi|T|\psi>, \tag{5.37}$$

subject to the constraint that the wave function ψ yield the correct density $\rho(\mathbf{r})$ at each point in space. This can be achieved by introducing a Lagrange multiplier $v(\mathbf{r})$ for each point in space and minimizing

$$<\psi|T|\psi> + \int v(\mathbf{r})\rho(\mathbf{r})\mathrm{d}\mathbf{r} =<\psi|T + V|\psi> \tag{5.38}$$

with respect to arbitrary variations of the (antisymmetric) wave function ψ, where V is given by Equation 5.5. But this is just the ground state energy of the Hamiltonian

$$H_0 = T + V. \tag{5.39}$$

As the kinetic energy T is a sum of one-electron operators, the many-electron wave function ψ can be constructed as a normalized Slater determinant of eigenfunctions of the equation,

$$[-\tfrac{1}{2}\nabla^2 + v(\mathbf{r})]|\varphi_k(\mathbf{r})> = \varepsilon_k|\varphi_k(\mathbf{r})>, \tag{5.40}$$

with the N lowest eigenvalues. This one-electron Schrödinger-like equation was derived in 1965 by Kohn and his postdoctoral fellow, Lu Sham, on Kohn's return to San Diego [22]. The φ_k are now known as *Kohn–Sham orbitals*. The ground state energy is the sum of those N lowest eigenvalues and

$$T_0[\rho(\mathbf{r})] = \sum_{k=1}^{N} \left\langle \varphi_k \left| -\frac{1}{2}\nabla^2 \right| \varphi_k \right\rangle \tag{5.41}$$

can be considered to be the kinetic energy of a hypothetical noninteracting system (with Hamiltonian H_0). Here, the external potential V acts as a set of Lagrange parameters to satisfy the constraint that the Kohn–Sham orbital densities sum up to give the true electron density of the interacting system:

$$\rho(\mathbf{r}) = \sum_{k=1}^{N} |\varphi_k(\mathbf{r})|^2. \tag{5.42}$$

Adding the interelectron repulsion operator U gives the energy functional for the interacting system:

$$E[\rho(\mathbf{r})] = T_0[\rho(\mathbf{r})] + \int v(\mathbf{r})\rho(\mathbf{r})d\mathbf{r} + \frac{1}{2}\int d\mathbf{r} \int d\mathbf{r}'\rho(\mathbf{r})\rho(\mathbf{r}')/|\mathbf{r}-\mathbf{r}'| + E_{xc}, \tag{5.43}$$

where we have used Equations 5.34 and 5.35, and E_{xc} is the exchange-correlation energy:

$$E_{xc} = \frac{1}{2}\int d\mathbf{r} \int d\mathbf{r}'\rho(\mathbf{r})\frac{\rho_{XC}(\mathbf{r},\mathbf{r}')}{|\mathbf{r}-\mathbf{r}'|} \tag{5.44}$$

defined in terms of the exchange-correlation hole $\rho_{XC}(\mathbf{r},\mathbf{r}')$ introduced in Section 5.3. The Kohn–Sham Equation 5.40 may also be rewritten in terms of a Kohn–Sham effective potential $v_{KS}(\mathbf{r})$ as:

$$[-½\nabla^2 + v_{KS}(\mathbf{r})]|\varphi_k(\mathbf{r})> = \varepsilon_k|\varphi_k(\mathbf{r})>, \tag{5.45}$$

where

$$v_{KS}(\mathbf{r}) = v(\mathbf{r}) + v_e(\mathbf{r}) + v_{xc}(\mathbf{r}), \tag{5.46}$$

$$v_e(\mathbf{r}) = \int d\mathbf{r}'\frac{\rho(\mathbf{r}')}{|\mathbf{r}-\mathbf{r}'|} \tag{5.47}$$

is the electrostatic potential, and $v_{xc}(\mathbf{r})$ is the exchange-correlation potential [23]:

$$v_{xc}(\mathbf{r}) = \frac{\delta E_{XC}}{\delta\rho(\mathbf{r})}. \tag{5.48}$$

The solution to the interacting many-electron problem thus reduces to finding a good approximation to the unknown exchange-correlation term, followed by solving the set of one-electron Kohn–Sham Equations 5.45. Kohn and Sham thus obtained a practical recipe for obtaining the electron density and the total electronic energy, formulating a theory going beyond the Hartree–Fock treatment—including electron correlation—at the cost of solving the one-electron Equation 5.45. This work was cited in the award of the Nobel Prize for chemistry in 1998 [24] to Walter Kohn, now at the University of California at Santa Barbara.

The simplest approximation (the Local density approximation [25]) is obtained by writing $E_{xc}(\mathbf{r})$ as a functional of only the local density $\rho(\mathbf{r})$ at $\mathbf{r}$:

$$E_{xc}^{LDA}[\rho] = \int d\mathbf{r} e_{xc}^{LDA}[\rho(\mathbf{r})]. \tag{5.49}$$

We can also explicitly introduce spins to give the Local spin density approximation (LSDA):

$$E_{xc}^{LSDA}[\rho_{\uparrow}, \rho_{\downarrow}] = \int d\mathbf{r} e_{xc}^{LSDA}[\rho_{\uparrow}(\mathbf{r}), \rho_{\downarrow}(\mathbf{r})], \tag{5.50}$$

where $e_{xc}{}^{LSDA}$ is the exchange-correlation energy density of a uniform electron gas with spin densities $\rho_{\uparrow}(\mathbf{r})$ and $\rho_{\downarrow}(\mathbf{r})$ equal to their local atomic or molecular values. Considering its simplicity, LSDA performs remarkably well for many atoms, molecules, and solids, and it has been widely used by solid-state physicists. Thus LSDA gives accurate geometries and charge densities for most systems. Inspite of a systematic overbinding, binding energies obtained with LSDA are often better than 1 eV. Calculated vibrational frequencies are often accurate to within 10–20%. In general, LSDA results are much better than those obtained with the Hartree–Fock approximation. This "unreasonable" effectiveness of the local density approximation is somewhat surprising, as the electron density distribution in an atom or small molecule is quite far from that of a uniform electron gas! The successes of LSDA can be attributed to various exact properties of its exchange-correlation hole, such as the sum rule, Equation 5.36. However, there are serious problems with LSDA that render it of limited use to problems of interest to chemists. Thus, for instance, negative ions are unbound in LSDA! LSDA overestimates the binding energy of the F_2 molecule by 100% and that of CO_2 by as much as 84 kcal/mol. Owing to rapid density variations within the interiors of atoms, atoms and small molecules represent very severe tests on the quality of density functional approximations. In general, LSDA predicts wrong dissociation limits for many molecules, incorrect ground states for many atoms and is severely inadequate for thermochemistry [26]. Furthermore, in contrast to methods such as configuration interaction, LSDA—and even more sophisticated density functionals—fail in situations where near-degeneracies are encountered (the so-called nondynamical correlation).

A possible improvement on LSDA is to introduce dependence on the local density and its gradients:

$$E_{xc}^{GGA}[\rho_\uparrow, \rho_\downarrow, \nabla\rho_\uparrow, \nabla\rho_\downarrow] = \int d\mathbf{r}\ e_{xc}^{GGA}[\rho_\uparrow(\mathbf{r}),\ \rho_\downarrow(\mathbf{r}),\ \nabla\rho_\uparrow(\mathbf{r}),\ \nabla\rho_\downarrow(\mathbf{r})]. \tag{5.51}$$

Such approximations are known as *Generalized gradient approximations* (GGAs). These are often used in conjunction with a number of adjustable parameters, which are determined by fitting known results of high accuracy for diverse systems. Introducing dependence on the Laplacian of the density and its gradients leads to meta-GGA.

Other methods of going beyond LSDA are based on obtaining an improved description of the exchange-correlation hole. John Perdew [27, 28] showed that while E_{xc}^{LDA} obeys the sum rule, Equation 5.36, straightforward addition of gradient terms to LSDA results in an exchange-correlation hole that does not satisfy the sum rule. Another exact property that should be required of the exchange-correlation energy functional is the asymptotic condition far from the nuclei, such as in the exponential tails of atomic and molecular electron distributions, where it follows from Equation 5.44 and hole normalization, Equation 5.36, that

$$e_{xc}(\mathbf{r}) \rightarrow -\frac{\rho(\mathbf{r})}{2\mathbf{r}} \quad \text{as } \mathbf{r} \rightarrow \infty. \tag{5.52}$$

This asymptotic condition is not satisfied by LSDA, but Axel Becke in 1988 designed a gradient correction for the exchange ($E_x{}^{B88}$) that satisfied this requirement [29]. Gradient corrections for correlation are important for properties that involve electron nonconserving processes, such as ionization energies, and in 1991, Perdew and Wang [30–34] designed such a parameter-free gradient-corrected correlation functional (E_c^{PW91}). These functionals reduced the errors associated with DFT calculations by an order of magnitude from LSDA.

Now exchange energies are known to be much larger than correlation energies and we already know, from Hartree-Fock theory, how to treat exchange exactly. This immediately suggests a hybrid method, using the exact Hartree–Fock exchange (E_x^{HF}) in combination with a density functional treatment for the correlation component. Unfortunately, this straightforward approach fails miserably and results are generally worse than LSDA, owing to fortuitous cancellation of errors between the exchange and correlation components of the energy in LSDA. So in 1993, Becke proposed using half of the exact Hartree-Fock exchange, leaving the other half together with the correlation energy to be determined by a GGA approximation [35]. Another hybrid scheme (B3PW91) suggested by him involved three semiempirical parameters a_0, a_x and a_c determined by fitting experimental data, and an exchange-correlation energy functional of the form:

$$E_{\text{xc}}^{\text{B3PW91}} = E_{\text{xc}}^{\text{LSDA}} + a_0(E_{\text{x}}^{\text{HF}} - E_{\text{x}}^{\text{LSDA}}) + a_{\text{x}}\Delta E_{\text{x}}^{\text{B88}} + a_{\text{c}}\Delta E_{\text{c}}^{\text{PW91}}, \tag{5.53}$$

where Δ signifies the gradient part of the corresponding functional, and the exact Hartree–Fock exchange E_{x}^{HF} is evaluated using the Kohn–Sham orbitals [36]. The most widely used density functional for chemical applications is B3LYP, which has the same form as mentioned above, but with the gradient part of the correlation energy replaced by one prescribed by Chengteh Lee, Weitao Yang, and Robert Parr (LYP) [37]. The success of the hybrid schemes B3PW91 and B3LYP underscores the importance of nonlocal exchange—beyond that which can be accounted for by local density gradients alone [38–40]. A discussion on more recent density functionals can be found in Chapter 6.

5.5 A MATTER OF PHASE

The unique mapping between the density $\rho(\mathbf{r})$ and the potential $v(\mathbf{r})$ shown by Hohenberg and Kohn does not apply to the case of an infinite periodic insulator in an electric field. Application of a uniform electric field to a system with a periodic potential does not permit a ground state solution, because a translation against the direction of the field by a whole number of lattice constants will always result in a decrease in the electronic energy. This impossibility of a ground state in the presence of a finite electric field invalidates the Hohenberg–Kohn proof of DFT for a periodic system in an electric field. The clue to extending DFT to such cases lies in the phase associated with a quantum wave function.

In 1963, Herzberg and Longuet-Higgins predicted a change of sign of the electronic wave function [41] around a closed loop of structure deformations encircling a point of Jahn–Teller degeneracy [42] between two electronic states: "*If the wave function of a given electronic state changes sign when transported adiabatically around a closed loop in nuclear configuration space then the state must become degenerate with another at some point within the loop. This sign reversal condition is necessary and sufficient to establish the existence of an intersection*" [41]. A few years earlier, Yariv Aharonov and David Bohm [43] had discovered that electromagnetic potentials play a far more fundamental role in the quantum theory than in classical mechanics. In classical mechanics, it is the electric and magnetic fields that are the fundamental quantities and electromagnetic potentials are introduced merely for convenience. Aharonov and Bohm found that quantum mechanics predicts physical effects of potentials on charged particles even in regions where all the fields, and hence all forces on the particles, are zero. This can be seen by passing an electron beam through a beam splitter (Fig. 5.1). The two resultant beams are allowed to pass on opposite sides (1 and 2) of a very tightly wound solenoid (so that the magnetic field $\mathbf{H}$ is essentially confined within the solenoid) and then recombined. The electromagnetic vector

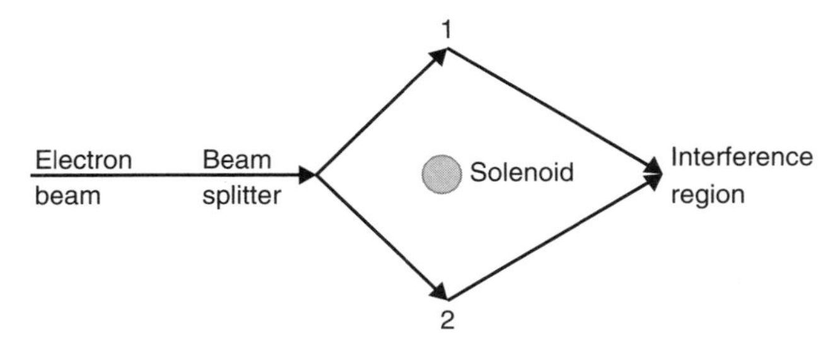

Figure 5.1 Schematic setup for the Aharonov–Bohm effect [43], demonstrating the fundamental role of electromagnetic potentials in the quantum theory. An electron beam is split into two by a beam splitter; the two resultant beams pass on opposite sides (1 and 2) of a tightly wound solenoid and are then recombined. Because of the nonzero magnetic flux through any closed loop around the solenoid, there is a nontrivial and experimentally detectable phase difference between the two beams 1 and 2, even in the absence of any significant magnetic field along the paths traversed by the electrons.

potential **A**, however, cannot be zero everywhere outside the solenoid because the total magnetic flux through any closed loop around the solenoid is a constant:

$$\oint \mathbf{A}.\mathbf{d}x = \mathbf{H}.\mathrm{ds} = \Phi. \tag{5.54}$$

Along each path 1 or 2, the wave function of the electron will acquire a phase factor $\exp\{-\mathrm{ie}/c\hbar \int_1 \mathbf{A}.\mathbf{dx}\}$ and $\exp\{-\mathrm{ie}/c\hbar \int_2 \mathbf{A}.\mathrm{dx}\}$, so that the phase difference between the two beams is

$$\Delta \mathrm{S}/\hbar = -\frac{\mathrm{e}}{c\hbar} \oint \mathbf{A}.\mathrm{d}x. \tag{5.55}$$

The Aharonov–Bohm effect was experimentally confirmed soon thereafter by Chambers [44], but a similar effect had been observed in the context of optics a few years earlier by Pancharatnam [45]. It was only in 1979 that the connection to molecular physics was made, when Alden Mead and Donald Truhlar [46] realized that the sign change of Herzberg and Longuet-Higgins is a molecular analog of the Aharonov–Bohm effect. The analog of the magnetic vector potential in the moecular situation is the diagonal Born vector term introduced in Chapter 3:

$$\mathbf{A}_{ii} = i < \psi_i | \nabla_{\mathbf{R}} | \psi_i >, \tag{5.56}$$

where **R** is a nuclear coordinate and ψ_i the electronic wave function in state i. The sign change of Herzberg and Longuet-Higgins is a manifestation of what is now known as *the geometric phase* or *Berry phase* [47–56]. The general nature of this phenomenon was discovered in 1984 by Michael Berry [47, 48] and generalized further [49–52] to encompass nonadiabatic, noncyclic and, even non-Abelian situations. The Kramers degeneracy between up spin and down spin states is a molecular manifestation of a non-Abelian geometric phase. In the general context, the geometric phase is given by:

$$\gamma = \int \mathbf{A}_{ii}.d\mathbf{R}. \tag{5.57}$$

In a Jahn–Teller molecule, where two electronic states are degenerate at a point in nuclear configuration space, there is thus a phase change of π acquired by the electronic wave function around the point of degeneracy, and a compensating phase change acquired by the nuclear wave function, such that the total wave function remains single valued. The effects of this geometric phase would be manifested as interference between trajectories passing on either side of the Jahn–Teller intersection. This effect was confirmed experimentally in 1986 through the observation of fractional quantization in the molecular pseudorotation (a series of molecular deformations in a closed loop around the degeneracy) of Na_3 [53]. In addition to the geometric phase in the molecular situation, Mead and Truhlar [46] predicted a further phase factor due to permutation of identical nuclei.

For a path 1, 2, ..., M in configuration space, the geometric phase may be rewritten as:

$$\begin{aligned}\gamma &= -\text{Im. log} \prod_{s=1}^{M} \langle \psi(s)|\psi(s+1)\rangle \\ &= -\text{Im. log Tr.} \prod_{s=1}^{M} \rho'(s),\end{aligned} \tag{5.58}$$

where $\hat{\rho}(s) = |\psi(s)> <\psi(s)|$ is the density operator. The geometric phase formalism has been applied to develop a density polarization functional theory of insulators [57–63]. Here, knowledge of both the change in periodic density and the change in polarization is required to obtain the change in periodic potential and the change in the electric field. The exchange-correlation functional now has an explicit dependence on the macroscopic polarization, which induces an exchange-correlation electric field. The macroscopic polarization is defined as the dipole moment induced in the material in response to an applied electric field and is conventionally related to the total surface charge. In the new theory the macroscopic polarization is a bulk property, the first moment of the electron distribution, and can be defined in terms of the geometric phase:

$$\begin{aligned}P &= -\text{Im. log} < \psi|e^{2\pi i\hat{\mathbf{R}}}|\psi > \\ &= i\int_{\text{BZ}} < \psi|\nabla_{\mathbf{R}}|\psi > .d\mathbf{R} = \gamma,\end{aligned} \tag{5.59}$$

where $\hat{\mathbf{R}}$ is the position operator, the integral is over the Brillouin zone and volume normalization has been assumed.

For conductors, the excitation gap goes to zero as the size of the system becomes large, while for insulators it remains finite. Walter Kohn showed in

1964 [64] that this difference is also reflected in the ground-state many-body wave function, which is delocalized for conductors and localized for insulators. The insulating state thus represents a rather special organization of electrons in the ground state, which can be probed by the localization tensor or the second moment of the electron distribution [65]:

$$<\mathbf{r}_i\mathbf{r}_j> = \frac{1}{N}\int d\mathbf{r}\int d\mathbf{r}'(\mathbf{r}-\mathbf{r}')_i(\mathbf{r}-\mathbf{r}')_j\rho(\mathbf{r})\rho_{XC}(\mathbf{r},\mathbf{r}'), \tag{5.60}$$

This is qualitatively different in insulators and in conductors. In insulators, the integral converges for large $|\mathbf{r}-\mathbf{r}'|$ and the localization tensor is finite, whereas in conductors, the integral diverges. This is a measure of what Kohn has called the "near-sightedness" of the electron distribution [66–68]. As we have seen in Chapter 4, it is this "near-sightedness" of electronic matter that is also responsible for the approximate transferability of atomic and functional group properties between molecules.

ACKNOWLEDGMENT

I thank the Institute of Mathematical Sciences, Chennai, India, for hospitality during the preparation of this manuscript.

REFERENCES

1. Hohenberg P, Kohn W. Phys Rev 1964;136:B864–B871.
2. Riess J, Münch W. Theor Chem Acta 1981;58:295–300.
3. Mezey PG. Mol Phys 1990;96:169–178.
4. Mezey PG. J Chem Inf Comput Sci 1999;39:224–230.
5. Gilbert TL. Phys Rev B 1975;12:2111–2120.
6. Levy M. Proc Natl Acad Sci U S A 1979;76:6062–6065.
7. Levy M. Phys Rev A 1982;26:1200–1208.
8. Dreizler RM, de Providencia J. *Density Functional Methods in Physics*. New York, London: Plenum Press; 1985.
9. Parr RG, Donnelly RA, Levy M, Palke WE. J Chem Phys 1978;68:3801–3807.
10. Parr RG, Yang W. *Density-Functional Theory of Atoms and Molecules*. New York: Oxford University Press; 1989.
11. Mulliken RS. J Chem Phys 1934;2:782–793.
12. Pariser R. J Chem Phys 1953;21:568–569.
13. Sanderson RT. *Chemical Bonds and Bond Energy*. 2nd ed. New York: Academic Press; 1976.
14. Gibbs JW. *Collected Works*. New York: Longmans Green; 1931.
15. Pearson RG. *Hard and Soft Acids and Bases*. Stroudsburg (PA): Dowden, Hutchinson & Ross; 1973.

16. Parr RG, Pearson RG. J Am Chem Soc 1983;150:7512–7516.
17. Fukui K, Yonezawa T, Shingu H. J Chem Phys 1952;20:722–725.
18. Fukui K, Yonezawa T, Nagata C, Shingu H. J Chem Phys 1954;22:1433–1442.
19. Fukui K. *Theory of Orientation and Stereoselection*. Berlin: Springer-Verlag; 1975.
20. Fukui K. Science 1987;218:747–754.
21. Chattaraj PK, editor. *Chemical Reactivity Theory: A Density Functional View*. Boca Raton: CRC Press; 2009.
22. Kohn W, Sham LJ. Phys Rev 1965;140:A1133–A1138.
23. Jones RO. Introduction to DFT and exchange-correlation energy functionals. Available at http://www2.fz-juelich.de/nic-series/volume31/jones.pdf. Accessed 2012.
24. Frängsmyr T, editor. *Les Prix Nobel. The Nobel Prizes 1998*. Stockholm: Nobel Foundation; 1999.
25. Dahl JP, Avery J, editors. *Local Density Approximation in Quantum Chemistry and Solid State Physics*. New York, London; Plenum Press; 1984.
26. von Barth U. Phys Scr 2004;T109:9–39.
27. Perdew JP. Phys Rev Lett 1985;55:1665.
28. Perdew JP. Physica B 1991;172:1.
29. Becke A. Phys Rev A 1988;38:3098–3100.
30. Perdew JP. In: Ziesche P, Eschrig H, editors. *Electronic Structure of Solids*. Berlin: Akademie Verlag; 1991.
31. Perdew JP, Chevary JA, Vosko SH, Jackson KA, Pederson MR, Singh DJ, Fiolhais C. Phys Rev B 1992;46:6671–6687; 1993;48:4978.
32. Perdew JP, Burke K, Wang Y. Phys Rev B 1996;54:16533–16539.
33. Burke K, Perdew JP, Wang Y. In: Dobson JF, Vignale G, Das MP, editors. *Electronic Density Functional Theory: Recent Progress and New Directions*. New York: Plenum Press; 1998.
34. Fiolhais C, Nogueira F, Marques M, editors. *A Primer in Density Functional Theory*. Berlin, Heidelberg, New York: Springer-Verlag; 2003.
35. Becke A. J Chem Phys 1993;98:1372–1377.
36. Becke A. J Chem Phys 1993;98:5648–5652.
37. Lee C, Yang W, Parr RG. Phys Rev B 1988;37:785–789.
38. Koch W, Holthausen MC. *A Chemist's Guide to Density Functional Theory*. 2nd ed. Weinheim: Wiley-VCH; 2002.
39. March NH. *Electron Density Theory of Atoms and Molecules*. London: Academic Press; 1992.
40. Dreizler R, Gross E. *Density Functional Theory*. New York: Plenum Press; 1995.
41. Herzberg G, Longuet-Higgins HC. Discuss Faraday Soc 1963;35:77–82.
42. Jahn H, Teller E. Proc R Soc London, Ser A 1937;161:220–235.
43. Aharonov Y, Bohm D. Phys Rev 1959;115:485–494.
44. Chambers RG. Phys Rev Lett 1960;5:3–5.
45. Pancharatnam S. Proc Ind Acad Sci A 1956;44:247–262.
46. Mead CA, Truhlar DG. J Chem Phys 1979;70:2284–2296.
47. Berry MV. Proc R Soc London, Ser A 1984;392:45–57.

48. Simon B. Phys Rev Lett 1983;51:2167–2170.
49. Wilczek F, Zee A. Phys Rev Lett 1984;52:2111–2114.
50. Aharonov Y, Anandan J. Phys Rev Lett 1987;58:1593–1596.
51. Samuel J, Bhandari R. Phys Rev Lett 1988;60:2339–2342.
52. Mead CA. Phys Rev Lett 1987;59:161–164.
53. Delacrétaz G, Grant ER, Whetten RL, Wöste L, Zwanziger JW. Phys Rev Lett 1986;56:2596–2601.
54. Shapere A, Wilczek F. *Geometric Phases in Physics*. Singapore: World Scientific; 1989.
55. Zwanziger JW, Koenig M, Pines A. Ann Rev Phys Chem 1990;41:601–645.
56. Mead CA. Rev Mod Phys 1992;64:51–85.
57. King-Smith RD, Vanderbilt D. Phys Rev B 1993;47(3):1651–1654.
58. Vanderbilt D, King-Smith RD. Phys Rev B 1993;48:4442–4445.
59. Ortiz G, Martin RM. Phys Rev B 1994;49(20):14202–14210.
60. Resta R. Rev Mod Phys 1994;66(3):899–915.
61. Gonze X, Ghosez Ph, Godby RW. Phys Rev Lett 1995;74(20):4035–4038; 1997;78(2):294–297.
62. Martin RM, Ortiz G. Phys Rev B 1997;56(3):1124–1140.
63. Resta R. Phys Rev Lett 1996;77(11):2265–2267.
64. Kohn W. Phys Rev 1964;133:A171–A181.
65. Resta R. J Chem Phys 2006;124:104104.
66. Kohn W, Yaniv A. Proc Natl Acad Sci U S A 1978;75(11):5270–5272.
67. Kohn W. Phys Rev Lett 1996;76(17):3168–3171.
68. Prodan E, Kohn W. Nearsightedness of Electronic Matter; 2005. Available at http://arxiv.org/abs/cond-mat/0503124. Accessed 2010, 2012.

6

DENSITY-FUNCTIONAL APPROXIMATIONS FOR EXCHANGE AND CORRELATION

Viktor N. Staroverov

6.1 THE CHALLENGE OF DENSITY-FUNCTIONAL THEORY

Density-functional theory (DFT) is based on two pivotal theorems due to Hohenberg and Kohn [1]. The first theorem states that the ground-state density $\rho(\mathbf{r})$ of a system of electrons uniquely determines the Hamiltonian and hence all properties that can be derived from it. Using mathematical language, we can say that the total electronic energy of the system is a functional of the electron density

$$E = E[\rho]. \tag{6.1}$$

The second Hohenberg–Kohn theorem demonstrates that the exact ground-state density and energy of the system can be found by minimizing the functional $E[\rho]$ over all admissible densities. The task of minimizing $E[\rho]$ amounts to solving the many-electron Schrödinger equation but, on the face of it, appears much simpler. Even a vague appreciation of the immense complexity of the Schrödinger equation makes one suspect that it cannot be tamed so easily and that the almost miraculous solution of the electronic structure problem by DFT must come at a price. At least, there must be a catch.

There is actually not one catch, but two. First, the exact functional $E[\rho]$ is not known and, some believe, is so complicated that it is practically unknowable.

A Matter of Density: Exploring the Electron Density Concept in the Chemical, Biological, and Materials Sciences, First Edition. Edited by N. Sukumar.

The second catch is that $E[\rho]$, even if it *were* known, is not explicit, meaning that the exact mapping from ρ to E cannot in general be written down as a formula with E on the left-hand side and ρ on the right. Not all is lost, however, since we are free to approximate $E[\rho]$ with expressions involving standard functions and usual mathematical operations. Development of approximate density functionals that yield accurate electronic energies for the widest possible range of systems and properties is the chief preoccupation of DFT.

As of 2011, there are hundreds of density-functional approximations to choose from. Most of them perform remarkably well for certain types of problems and fail for others. For example, the B3LYP and PBE functionals are very good at predicting structural and thermodynamical properties but not for charge-transfer excitation energies, barriers of chemical reactions, polarizabilities, and noncovalent interactions. Sometimes, approximate functionals are designed to perform well for a particular property. However, this works like a see-saw: improvement for one target property often results in deterioration for others. The great proliferation of approximate density functionals and their uneven performance are in part responsible for certain skepticism toward DFT as a method.

In fairness to DFT, one should always keep in mind that practical computational chemistry never deals with the exact density functional but only with density-functional *approximations*. If we knew the exact functional, then every DFT calculation would be exact. When we say "DFT fails," we mean that the density-functional approximation we chose to use fails to give the correct prediction. Such failures are not surprising and even should be expected, given how simple some approximations are. What *is* surprising is that compact closed-form density-to-energy expressions developed by theorists work as well as they do.

There are currently two views on the status of DFT. One view is that theorists have done everything they could, but the problem of approximating the exact functional is so hard that the hopes of making further progress may be fading. This sentiment is sometimes felt by the users who have been growing impatient with the incremental progress of density functionals since the early 1990s when functionals such as B3LYP entered computational chemistry and completely transformed it. One should be reminded, however, that DFT has been in a similar position before: the significance of the Hohenberg–Kohn theorems was realized back in 1964, but it took two decades to understand the limitations of the early density-functional approximations and develop density functionals that were usefully accurate for chemical applications. Likewise, the limitations of present-day DFT only reflect the inadequacies of density-functional approximations that have been invented so far. Over the past decade, theorists have been busy trying to understand the reasons for successes and failures of currently available functionals and made great strides in this regard. As a result, there is a basis for an optimistic view that DFT is ripe for a "paradigm shift," which will eventually lead to qualitatively better functionals.

The purpose of this chapter is to explain the inner workings of density-functional approximations and to give a sense of where DFT is heading in the near future. For a more technical account of some of the older topics discussed in this chapter, the reader is referred to Reference 2. To keep things simple, we write many equations in this chapter in the form applicable only to closed-shell ("spin-unpolarized") systems. For spin-polarized systems, where the spin-up and spin-down densities are not equal, some modifications may be required (see Section 8.2 in Reference 3). This convention eliminates the need to include spin subscripts and sums over spins. Spin-specific quantities are discussed only when necessary.

6.2 EXCHANGE AND CORRELATION FUNCTIONALS

The starting point for approximating the electronic energy functional is to think of $E[\rho]$ as a sum of several terms. The idea is to identify those terms that are known exactly, define others in some convenient way, and then focus on the only unknown term that remains. This is precisely what Kohn and Sham [1] did by writing the total energy functional as

$$E[\rho] = T_s[\rho] + V[\rho] + U[\rho] + E_{xc}[\rho]. \tag{6.2}$$

In Equation (6.2),

$$T_s[\rho] = -\frac{1}{2}\sum_k^{\text{occ.}} \int \phi_k^*(\mathbf{r})\nabla^2\phi_k(\mathbf{r})\,d\mathbf{r} \tag{6.3}$$

is the kinetic energy of a hypothetical system of noninteracting electrons whose total ground-state density is exactly equal to $\rho(\mathbf{r})$, and $\phi_k(\mathbf{r})$ are the so-called Kohn–Sham orbitals occupied by these electrons, such that

$$\rho(\mathbf{r}) = \sum_k^{\text{occ.}} |\phi_k(\mathbf{r})|^2. \tag{6.4}$$

The symbol $\sum_k^{\text{occ.}}$ means that each term in the sum must be included as many times as there are electrons occupying the orbital ϕ_k (one, two, or zero). The functional

$$V[\rho] = \int \rho(\mathbf{r})v(\mathbf{r})\,d\mathbf{r} \tag{6.5}$$

is the electrostatic energy of the electron density interacting with the external potential $v(\mathbf{r})$, whereas

$$U[\rho] = \frac{1}{2}\int d\mathbf{r}_1 \int d\mathbf{r}_2\, \frac{\rho(\mathbf{r}_1)\rho(\mathbf{r}_2)}{|\mathbf{r}_1 - \mathbf{r}_2|} \tag{6.6}$$

is the electrostatic energy of $\rho(\mathbf{r})$ interacting with itself. The last term, $E_{xc}[\rho]$, incorporates everything else and is called the *exchange-correlation* energy. It is the only term that is unknown. In a way, the Kohn–Sham method packs all the complexity of the total energy functional into the exchange-correlation functional. But this is a clever reshuffle because approximating a part (E_{xc}) is safer than directly approximating the whole (E).

The first step in tackling the exchange-correlation functional involves the same trick as for $E[\rho]$: we divide $E_{xc}[\rho]$ into two parts, one large and one small, in such a way that the large part can be defined and computed exactly. These two parts are called, respectively, *exchange* and *correlation functionals*

$$E_{xc}[\rho] = E_x[\rho] + E_c[\rho]. \tag{6.7}$$

For closed-shell systems, where each Kohn–Sham orbital is doubly occupied, the exchange part is *defined* exactly by the expression

$$E_x^{\text{exact}}[\rho] = -\sum_{k,l=1}^{N/2} \int d\mathbf{r}_1 \int d\mathbf{r}_2 \, \frac{\phi_k(\mathbf{r}_1)\phi_k^*(\mathbf{r}_2)\phi_l^*(\mathbf{r}_1)\phi_l(\mathbf{r}_2)}{|\mathbf{r}_1 - \mathbf{r}_2|}. \tag{6.8}$$

This definition is borrowed from the closed-shell Hartree–Fock theory, where an equation identical to Equation 6.8 represents the Hartree–Fock exchange energy (see Section 2.3.5 in Reference 4). The functional $E_x^{\text{exact}}[\rho]$ is an *implicit* functional of the density: it depends on ρ through the Kohn–Sham orbitals that are related to ρ by Equation 6.4.

In atoms and molecules near their equilibrium geometries, the correlation energy E_c is roughly an order of magnitude smaller than the exchange energy E_x. We seem to be making progress: instead of approximating the total energy we now need to approximate only a relatively small part, E_c. Yet anyone who has ever run DFT calculations knows that the exchange energy is usually approximated. This brings up the question: why would one want to use an approximate functional for exchange when an exact formula is readily available? The short answer is that the pairing of E_x^{exact} with standard correlation functionals gives poor accuracy in calculations of most properties of interest. In order to understand how this comes about and why theorists work so hard to approximate something that is already known, we need to invoke the concept of exchange and correlation holes.

As explained in Section 1.3.5 of Reference 5, the exchange-correlation energy can be written *exactly* as

$$E_{xc}[\rho] = \frac{1}{2} \int d\mathbf{r}_1 \, \rho(\mathbf{r}_1) \int d\mathbf{r}_2 \, \frac{\rho_{xc}(\mathbf{r}_1, \mathbf{r}_2)}{|\mathbf{r}_1 - \mathbf{r}_2|}, \tag{6.9}$$

where $\rho_{xc}(\mathbf{r}_1, \mathbf{r}_2)$ is a function called the (coupling-constant-averaged) *exchange-correlation hole density*. Equation 6.9 is physically revealing: it suggests that we think of the exchange-correlation energy as coulombic interaction between an electron at $\mathbf{r}_1$ and the surrounding exchange-correlation hole charge $\rho_{xc}(\mathbf{r}_1, \mathbf{r}_2)$. Note that hole charge at $\mathbf{r}_2$ is not static but depends on the current position of the electron $\mathbf{r}_1$—as if the hole were riding along with the electron.

The exchange-correlation hole may be subdivided into exchange and correlation holes

$$\rho_{\mathrm{xc}}(\mathbf{r}_1, \mathbf{r}_2) = \rho_{\mathrm{x}}(\mathbf{r}_1, \mathbf{r}_2) + \rho_{\mathrm{c}}(\mathbf{r}_1, \mathbf{r}_2), \tag{6.10}$$

so we can write for the exchange functional

$$E_{\mathrm{x}}[\rho] = \frac{1}{2}\int d\mathbf{r}_1\, \rho(\mathbf{r}_1) \int d\mathbf{r}_2\, \frac{\rho_{\mathrm{x}}(\mathbf{r}_1, \mathbf{r}_2)}{|\mathbf{r}_1 - \mathbf{r}_2|}. \tag{6.11}$$

The analogous expression for the correlation functional is

$$E_{\mathrm{c}}[\rho] = \frac{1}{2}\int d\mathbf{r}_1\, \rho(\mathbf{r}_1) \int d\mathbf{r}_2\, \frac{\rho_{\mathrm{c}}(\mathbf{r}_1, \mathbf{r}_2)}{|\mathbf{r}_1 - \mathbf{r}_2|}. \tag{6.12}$$

By comparing Equation 6.11 with Equation 6.8 we see that the exact exchange hole for a closed-shell system is

$$\rho_{\mathrm{x}}^{\mathrm{exact}}(\mathbf{r}_1, \mathbf{r}_2) = -\frac{2}{\rho(\mathbf{r}_1)} \sum_{k,l=1}^{N/2} \phi_k(\mathbf{r}_1)\phi_k^*(\mathbf{r}_2)\phi_l^*(\mathbf{r}_1)\phi_l(\mathbf{r}_2), \tag{6.13}$$

where ϕ_k are the occupied Kohn–Sham orbitals. The exact correlation hole is, of course, not known.

It turns out [6] that the exact exchange hole in a molecule is delocalized, meaning that for a given position of the reference electron $\mathbf{r}_1$, the plot of $\rho_{\mathrm{x}}(\mathbf{r}_1, \mathbf{r}_2)$ as a function of $\mathbf{r}_2$ has deep minima at other nuclei, no matter how remote. By contrast, the *total* exchange-correlation hole, $\rho_{\mathrm{xc}}(\mathbf{r}_1, \mathbf{r}_2)$, is typically localized around the reference electron. This implies that the exact correlation hole, $\rho_{\mathrm{c}}(\mathbf{r}_1, \mathbf{r}_2)$, must also be highly delocalized in order to cancel out the nonlocality of the exact exchange hole. Thus, if we want to combine the exact exchange functional with a density-functional approximation for correlation, we need to devise a very sophisticated, highly nonlocal functional.

For a long time, all attempts to marry the exact exchange expression with an approximate correlation functional were defeated, although recently there has been some progress, which we will discuss toward the end of this chapter. A simpler, pragmatic alternative is to abandon the exact exchange functional and use instead an approximation that is based on a localized hole and so is compatible with an approximate correlation functional. Of course, by giving up exact exchange in favor of approximations, one introduces an error into $E_{\mathrm{x}}[\rho]$. Fortunately, this error tends to be canceled out by a similar opposite-sign error in the approximation for $E_{\mathrm{c}}[\rho]$. This built-in cancellation of errors has proved to be a very fruitful idea, and it was the principal reason for the tremendous success of exchange-correlation functionals developed in the 1980s and 1990s.

6.3 INGREDIENTS AND TECHNIQUES FOR CONSTRUCTING DENSITY FUNCTIONAL APPROXIMATIONS

Development of density-functional approximations is a bold enterprise with relatively few strict guidelines. This means that one can be creative and try different routes. In fact, it is the absence of any mechanical prescriptions for systematic improvement of approximate functionals that makes DFT such an interesting subject.

The central objective of Kohn–Sham DFT is to come up with accurate approximations to the exact exchange-correlation functional. These approximations are usually cast in the form of integral expressions of the type

$$E_{\mathrm{xc}}[\rho] = \int \mathrm{e}_{\mathrm{xc}}(\rho, \ldots)\, d\mathbf{r}, \tag{6.14}$$

where e_{xc} is some function of $\rho(\mathbf{r})$ and other density-dependent ingredients. Since the dimension of this quantity is $\frac{\text{energy}}{\text{volume}}$, e_{xc} is called the *exchange-correlation energy density*.

The most common ingredients of e_{xc} are the modulus of the gradient of the density

$$g = |\nabla\rho|, \tag{6.15}$$

the Laplacian of the density

$$l = \nabla^2\rho, \tag{6.16}$$

the Kohn–Sham (noninteracting) kinetic energy density

$$\tau = \frac{1}{2}\sum_{k}^{\text{occ.}} |\nabla\phi_k|^2, \tag{6.17}$$

the (closed-shell) exact exchange energy density

$$\mathrm{e}_{\mathrm{X}}^{\mathrm{exact}}(\mathbf{r}_1) = -\sum_{k,l=1}^{N/2} \int \frac{\phi_k(\mathbf{r}_1)\phi_k^*(\mathbf{r}_2)\phi_l^*(\mathbf{r}_1)\phi_l(\mathbf{r}_2)}{|\mathbf{r}_1 - \mathbf{r}_2|}\, d\mathbf{r}_2, \tag{6.18}$$

which is just the inner integral of Equation 6.8, and the paramagnetic current density, defined in atomic units by

$$\mathbf{j} = \frac{1}{2i}\sum_{k}^{\text{occ.}} \left(\phi_k^*\nabla\phi_k - \phi_k\nabla\phi_k^*\right). \tag{6.19}$$

Observe that in the last equation, the expression in parentheses is purely imaginary, so that $\mathbf{j}$ itself is always real. Obviously, if the Kohn–Sham orbitals are real, the current density is zero.

Both g and l depend on ρ explicitly, whereas τ, $\mathrm{e}_{\mathrm{x}}^{\mathrm{exact}}$, and $\mathbf{j}$ cannot be written entirely in terms of ρ, although they are uniquely determined by it. Accordingly, density-functional approximations of the type

$$E_{\mathrm{xc}}[\rho] = \int \mathrm{e}_{\mathrm{xc}}(\rho, g, l)\, \mathrm{d}\mathbf{r} \tag{6.20}$$

are called *explicit*, whereas functionals of the type

$$E_{\mathrm{xc}}[\rho] = \int \mathrm{e}_{\mathrm{xc}}(\rho, g, \tau, \mathrm{e}_{\mathrm{x}}^{\mathrm{exact}}, \ldots)\, \mathrm{d}\mathbf{r} \tag{6.21}$$

are called *implicit*. Orbital-dependent functionals [7] are the most practically important type of implicit density functionals.

The ingredients g, l, τ, and $\mathbf{j}$ are called *semilocal* because they depend on the value of ρ or ϕ_k at $\mathbf{r}$ and/or in an infinitesimal neighborhood of $\mathbf{r}$. The exact exchange energy density $\mathrm{e}_{\mathrm{x}}^{\mathrm{exact}}$ is different in this respect because it depends on values of all ϕ_k everywhere, as reflected in the integration over $\mathbf{r}_2$. Such ingredients are said to be *nonlocal*. Semilocal density-functional approximations are those that involve one or more semilocal ingredients.

A significant portion of the vocabulary of modern DFT was developed by John Perdew in reference to a systematic approach called *Jacob's ladder of density-functional approximations* [8]. In this classification, density-functional approximations that are constructed using the electron density ρ and no other ingredients represent rung 1 of the ladder and are termed local density approximations (LDA)

$$E_{\mathrm{xc}}^{\mathrm{LDA}}[\rho] = \int \mathrm{e}_{\mathrm{xc}}(\rho)\, \mathrm{d}\mathbf{r}. \tag{6.22}$$

Approximations where e_{xc} depends on ρ and g represent rung 2 and are called generalized-gradient approximations (GGA)

$$E_{\mathrm{xc}}^{\mathrm{GGA}}[\rho] = \int \mathrm{e}_{\mathrm{xc}}(\rho, g)\, \mathrm{d}\mathbf{r}. \tag{6.23}$$

Rung 3 approximations depend, in addition to ρ and g, on l and/or τ and are called meta-GGAs (MGGA)

$$E_{\mathrm{xc}}^{\mathrm{MGGA}}[\rho] = \int \mathrm{e}_{\mathrm{xc}}(\rho, g, l, \tau)\, \mathrm{d}\mathbf{r}. \tag{6.24}$$

The functionals of rung 4 involve dependence on a nonlocal ingredient, the exact exchange energy density, and are termed hyper-GGAs (HGGA),

$$E_{\rm xc}^{\rm HGGA}[\rho] = \int {\rm e}_{\rm xc}(\rho, g, l, \tau, {\rm e}_{\rm x}^{\rm exact})\, d\mathbf{r}. \tag{6.25}$$

Approximations of rungs 1 through 4 involve only occupied Kohn–Sham orbitals. There is also a fifth rung where one finds approximations that involve occupied *and* virtual Kohn–Sham orbitals.

The historical development of density-functional approximations for exchange correlation may be regarded as the process of climbing Jacob's ladder or as a story of passing the following milestones:

1. Analysis of exactly solvable models and introduction of various local density approximations.
2. Development of GGAs and meta-GGAs by bringing into play semilocal ingredients and by grafting selected properties of the exact functional.
3. Introduction of exact exchange into semilocal functionals (hybrid DFT).
4. Empirical construction (fitting).
5. Development of nonlocal correlation functionals compatible with exact exchange.

Most density functionals that are currently in use fall into groups 1 through 4, while functionals of group 5 are still at experimental stage. The rest of this chapter offers a close look at various strategies of devising density-functional approximations.

6.4 NONEMPIRICAL DERIVATION AND LOCAL DENSITY MODELS

In an ideal world, we might be able to derive the exact exchange-correlation functional from first principles. In reality, we have to settle for less. One possible strategy is to obtain the exact functional for a solvable model system and hope that the same expression will work well in general. To illustrate this approach, let us consider a trivial example of one electron in an external potential $v(\mathbf{r})$. The Schrödinger equation for this system is identical with the Kohn–Sham equation

$$\left[-\frac{1}{2}\nabla^2 + v(\mathbf{r})\right]\phi(\mathbf{r}) = E\phi(\mathbf{r}), \tag{6.26}$$

where $\phi(\mathbf{r})$ is the exact wavefunction and simultaneously the exact Kohn–Sham orbital. Suppose that $\phi(\mathbf{r})$ is normalized and real. (If ϕ is complex, it can always be made real as explained in Section 2.2 of Reference 9.) Since there is only one electron in this system, the density is just $\rho = \phi^2$, so $\phi = \rho^{1/2}$. Let us multiply Equation 6.26 from the left by $\phi(\mathbf{r})$, integrate over $\mathbf{r}$, and write the result as

$$E[\rho] = -\frac{1}{2}\int \rho^{1/2}(\mathbf{r})\nabla^2\rho^{1/2}(\mathbf{r})\,d\mathbf{r} + \int \rho(\mathbf{r})v(\mathbf{r})\,d\mathbf{r}. \tag{6.27}$$

This is clearly an explicit density functional, and it is exact for any one-electron system. One should not be surprised, however, that this functional gives dismal results for many-electron systems.

Although Equation 6.27 is useless for practical purposes, it tells us something about the true functional. First, for any one-electron system with a constant external potential, the true $E[\rho]$ should reduce to Equation 6.27. Second, the fact that Equation 6.27 is exact for some systems but not for others suggests that the true $E[\rho]$ and hence $E_{xc}[\rho]$ cannot be written as a single analytic expression valid for all electron numbers. When a second electron is added to the system, the true $E[\rho]$ must switch discontinuously from Equation 6.27 to something else. Such sudden switching is not a property of analytic functionals.

Another model system that gives rise to a more useful nonempirical functional is a *uniform electron gas*, also called the *jellium model*. The uniform electron gas is a system of many interacting electrons moving in the field of a uniform positive background charge of the same density as the averaged electron density. The latter requirement ensures overall electric neutrality. The total volume of this system is assumed to be large but finite, so that Kohn–Sham orbitals can be normalized. For a uniform electron gas, $\rho(\mathbf{r}) = \text{const}$.

One can show (see, for instance, Section 6.1 in Reference 3) that for a clot of spin-unpolarized uniform electron gas of volume V the exchange energy is given *exactly* by the expression

$$E_{\mathrm{x}}^{\mathrm{LDA}}[\rho] = -C_{\mathrm{x}}\int \rho^{4/3}(\mathbf{r})\,d\mathbf{r}, \tag{6.28}$$

where $C_{\mathrm{x}} = (3/4)(3/\pi)^{1/3} \approx 0.73856$ and the integration is over V. The exact correlation functional for a uniform electron gas is not known (except in the high and low density limits), but the correlation energy of this system has been studied numerically and parametrized in the form of analytic functionals such as [10]

$$E_{\mathrm{c}}^{\mathrm{LDA}}[\rho] = -A\int \rho(1+\alpha_1 r_s)\ln\left[1 + \frac{1}{A(\beta_1 r_s^{1/2} + \beta_2 r_s + \beta_3 r_s^{3/2} + \beta_4 r_s^2)}\right]d\mathbf{r}, \tag{6.29}$$

where $r_s = (3/4\pi\rho)^{1/3}$ and A, α_1, β_1, β_2, β_3, and β_4 are fixed parameters.

In real atoms and molecules, the electron density is far from uniform (it is approximately piecewise exponential), so Equations 6.28 and 6.29 are no longer exact. Despite this, the sum of Equations 6.28 and 6.29 gives a reasonably accurate approximation to the true exchange-correlation energy. The LDA predicts fairly accurate bond lengths and lattice constants but severely overestimates atomization energies of molecules and solids. For comparison, the Hartree–Fock method, which is computationally *more* expensive than the LDA, predicts bond lengths much less accurately than LDA and underestimates atomization energies with a mean absolute error that is twice as large as the overbinding error of LDA. This is remarkable: a basic DFT method outperforms a basic wavefunction method. Good as LDA is, it is still not good enough for most chemical applications. As we shall see in the following section, attempts to derive exact density functionals for nonuniform densities by formal density-gradient expansions do not yield better general-purpose approximations. This compels one to seek different, less formulaic procedures for going beyond LDA.

One way to improve the LDA is to relax the requirement that this functional be exact for a uniform electron gas and instead demand better performance for chemically relevant systems. For the exchange component, this can be achieved by treating the constant C_x in Equation 6.28 as an empirical parameter—the technique is known as *Slater's $X\alpha$ method* [11]. For correlation, one can start with some LDA expression and reparametrize it by fitting to the exact correlation energies of a few atoms. This strategy is represented by the Brual–Rothstein functional [12]. The gains in accuracy made in this manner, however, are modest.

A third method for deriving density functionals is to start with a model for the coupling-constant-averaged exchange-correlation hole, $\rho_{xc}(\mathbf{r}_1, \mathbf{r}_2)$. Once the hole is specified, we insert it into Equation 6.9 and integrate over $\mathbf{r}_2$ to obtain a density functional. For a density functional that is not explicitly derived from an exchange-correlation hole, one assumes that there is a model hole underlying it. The implied hole may be hard or even impossible to recover from a given $E_x[\rho]$ or $E_c[\rho]$, but it strongly influences the performance of the functional. Unfortunately, approximation of exchange-correlation hole densities is as difficult as direct approximation of functionals themselves, so this method does not by itself lead to more accurate results.

6.5 SEMILOCAL FUNCTIONALS BEYOND THE LOCAL DENSITY APPROXIMATION

The most natural way to account for the nonuniformity of electron density in atoms and molecules is to construct an approximate functional in terms of ρ and its gradient $\nabla\rho$ or, rather, the gradient norm $|\nabla\rho|$. Because density-functional approximations must satisfy certain dimensionality requirements, it is convenient to make the energy density e_{xc} depend on $|\nabla\rho|$ through the so-called reduced density gradient,

$$s = \frac{|\nabla\rho|}{\rho^{4/3}}. \qquad (6.30)$$

The reduced gradient s is a dimensionless quantity since the dimensions of ρ and $|\nabla\rho|$ are length^{-3} and length^{-4}, respectively. Now one can attempt to improve on the LDA by devising a functional of the form

$$E_{xc}[\rho] = \int e_{xc}^{LDA}(\rho)\left[1 + \mu(\rho)s^2 + \ldots\right] d\mathbf{r}, \tag{6.31}$$

where $\mu(\rho)$ is a function of the density that reduces to a constant for the exchange component. Approximations of this type are called *density-gradient expansions*. The coefficients of the lowest powers of s in Equation 6.31 can be rigorously derived for two extreme cases: the slowly varying density limit and the high-density limit [5]. Since the leading gradient correction terms are nonempirical, one might assume that Equation 6.31 cannot be worse than the LDA. But DFT often confounds expectations. It turns out that truncated density-gradient expansions are *less* accurate than the LDA for atoms and molecules. In particular, addition of the $\mu(\rho)s^2$ term to the LDA energy density makes total correlation energies positive [5], which is an unphysical result.

The failure of truncated density-gradient expansions for $E_{xc}[\rho]$ was analyzed and explained by Perdew and coworkers [5]. They showed that the exchange-correlation hole underlying the second-order gradient expansion exhibits spurious undamped oscillations as $|\mathbf{r}_1 - \mathbf{r}_2| \to \infty$ and so violates two important conditions, namely, the negativity constraint for the exchange hole charge

$$\rho_x(\mathbf{r}_1, \mathbf{r}_2) < 0, \tag{6.32}$$

and the requirement that the exchange-correlation hole charge be normalized to -1 for every reference point $\mathbf{r}_1$

$$\int \rho_{xc}(\mathbf{r}_1, \mathbf{r}_2)\, d\mathbf{r}_2 = -1. \tag{6.33}$$

The incorrect behavior of the function $\rho_x(\mathbf{r}_1, \mathbf{r}_2)$ associated with second-order truncated density-gradient expansions translates via Equation 6.9 into large errors in energy for real atoms and molecules.

Another problem with truncated density-gradient expansions is that the corresponding exchange potential, $v_x(\mathbf{r}) = \delta E_x[\rho]/\delta\rho(\mathbf{r})$, has a pathological divergence in the exponential density tails found in all atomic and molecular charge distributions. This divergence is caused by the density-gradient correction term that is proportional to $\rho^{1/3}s^2$ and so diverges asymptotically for an exponential density. To see this, we substitute $\rho(r) = e^{-br}$ into Equation 6.30 and obtain

$$s = \frac{|\nabla\rho|}{\rho^{4/3}} = \frac{|\partial\rho/\partial r|}{\rho^{4/3}} = \frac{be^{-br}}{e^{-4br/3}} = be^{br/3}. \tag{6.34}$$

This shows that $\rho^{1/3}s^2 \sim e^{br/3} \to \infty$ as $r \to \infty$.

In order to remedy the unphysical behavior of the exchange hole and exchange potential associated with density-gradient expansions, Perdew, Becke, and others proposed to replace the truncated series in square brackets in Equation 6.31 with a damping function $F_{\text{xc}}(\rho, s)$, such that it remains finite as $r \to \infty$. This leads to density-functional approximations of the form

$$E_{\text{xc}}[\rho] = \int e_{\text{xc}}^{\text{LDA}}(\rho) F_{\text{xc}}(\rho, s)\, d\mathbf{r}, \tag{6.35}$$

which are called *GGA*s. The analytic form of the function F_{xc} varies from case to case. For example, Becke's exchange functional of 1986 (B86) [13] and the Perdew–Burke–Ernzerhof (PBE) GGA [14] employ damping functions of the form

$$F_{\text{x}}^{\text{PBE}}(s) = 1 + \frac{as^2}{1 + bs^2}, \tag{6.36}$$

whereas Becke's exchange functional of 1988 (B88) uses

$$F_{\text{x}}^{\text{B88}}(s) = 1 + \frac{as^2}{1 + bs \ln(s + \sqrt{1 + s^2})}. \tag{6.37}$$

In both cases, a and b are functional-specific constants that are either determined from known exact properties of $E_{\text{x}}[\rho]$ or are fitted to experimental data. GGAs for the correlation energy have a more complicated form but also use damping functions to ensure that the correlation energy density has proper behavior in various physically relevant limits.

After GGA were perfected by the late 1980s, they were found to perform not only much better than the LDA but also quite well relative to medium-level wavefunction methods. The latter fact is especially significant if we recall that GGAs have a much lower computational cost than wavefunction methods. As soon as all that came to light around 1991, many quantum chemists who had been previously skeptical about DFT finally became converts.

6.6 CONSTRAINT SATISFACTION

Although we do not know the exact exchange-correlation functional, we do know quite a few of its mathematical properties. Suppose we identify several such properties, adopt them as constraints, and then construct a density-functional approximation that satisfies those constraints. With respect to these mathematical properties, the resulting approximation will mimic the exact functional. We might also expect that the more properties our approximation shares with the exact functional, the more accurate and transferable it will be. This strategy of density-functional design, called *constraint satisfaction* [15], has produced some of the most successful density-functional approximations available today.

What properties of the exact functional are known? First of all, we know that for any admissible electron density, the exact exchange energy is strictly negative

$$E_{\mathrm{x}}[\rho] < 0, \tag{6.38}$$

while the exact correlation energy is nonpositive

$$E_{\mathrm{c}}[\rho] \leq 0. \tag{6.39}$$

The equality in Equation 6.39 holds for all one-electron systems and only for such systems. Lieb and Oxford [16] showed that the exchange-correlation energy in Coulombic systems of electrons is also bounded from below

$$E_{\mathrm{x}}[\rho] \geq E_{\mathrm{xc}}[\rho] \geq -C \int \rho^{4/3}(\mathbf{r})\, d\mathbf{r}, \tag{6.40}$$

where $C = 1.68$.

For any one-electron density $\rho_1(\mathbf{r})$, the exact $E_{\mathrm{x}}[\rho]$ cancels out the spurious Coulomb self-repulsion energy. This means that for any one-electron density ρ_1, the exact functionals should satisfy the relations

$$E_{\mathrm{xc}}[\rho_1] = E_{\mathrm{x}}[\rho_1] = -U[\rho_1], \tag{6.41}$$

where $U[\rho_1]$ is given by Equation 6.6 with $\rho = \rho_1$. Notice that when this constraint applies, the Kohn–Sham functional of Equation 6.2 correctly reduces to the exact one-electron density functional of Equation 6.27.

For uniform electron densities, every exchange density-functional approximation should reduce to the known exact expression for a uniform electron gas,

$$E_{\mathrm{x}}[\rho] = E_{\mathrm{x}}^{\mathrm{LDA}}[\rho] \quad \text{if} \quad \rho(\mathbf{r}) = \text{const}, \tag{6.42}$$

where $E_{\mathrm{x}}^{\mathrm{LDA}}[\rho]$ is given by Equation 6.28.

Mel Levy [17] deduced many properties of the exact exchange and correlation functionals under various coordinate scaling transformations of the density. The most important of these transformations is the *uniform* scaling of the density, defined by

$$\rho_\lambda(\mathbf{r}) = \lambda^3 \rho(\lambda \mathbf{r}), \tag{6.43}$$

where λ is a constant. The name "uniform" refers to the fact that all three Cartesian components of $\mathbf{r} = (x, y, z)$ are scaled by the same λ. As λ is varied, the density either contracts or becomes more diffuse, but the integral of $\rho_\lambda(\mathbf{r})$ over the entire space remains independent of λ

$$\begin{aligned}\int \rho_\lambda(\mathbf{r})\,\mathrm{d}\mathbf{r} &= \int \lambda^3 \rho(\lambda\mathbf{r})\,\mathrm{d}\mathbf{r} \\ &= \int \mathrm{d}x \int \mathrm{d}y \int \mathrm{d}z\, \lambda^3 \rho(\lambda x, \lambda y, \lambda z) \\ &= \int \mathrm{d}(\lambda x) \int \mathrm{d}(\lambda y) \int \mathrm{d}(\lambda z)\, \rho(\lambda x, \lambda y, \lambda z) \\ &= \int \mathrm{d}x' \int \mathrm{d}y' \int \mathrm{d}z'\, \rho(x', y', z') = \int \rho(\mathbf{r}')\,\mathrm{d}\mathbf{r}' = N. \end{aligned} \tag{6.44}$$

The key property of the exact exchange functional is that it obeys the simple scaling law

$$E_\mathrm{x}[\rho_\lambda] = \lambda E_\mathrm{x}[\rho]. \tag{6.45}$$

The exact correlation functional does not have a simple scaling behavior, but it is known that

$$\lim_{\lambda\to\infty} E_\mathrm{c}[\rho_\lambda] > -\infty. \tag{6.46}$$

It is also known that in a finite many-electron system, the true exchange-correlation potential $v_\mathrm{xc}(\mathbf{r})$, defined as the functional derivative of $E_\mathrm{xc}[\rho]$ with respect to ρ, has the following asymptotic behavior:

$$v_\mathrm{xc}(\mathbf{r}) \equiv \frac{\delta E_\mathrm{xc}[\rho]}{\delta \rho(\mathbf{r})} \xrightarrow[r\to\infty]{} -\frac{1}{r}. \tag{6.47}$$

The asymptotic behavior of the exchange-correlation energy density is as follows:

$$\mathrm{e}_\mathrm{xc}(\mathbf{r}) \xrightarrow[r\to\infty]{} -\frac{\rho(r)}{2r}. \tag{6.48}$$

The list can be continued, but the message is clear: (i) density-functional approximations should reproduce known properties of the exact exchange-correlation functional and (ii) any approximation that violates a known exact constraint should be suspect. To illustrate the method of constraint satisfaction, we will explain how it was used to eliminate one embarrassing artifact of early density-functional approximations.

The hydrogen atom is one of the few systems of chemical interest for which the Schödinger equation can be solved analytically. The exact ground-state density of the H atom is $\rho(\mathbf{r}) = \frac{1}{\pi}\mathrm{e}^{-2r}$ and the corresponding exact total energy is $E = -\frac{1}{2}$ hartree. The LDA and most GGAs fail to give these results because these functionals incorrectly predict nonzero correlation energies for one-electron systems, in violation of the constraint

$$E_c[\rho_1] = 0, \tag{6.49}$$

where ρ_1 is a one-electron density. For the same reason, LDA and GGA give nonzero correlation energies for other one-electron systems such as H_2^+. One notable exception is the Lee–Yang–Parr (LYP) correlation GGA in which the correlation energy density is proportional to the product of spin-up and spin-down densities, $\rho_\alpha \rho_\beta$. As a result, LYP predicts $E_c = 0$ for *any* N-electron system where all electrons have parallel spins. That is, LYP happens to be correct for $N = 1$ but is wrong for $N \geq 2$.

To satisfy the constraint of Equation 6.49, Becke devised an indicator function that distinguishes one-electron densities from all others. This function is based on certain properties of the kinetic energy density $\tau(\mathbf{r})$ and its interplay with other density-functional ingredients. To understand Becke's reasoning, we consider the quantity

$$\tau_W = \frac{1}{8} \frac{|\nabla \rho|^2}{\rho}, \tag{6.50}$$

called the *Weizsäcker gradient correction* to the Thomas–Fermi kinetic energy density. The property of τ_W that we need is the following double inequality

$$0 \leq \tau_W \leq \tau - \frac{1}{2} \frac{|\mathbf{j}|^2}{\rho}, \tag{6.51}$$

where $\mathbf{j}$ is the current density defined by Equation 6.19. The first part of this inequality, $\tau_W \geq 0$, is obvious from the definition of τ_W. Proof of the second part of Equation 6.51 requires some work.

Let us consider first closed-shell systems. For such systems, the gradient of the density is given by

$$\nabla \rho = \nabla \left(2 \sum_{k=1}^{N/2} \phi_k^* \phi_k \right) = 2 \sum_{k=1}^{N/2} \left(\phi_k^* \nabla \phi_k + \phi_k \nabla \phi_k^* \right). \tag{6.52}$$

Here, $2 \sum_{k=1}^{N/2} \phi_k^* \nabla \phi_k$ is a complex-valued vector quantity which we can rewrite as

$$2 \sum_{k=1}^{N/2} \phi_k^* \nabla \phi_k = \sum_{k=1}^{N/2} (\phi_k^* \nabla \phi_k + \phi_k \nabla \phi_k^*) + \sum_{k=1}^{N/2} (\phi_k^* \nabla \phi_k - \phi_k \nabla \phi_k^*) = \frac{1}{2} \nabla \rho + i\mathbf{j}, \tag{6.53}$$

where we used Equations 6.52 and 6.19 (the latter without the factor of $\frac{1}{2}$ because we are summing over $\frac{N}{2}$ orbitals). Since $\frac{1}{2}\nabla\rho$ and $\mathbf{j}$ are always real, we can think of them, respectively, as the real and imaginary parts of $2\sum_{k=1}^{N/2}\phi_k^*\nabla\phi_k$. Since for $z = x + iy$ we have $|z|^2 = [\mathrm{Re}(z)]^2 + [\mathrm{Im}(z)]^2$, we can write

$$\left|2\sum_{k=1}^{N/2}\phi_k^*\nabla\phi_k\right|^2 = 4\left|\sum_{k=1}^{N/2}\phi_k^*\nabla\phi_k\right|^2 = \frac{1}{4}|\nabla\rho|^2 + |\mathbf{j}|^2 = 2\rho\tau_W + |\mathbf{j}|^2. \quad (6.54)$$

But according to the Cauchy–Schwarz inequality

$$4\left|\sum_{k=1}^{N/2}\phi_k^*\nabla\phi_k\right|^2 \leq 4\left(\sum_{k=1}^{N/2}|\phi_k|^2\right)\left(\sum_{k=1}^{N/2}|\nabla\phi_k|^2\right) = 2\rho\tau. \quad (6.55)$$

Comparing Equations 6.54 and 6.55 we see that $2\rho\tau_W + |\mathbf{j}|^2 \leq 2\rho\tau$ or, equivalently,

$$\tau_W \leq \tau - \frac{1}{2}\frac{|\mathbf{j}|^2}{\rho}. \quad (6.56)$$

This concludes the proof of Equation 6.51. Note that for real orbitals, where $\mathbf{j}$ is identically zero, Equation 6.51 reduces to

$$0 \leq \tau_W \leq \tau. \quad (6.57)$$

The next step is an important observation that the equality in Equation 6.55 holds only if the number of occupied Kohn–Sham orbitals is one. In this case,

$$\tau = \tau_W + \frac{1}{2}\frac{|\mathbf{j}|^2}{\rho} \quad (6.58)$$

or simply $\tau = \tau_W$ if the orbital is real.

For spin-polarized system (when $\rho_\alpha \neq \rho_\beta$), Equation 6.51 branches into two separate inequalities, one for each spin

$$0 \leq \frac{|\nabla\rho_\sigma|^2}{8\rho_\sigma} \leq \tau_\sigma - \frac{1}{2}\frac{|\mathbf{j}_\sigma|^2}{\rho_\sigma}, \quad (6.59)$$

where $\sigma = \alpha$ or β. The quantities ρ_σ, τ_σ, and $\mathbf{j}_\sigma$ are given by equations similar to Equations 6.4, 6.17, and 6.19 in which only singly-occupied σ-spin orbitals are included. Again, if only one σ-spin orbital is occupied, the second inequality in Equation 6.59 becomes a strict equality

$$\tau_\sigma = \frac{1}{8}\frac{|\nabla\rho_\sigma|^2}{\rho_\sigma} + \frac{1}{2}\frac{|\mathbf{j}_\sigma|^2}{\rho_\sigma}. \quad (6.60)$$

Following Becke [18, 19], we now introduce the function

$$\eta_\sigma = \frac{1}{\tau_\sigma}\left(\tau_\sigma - \frac{1}{8}\frac{|\nabla\rho_\sigma|^2}{\rho_\sigma} - \frac{1}{2}\frac{|\mathbf{j}_\sigma|^2}{\rho_\sigma}\right). \quad (6.61)$$

As explained above, $\eta_\sigma(\mathbf{r})$ vanishes identically for any one-electron system and is strictly positive in systems that contain two or more σ-spin electrons. Consider now the meta-GGA correlation functional

$$E_c[\rho] = \int \left[e_c^{\alpha\beta}(\mathbf{r}) + \sum_\sigma e_c^{\sigma\sigma}(\mathbf{r})\eta_\sigma(\mathbf{r}) \right] d\mathbf{r}, \quad (6.62)$$

where $e_c^{\alpha\beta}$ and $e_c^{\sigma\sigma}$ are some GGA-type expressions for the opposite-spin and parallel-spin correlation energy densities, respectively. Because of the presence of $\eta_\sigma(\mathbf{r})$ in Equation 6.62, every functional of this form will correctly yield zero for the $\sigma\sigma$-spin correlation energy in any system with a single σ-spin electron and a nonzero energy in any system with two or more σ-spin electrons. This is now a standard trick for constructing correlation functionals that are free from the one-electron self-interaction error. Density-functional approximations that use it include Bc88 [20], Bc95 [21], B98 [22], τ-HCTH [23], TPSS [24], VS98 [25], M06 [26], and others.

Although constraint satisfaction is currently the most rigorous practical method of constructing density-functional approximations, it has its limitations. Enforcement of any particular constraint does not by itself guarantee that the resulting functional will be better. This is because by imposing one known constraint we may unwittingly violate other—unknown—constraints that may be more important. In fact, better performance is sometimes achieved when an exact constraint is relaxed. For example, any GGA can and should reduce to the LDA functional of Equation 6.28 when $\rho(\mathbf{r}) = \text{const}$ because LDA is the proper functional for a uniform density. Some of the most successful density functionals in chemistry sacrifice this property in favor of better performance for nonuniform densities. In particular, the LYP correlation functional in not exact for a uniform electron gas, yet predicts highly accurate correlation energies for atoms. BLYP, B3LYP, and other exchange-correlation functionals that include LYP also fail to yield correct energies for a uniform electron gas, but this has little effect on their performance in chemical applications.

Another example of beneficial and even intentional constraint violation involves GGA functionals. For a slowly varying density (i.e., for $s \to 0$), any exchange GGA should reproduce the known low-order terms in the exact density-gradient expansion:

$$E_x[\rho] = -C_x \int \rho^{4/3}(1 + \mu s^2 + \ldots)\, d\mathbf{r}, \quad (6.63)$$

where the theoretical value of μ is $10/81 \approx 0.1235$. When this constraint is enforced, GGAs predict accurate bond lengths in molecules and lattice constants in solids but give poor atomization energies. Perdew and coworkers [27] showed that a GGA can produce accurate atomization energies only if it strongly violates Equation 6.63 and has an enhanced gradient dependence. The PBE GGA in particular was designed to give accurate atomization energies, and so it has $\mu = 0.2195$, which is about twice as large as required by the density-gradient expansion. All attempts to construct an accurate GGA face the dilemma [28]: using the theoretical value of μ leads to accurate bond lengths but yields poor atomization energies; increasing μ improves atomization energies but worsens bond lengths and lattice constants. Being a very restrictive form, GGAs cannot simultaneously perform well for both properties. Thus, for calculations of atomization energies, one should use the PBE GGA or its hybrid versions with $\mu = 0.2195$. For bulk properties of solids, one should use a modified version called *PBEsol* (PBE revised for solids) that restores the nonempirical value $\mu = 10/81$.

6.7 THE COMEBACK OF EXACT EXCHANGE: GLOBAL AND LOCAL HYBRIDS

A decade ago, Peter Gill [29] published an "obituary" for DFT in the *Australian Journal of Chemistry*. According to his account, DFT was born in 1927 and passed away in 1993. The cause of her demise was an unsuccessful operation performed on her by "an eminent Canadian surgeon," a follower of Dr. Frankenstein, who attempted to cure DFT by blending her with wavefunction theory into a "grisly hybrid." It would be instructive for us here to understand what prompted the famous surgeon to recommend such a drastic treatment.

As we discussed earlier, semilocal correlation functionals do not work well in combination with the exact exchange functional of Equation 6.8, but good performance is easily achieved if both exchange and correlation approximations are semilocal. In 1993, however, Becke showed [30] that one can go beyond the accuracy of GGAs by representing the exchange contribution with a *mixture* of the exact exchange functional and a semilocal approximation. This discovery led to many so-called *hybrid* functionals such as B3PW91, B3LYP, and PBEh.

The basic form of hybrid functionals is

$$E_{\mathrm{xc}} = aE_{\mathrm{x}}^{\mathrm{exact}} + (1-a)E_{\mathrm{x}} + E_{\mathrm{c}}, \tag{6.64}$$

where E_{x} and E_{c} are some semilocal density-functional approximations and a $(0 \le a \le 1)$ is a universal parameter called *a mixing fraction*. The value of a is usually determined by empirical fitting of Equation 6.64 to reproduce experimental atomization energies, exact nonrelativistic energies, reaction barrier heights, and other data. Fitting to atomization energies typically gives $a \approx 0.2$ for GGAs and $a \approx 0.1$ for meta-GGAs, while fitting to reaction barrier heights yields $a \approx 0.5$.

Mixing exact and approximate exchange functionals is not an empirical cookbook recipe. The hybrid scheme has a theoretical underpinning that not only explains why hybrid functionals work better than GGAs but also predicts the optimal value of a in various situations [31].

From the point of view of computational chemists, hybrid functionals were a smashing success because they represented the first quantum-mechanical method that was simultaneously accurate, reliable, and computationally cheap. Ironically, it is the "grisly hybrids" that made DFT so effective and popular.

The term "hybrid" in relation to functionals such as B3LYP is now often used with the qualifier *global* to indicate that the value of a in Equation 6.64 is position-independent. This can be emphasized by rewriting Equation 6.64 in terms of energy densities

$$E_{\mathrm{xc}} = \int \left[a e_{\mathrm{x}}^{\mathrm{exact}}(\mathbf{r}) + (1-a)e_{\mathrm{x}}(\mathbf{r}) + e_{\mathrm{c}}(\mathbf{r})\right] d\mathbf{r}. \tag{6.65}$$

The fact that the optimal value of a has large system-dependent variations suggests a generalization of Equation 6.65 by turning the mixing fraction a into a function of $\mathbf{r}$

$$E_{\mathrm{xc}} = \int \left\{a(\mathbf{r}) e_{\mathrm{x}}^{\mathrm{exact}}(\mathbf{r}) + [1-a(\mathbf{r})]e_{\mathrm{x}}(\mathbf{r}) + e_{\mathrm{c}}(\mathbf{r})\right\} d\mathbf{r}, \tag{6.66}$$

Such forms are called *local hybrids*. In the local hybrid scheme, the objective is to devise a mixing fraction $a(\mathbf{r})$ that adapts to the local chemical environment. The basic requirements for the mixing fraction $a(\mathbf{r})$ are that it be restricted to the range of values between 0 and 1 and reduce to 1 for any one-electron density.

The first mixing fraction was suggested by Becke [19]

$$a(\mathbf{r}) = \frac{\tau_W(\mathbf{r})}{\tau(\mathbf{r})}, \tag{6.67}$$

and implemented in a local hybrid functional by Jaramillo et al. [32]. This choice gives accurate reaction barriers but produces disappointing results for atomization energies [32]. More recently, Kaupp and coworkers [33] constructed and implemented self-consistently several local hybrid functionals with various mixing fractions. One of those is given by

$$a(\mathbf{r}) = \sum_{m=1}^{M} b_m \left[\frac{\tau_W(\mathbf{r})}{\tau(\mathbf{r})}\right]^m, \tag{6.68}$$

where M is a small integer and b_m are fractional coefficients. Another is

$$a(\mathbf{r}) = \left[\frac{s(\mathbf{r})}{b + s(\mathbf{r})}\right]^2, \tag{6.69}$$

where s is the reduced gradient of Equation 6.30 and b is a positive parameter. It was found that a local hybrid functional using the mixing fraction of

Equation 6.68 with $M = 1$ and $b_1 \approx 0.5$ predicts simultaneously accurate atomization energies and reaction barriers.

In view of the resounding success of global hybrid functionals, the local hybrid scheme was initially thought to hold great promise. However, finding a mixing fraction $a(\mathbf{r})$ that would decisively beat global hybrid functionals proved more difficult than anticipated. As a result, the overall accuracy of the best local hybrid functionals proposed to date is not significantly higher than that of the global hybrid scheme with an optimal mixing constant. Attempts to develop better local hybrid approximations continue despite these setbacks.

6.8 THE BEST OF BOTH WORLDS: RANGE-SEPARATED HYBRIDS

Interaction of opposite-spin electrons at close range (small $r_{12} \equiv |\mathbf{r}_1 - \mathbf{r}_2|$) is adequately described by semilocal exchange-correlation approximations but not by the exact (Hartree–Fock-type) exchange functional. In fact, the Hartree–Fock method does not correlate the motion of electrons with opposite spins at all. That is why molecular properties for which short-range (SR) electron interactions are dominant (e.g., equilibrium geometries and atomization energies) are predicted by approximate DFT much better than by the Hartree–Fock method. Conversely, when two electrons are far apart (large r_{12}), their interaction is better described with the exact exchange functional than with semilocal density-functional approximations. Consequently, properties determined by long-range (LR) interactions (e.g., electronic Rydberg excitations, polarizabilities, and charge-transfer processes) require a large fraction of exact exchange (50% or more). The physical insight arising from these observations suggests a hybrid scheme in which SR interactions are treated by density-functional approximations while LR interactions are described by the exact exchange. This is precisely the idea of the so-called *range-separated* or *screened* hybrid functionals, and it proved to be one of the DFT's biggest successes of the past decade.

In the range-separated hybrid scheme, the electron–electron Coulomb repulsion operator is partitioned into a SR and a LR component

$$\frac{1}{r_{12}} = \underbrace{\frac{1 - f(r_{12})}{r_{12}}}_{\text{SR}} + \underbrace{\frac{f(r_{12})}{r_{12}}}_{\text{LR}}, \tag{6.70}$$

where $f(r_{12})$ is a "screening function" that satisfies the following requirements: (a) $0 \leq f \leq 1$, (b) $f \to 0$ when $r_{12} \to 0$, and (c) $f \to 1$ when $r_{12} \to \infty$. The SR component of a given exchange functional can be obtained by replacing the Coulomb operator $1/r_{12}$ in Equation 6.11 with its SR part to give

$$E_{\mathrm{x}}^{\mathrm{SR}}[\rho] = \frac{1}{2}\int d\mathbf{r}_1\, \rho(\mathbf{r}_1) \int d\mathbf{r}_2\, \frac{1 - f(r_{12})}{r_{12}} \rho_{\mathrm{x}}(\mathbf{r}_1, \mathbf{r}_2), \tag{6.71}$$

where $\rho_x(\mathbf{r}_1, \mathbf{r}_2)$ is the exchange hole density corresponding to the functional. Similarly, the LR exchange component of a functional may be defined by

$$E_x^{LR}[\rho] = \frac{1}{2}\int d\mathbf{r}_1\,\rho(\mathbf{r}_1)\int d\mathbf{r}_2\,\frac{f(r_{12})}{r_{12}}\rho_x(\mathbf{r}_1, \mathbf{r}_2). \tag{6.72}$$

For instance, the LR part of the exact exchange functional (whose exchange hole is given by Equation 6.13) is

$$E_x^{\text{exact,LR}}[\rho] = -\sum_{k,l=1}^{N/2}\int d\mathbf{r}_1\int d\mathbf{r}_2\,\phi_k(\mathbf{r}_1)\phi_k^*(\mathbf{r}_2)\frac{f(r_{12})}{r_{12}}\phi_l^*(\mathbf{r}_1)\phi_l(\mathbf{r}_2). \tag{6.73}$$

The two popular choices for the screening function are the exponential function

$$f(r_{12}) = 1 - e^{-\omega r_{12}}, \tag{6.74}$$

where ω is a positive constant, and the Gauss error function

$$f(r_{12}) = \text{erf}(\omega r_{12}) = \frac{2}{\sqrt{\pi}}\int_0^{\omega r_{12}} e^{-t^2}\,dt, \tag{6.75}$$

where ω is also a positive parameter. The error function is convenient in calculations employing Gaussian-type basis sets because all necessary two-electron integrals in this case can be evaluated efficiently.

To separate a functional into a LR and a SR parts by Equations 6.71 and 6.72, one needs the associated exchange hole. Aside from the exact exchange functional, exchange holes are known for only a handful of density-functional approximations such as LDA, Becke–Roussel [34], PBE, and TPSS. (In the case of LDA, the SR and LR parts can be derived in closed form [35, 36]; in the cases of PBE and TPSS, exchange holes were reverse-engineered from the corresponding functionals.) To circumvent this restriction, Hirao and coworkers [37, 38] proposed a different definition of the screened components, which does not require the exchange hole and so is applicable to any GGA.

Screened hybrid functionals that combine the LR part of exact exchange with the SR part of a semilocal density-functional approximation have been proposed by several researchers [37, 38]. In particular, Vydrov and Scuseria [39] combined the SR PBE exchange with the LR exact exchange into a LR-corrected PBE hybrid functional called *LC-ωPBE*. This functional is given by

$$E_{xc}^{\text{LC-}\omega\text{PBE}}(\omega) = E_x^{\text{exact,LR}}(\omega) + E_x^{\text{PBE,SR}}(\omega) + E_c^{\text{PBE}}, \tag{6.76}$$

where the recommended value of the screening parameter is $\omega = 0.40$ bohr^{-1}. LR-corrected functionals such as LC-ωPBE have excellent performance for a wider range of properties than other types of density-functional approximations.

A different way of combining SR and LR parts of exchange functionals has found use in condensed-matter physics. It had long been suspected that certain properties of solids should be better described with hybrid functionals than with semilocal approximations. Unfortunately, the exact exchange energy is difficult to evaluate accurately for metallic and weakly insulating solids using conventional techniques. This is due to the unphysically slow spatial decay of exact exchange interactions in systems with vanishing band gaps, which itself is a consequence of the essentially nonlocal character of the exact exchange energy density. To make hybrid DFT calculations on solids possible, Heyd, Scuseria, and Ernzerhof (HSE) [40] proposed to replace the LR portion of exact exchange in a global hybrid functional with a LR part of a semilocal density functional. This is equivalent to taking a semilocal functional and hybridizing the SR part of exchange. If the starting functional is the PBE GGA, this construction yields the HSE functional

$$E_{\mathrm{xc}}^{\mathrm{HSE}}(\omega) = aE_{\mathrm{x}}^{\mathrm{exact,SR}}(\omega) + (1-a)E_{\mathrm{x}}^{\mathrm{PBE,SR}}(\omega) + E_{\mathrm{x}}^{\mathrm{PBE,LR}}(\omega) + E_{\mathrm{c}}^{\mathrm{PBE}}, \quad (6.77)$$

where the parameter ω ($0 \leq \omega < \infty$) is adjusted to achieve the best possible accuracy for the problem of interest. Observe that smaller values of ω cause the mixing to be switched on at shorter interelectron distances. The HSE functional can be viewed as an interpolation between pure PBE and the global hybrid PBE functional (PBEh): When $a = 0.25$ and $\omega = 0$, HSE reduces to PBEh, while in the limit $\omega \to \infty$ it reduces to PBE. For solids, computational cost of HSE is much closer to that of PBE than of PBEh. The main practical advantage of the HSE hybrid is that it predicts much more accurate lattice constants and band gaps than any standard semilocal functional including LDA, PBE, and TPSS [41].

6.9 EMPIRICAL FITS

So far, we have discussed the methods of density-functional design that avoid empiricism as much as possible. New density-functional approximations were obtained either by rigorous derivations for exactly solvable models or by devising phenomenological mathematical expressions that were consistent with known properties of the exact functional. At some point in this process, it was necessary to introduce one or more parameters whose values were *a priori* unknown. These values were found by fitting computed properties to high-quality experimental data. It is because of this step that DFT is sometimes regarded as a semiempirical method.

Since there is no hope of deriving the exact exchange-correlation functional, while the method of constraint satisfaction is arduous and slow, it is hard to resist the pragmatism of fully empirical constructions. In the empirical approach, one starts by postulating a flexible analytic representation for the energy density and then tunes it by minimizing discrepancies between theoretical predictions and experimental observations. For example, on the basis of analysis of the density matrix expansion, Van Voorhis and Scuseria (VS98) [25] proposed parametrizing the exchange functional in the following form:

$$E_{\mathrm{x}}^{\mathrm{VS98}}[\rho] = \int \rho^{4/3} \left[\frac{b_0}{h(s,z)} + \frac{b_1 s^2 + b_2 z}{h^2(s,z)} + \frac{b_3 s^4 + b_4 s^2 z + b_5 z^2}{h^3(s,z)} \right] \mathrm{d}\mathbf{r}, \quad (6.78)$$

where s is defined by Equation 6.30, $z = \tau\rho^{-5/3} - C_F$ with $C_F = \frac{3}{5}(3\pi^2)^{2/3}$ and $h(s,z) = 1 + c(s^2 + z)$, whereas b_0, b_1, b_2, b_3, b_4, b_5, and c are adjustable parameters.

Optimization of empirical functionals can be carried out on several levels. On the first level, one optimizes linear and nonlinear parameters appearing in the expression for the energy density. On the second level, one writes the density-functional approximation in the form

$$E_{\mathrm{xc}}[\rho] = \int \sum_m a_m \mathrm{e}_{\mathrm{xc}}^{(m)}(\rho, s, \ldots)\, \mathrm{d}\mathbf{r}, \quad (6.79)$$

where $\mathrm{e}_{\mathrm{xc}}^{(m)}$ are various representations of the exchange-correlation energy density and a_m are adjustable empirical coefficients. All global and local hybrid functionals belong to this type. If desired, one may proceed even further to the third level, called *external optimization* [42], and consider a linear combination of several "model chemistries,"

$$E_{\mathrm{xc}}[\rho] = \sum_n \mathrm{d}_n E_{\mathrm{xc}}^{(n)}[\rho], \quad (6.80)$$

where the quantities $E_{\mathrm{xc}}^{(n)}[\rho]$ represent results of fully self-consistent Kohn–Sham calculations using different functionals; d_n are their weights fitted to a set of experimental data. Naturally, functionals that are optimized on two or three levels achieve a higher accuracy than functionals optimized on one level only.

The parameter optimization is usually accomplished by minimizing the root mean square (RMS) deviation of predictions from experiment

$$\mathrm{RMS} = \sqrt{\frac{\sum_s \sum_p (x_{sp}^{\mathrm{calc}} - x_{sp}^{\mathrm{exp}})^2}{N_x}}, \quad (6.81)$$

where x_{sp}^{calc} and x_{sp}^{exp} are, respectively, the calculated and experimental values of property p in system s, and N_x is total number of such data.

Most of the existing empirical density functionals are based on the analytic representations of exchange and correlation energy densities proposed, respectively, by Van Voorhis and Scuseria [25] and by Becke [43, 44]. These functionals include VS98 [25], Becke's exchange-correlation approximation of 1997 (B97) [43], the 1998 hybrid GGA [45], and hybrid meta-GGA [22] of Schmider and Becke, the GGA of Hamprecht, Cohen, Tozer, and Handy [46]

(HCTH), and its various reparametrizations. The most sophisticated empirical exchange-correlation functionals existing today are those of the Minnesota 2006 (M06) suite developed by Zhao and Truhlar [26, 47].

The M06 suite consists of four functionals: M06, M06-2X, M06-L, and M06-HF. All four have the same analytic form combining the functional forms of LDA, PBE, VS98, and B97 but differ by the values of more than 40 independent empirical parameters. The parameters are adjusted for optimal performance in four different types of chemical problems. M06 is a hybrid meta-GGA with 27% of exact exchange; it is designed to provide a consistently good accuracy for transition metals, main-group thermochemistry, medium range correlation energy, and barrier heights. M06-2X has twice as much exact exchange as M06 ($a = 0.54$; other parameters are reoptimized) and is trained to give the best possible performance for main-group compounds, valence and Rydberg electronic excitation energies, and noncovalent interactions. The M06-2X parametrization, however, is not good for transition metals. M06-L (where L stands for local) is a reparametrization of M06 with no exact exchange, dropped to enable application of the functional to very large and periodic systems. M06-L is the most accurate for transition metal compounds but not very accurate for reaction barrier heights that require a large fraction of exact exchange. Finally, M06-HF includes 100% of exact exchange to achieve good performance for charge-transfer excited states. Although M06 functionals contain many empirical parameters, they also respect several important exact constraints including the uniform electron gas limit (Eq. 6.42) and are free from the one-electron self-interaction error (Eq. 6.41).

Development of empirical density functionals requires large databases of accurate experimental data. Early empirical functionals were trained on relatively small test sets of atomization energies. By contrast, functionals of the M06 suite rely on a truly massive set of data that includes dozens of atomization energies, ionization potentials, and electron and proton affinities; bond dissociation energies, isomerization energies, and a variety of reaction barriers; hydrogen-bonded systems; charge-transfer, dipole-interaction, and $\pi - \pi$ stacking complexes; valence and Rydberg vertical excitation energies; and thermochemistry of transition metal reactions. The high flexibility combined with the unprecedented diversity of the training set enable the M06 functionals to predict chemical and physical properties with a reliability matching that of some high-level wavefunction methods.

6.10 CORRELATION FUNCTIONALS COMPATIBLE WITH EXACT EXCHANGE

Perhaps the most sophisticated density functionals constructed to date are Becke's nondynamical correlation functional of 2005 (B05) [48] and the 2008 hyper-GGA of Perdew, Staroverov, Tao, and Scuseria (PSTS) [49]. Both functionals use the exact exchange energy density of Equation 6.18 as an ingredient in the *correlation* part. This makes the correlation functional compatible with the exact exchange functional. The complexity of the B05 and PSTS functionals reflects not only the difficulty of the problem but also our growing understanding of the interplay between exchange and correlation.

The starting point for the B05 model is analysis of two types of electron correlation, called *dynamical* and *nondynamical (static)*. Dynamical correlation is due to close-range Coulombic interactions and so is essentially local in character. In systems where most of the correlation energy is dynamical, the exact exchange and correlation holes are both localized around the reference electron. Such systems include atoms, molecules near their equilibrium geometry, and the uniform electron gas. Semilocal density-functional approximations (LDA, GGA, meta-GGA) work well for such systems precisely because the LDA, GGA, meta-GGA exchange, and correlation holes are themselves localized. Nondynamical correlation arises in many-electron systems consisting of two of more fragments whose Coulombic interaction is weak or negligible. Each such fragment is effectively an independent system, so the exact exchange-correlation hole for electrons of any one fragment is contained entirely within that fragment. The exact exchange hole in such systems is split between all fragments, which means that the exact correlation hole must be delocalized as well. Semilocal density-functional approximation cannot recognize this delocalization and so they do not work well for systems with strong nondynamical correlation. A possible way to detect and account for nondynamical correlation is by using the real-space structure of the exact exchange hole as a diagnostic tool.

In an isolated hydrogen atom, the exact exchange hole around the reference electron is contained entirely within the vicinity of the nucleus and integrates to -1. In a highly stretched H_2 molecule, the exact exchange hole is divided between the two atoms. As a result, the effective normalization of the exact exchange hole around each H atom in stretched H_2 is only $-\frac{1}{2}$. According to Becke [50], an effective normalization of $-\frac{1}{2}$ means that the reference electron excludes less than one opposite-spin electron from its immediate vicinity, which raises the energy of each H atom in the stretched H_2 molecule. Therefore, Becke argued, the effective hole in each half of stretched H_2 needs to be deepened to repel electrons of opposite spin. The deepening of the effective exchange hole amounts to introducing *nondynamical correlation* and is modeled as follows:

$$\rho_{\mathrm{xc}}^{\alpha}(\mathbf{r}_1, \mathbf{r}_2) = \rho_{\mathrm{x}}^{\alpha}(\mathbf{r}_1, \mathbf{r}_2) + f_{\mathrm{c}}(\mathbf{r}_1)\rho_{\mathrm{x}}^{\beta}(\mathbf{r}_1, \mathbf{r}_2), \tag{6.82}$$

$$\rho_{\mathrm{xc}}^{\beta}(\mathbf{r}_1, \mathbf{r}_2) = \rho_{\mathrm{x}}^{\beta}(\mathbf{r}_1, \mathbf{r}_2) + f_{\mathrm{c}}(\mathbf{r}_1)\rho_{\mathrm{x}}^{\alpha}(\mathbf{r}_1, \mathbf{r}_2). \tag{6.83}$$

Here, $\rho_{\mathrm{x}}^{\alpha}$ and $\rho_{\mathrm{x}}^{\beta}$ are effective holes seen, respectively, by spin-up and spin-down electrons, while f_{c} is a position-dependent correlation parameter determined by two physical constraints: (i) $0 \leq f_{\mathrm{c}} \leq 1$; (ii) an exchange-correlation hole cannot contain more than one electron. The explicit form of this parameter proposed by

Becke is

$$f_{\mathrm{c}}(\mathbf{r}) = \min\left[\frac{1 - N_{\mathrm{x}}^{\alpha}(\mathbf{r})}{N_{\mathrm{x}}^{\beta}(\mathbf{r})}, \frac{1 - N_{\mathrm{x}}^{\beta}(\mathbf{r})}{N_{\mathrm{x}}^{\alpha}(\mathbf{r})}, 1\right], \tag{6.84}$$

where $N_{\mathrm{x}}^{\alpha}(\mathbf{r})$ and $N_{\mathrm{x}}^{\beta}(\mathbf{r})$ are position-dependent integrals of the exchange hole charge over the atomic region when the reference electron is at $\mathbf{r}$. The exchange-correlation energy is obtained by substituting the above expressions for the exchange-correlation holes into Eq. 6.9 to give

$$E_{\mathrm{xc}}[\rho] = \frac{1}{2} \sum_{\sigma=\alpha,\beta} \int d\mathbf{r}_1\, \rho_{\sigma}(\mathbf{r}_1) \int d\mathbf{r}_2\, \frac{\rho_{\mathrm{xc}}^{\sigma}(\mathbf{r}_1, \mathbf{r}_2)}{r_{12}}. \tag{6.85}$$

Since the exact exchange hole of Equation 6.13 cannot be integrated efficiently, the values of N_{x}^{α} and N_{x}^{β} in the B05 model are found using the approximate Becke–Roussel model exchange hole [34] instead of the exact $\rho_{\mathrm{x}}^{\sigma}(\mathbf{r}_1, \mathbf{r}_2)$. Even with this simplification, one still needs to solve numerically a complicated nonlinear equation for each point $\mathbf{r}$. Another obstacle is that the piecewise definition of $f_{\mathrm{c}}(\mathbf{r})$ by Equation 6.84 causes this function to have a discontinuous derivative that complicates self-consistent implementation of the B05 model. These difficulties were surmounted by Arbuznikov and Kaupp [51] and by Proynov and coworkers [52] who implemented the B05 functional in a fully self-consistent Kohn–Sham scheme, with minor modifications of Becke's original definitions. Although these researchers have so far reported only preliminary results, the numbers are encouraging: B05 does perform significantly better than B3LYP for difficult reaction barriers and gives excellent bond lengths [52].

The PSTS hyper-GGA also employs the exact exchange energy density to model nondynamical correlation. PSTS is essentially a local hybrid with a very complicated mixing fraction designed to interpolate between two extreme types of density regions for which the proper amount of exact exchange is known. The first type of density regions are called *normal*. These are the regions where the exact exchange-correlation hole is spatially localized around an electron and integrates to -1 over a narrow range. As we saw above, this is the situation where semiempirical density-functional approximations for exchange and correlation work very well because of mutual error cancellation. Therefore, in normal regions, the local fraction of exact exchange, $a(\mathbf{r})$, is designed to be small. "Abnormal" regions are those where the exact exchange-correlation hole is highly nonlocal and integrates to a value greater than -1 over the region. For example, all multicenter one-electron densities and regions with a fractional electron charge are abnormal in this sense. In abnormal regions, the mixing fraction $a(\mathbf{r})$ should be close to 1. For all intermediate situations, the mixing fraction adjusts the amount of exact exchange to some appropriate value between 0 and 1.

Construction of the PSTS mixing fraction is largely phenomenological and is guided by exact constraints. Nevertheless, the complexity of the problem requires a few empirical parameters. These parameters were determined by fitting to 97 molecular standard enthalpies of formation and 42 reaction barrier heights. Initial assessment of the PSTS functional showed that it performs much better than conventional global hybrids for reaction barrier heights, although there was no accuracy gain for atomization energies. As with the B05 functional, self-consistent implementation of the PSTS hyper-GGA is nontrivial and requires further simplifications [53].

6.11 CURRENT TRENDS AND OUTLOOK FOR THE FUTURE

One of the most fascinating topics in DFT that has come to prominence recently is the performance of density functionals for systems with fractional electron numbers and fractional spins [54–59]. It has been even argued [56] that all failures of present-day DFT can be understood by analyzing the errors of existing exchange-correlation approximations in such systems.

The story starts in 1982 when Perdew et al. [60] published a seminal paper in which they analyzed behavior of the exact density functional in systems with a fractional number of electrons. How can a system have a "fractional number of electrons?" As far as real atoms and molecules are concerned, electrons are of course indivisible, so the total number of electrons in a real chemical system is always an integer. What is meant by a system with a fractional electron number is a linear combination ("ensemble") of wavefunctions representing systems with different integer electron numbers.

Consider an example. When the internuclear distance in an H_2^+ molecule is stretched to infinity, the electron is physically localized either on one nucleus or on the other. Let the wavefunctions representing these two states, $H\cdots H^+$ and $H^+\cdots H$, be ϕ_L and ϕ_R, respectively. The wavefunctions ϕ_L and ϕ_R are degenerate ground-state eigenfunctions of the Hamiltonian. By the fundamental quantum-mechanical principle of linear superposition, any normalized linear combination of these wavefunctions, $c_L\phi_L + c_R\phi_R$, where $|c_L|^2 + |c_R|^2 = 1$, is also a valid ground-state solution of the Schrödinger equation. This includes a half-and-half combination with $|c_L|^2 = |c_R|^2 = \frac{1}{2}$ in which each of the atoms has only half an electron. In this sense, any fractional electron number $q = |c_L|^2$ is possible on the left atom, giving rise to the supermolecule $H^{1-q}\cdots H^{+q}$. We say that the region around each proton in stretched H_2 is a system with a fractional electron number (or a fractional charge). In practice, fractional charges are found not only at infinite nuclear separation but also in a moderately stretched H_2^+ molecule where the internuclear distance is a little greater than at equilibrium. Similarly, a molecular ion of the general formula A_2^+ can be viewed as a supersystem composed of two many-electron systems with fractional electron numbers.

Suppose now that we have a system with $N = J + q$ electrons, where J is a positive integer and $0 \leq q \leq 1$. Within Kohn–Sham DFT, the electron density of this system is constructed in accordance with the *Aufbau* principle, that is, by filling each of the J lowest energy Kohn–Sham spin-orbitals with one electron and placing the fraction q of an electron in the highest occupied molecular orbital (HOMO). The reason for using the *Aufbau* rule is because our system belongs to a supersystem that is supposed to be in the ground state. Thus, we write

$$\rho(\mathbf{r}) = \sum_{k=1}^{J} |\phi_k(\mathbf{r})|^2 + q|\phi_{\text{HOMO}}(\mathbf{r})|^2. \tag{6.86}$$

Perdew and coworkers [60] showed that, in general, the exact ground-state energy of a $(J + q)$-electron system is a linear combination of the ground-state energies of the J- and $(J + 1)$-electron systems

$$E(J + q) = (1 - q)E(J) + qE(J + 1), \qquad 0 \leq q \leq 1 \tag{6.87}$$

This means that the plot of the exact E as a function of q between J and $J + 1$ is a straight line.

It turns out that if we calculate the electronic energy using any existing density-functional approximation and then plot $E(J + q)$ as a function of q, the result will *not* be a straight line. Approximate density functionals are close to target at the end points J and $J + 1$ but fail to reproduce the straight line in between: the actual plot is a curve that is usually bent downward. This means that in systems such as H_2^+, application of the variational principle to approximate density functionals yields the maximally delocalized density ($H^{+1/2} \cdots H^{+1/2}$) whose energy is much lower than it should be. This artificial lowering of the energy in systems with fluctuating electron number is known as the *charge delocalization error*.

Similar analysis of spin-up and spin-down degeneracies in electrically neutral open-shell systems leads to the concept of fractional spin [56]. Consider an isolated hydrogen atom. In the absence of an external magnetic field, the ground state of this system is doubly degenerate: the spin-up eigenstate $\psi_\uparrow$ has the same energy as the spin-down eigenstate $\psi_\downarrow$, that is, $E_\uparrow = E_\downarrow$. Since the Hamiltonian is spin-independent, any normalized linear combination of these eigenfunctions, $c_\uparrow\psi_\uparrow + c_\downarrow\psi_\downarrow$, with $|c_\uparrow|^2 + |c_\downarrow|^2 = 1$ is also an eigenfunction of the Hamiltonian. Assuming that this linear combination is normalized, we can interpret it as a wavefunction of an H atom with a fraction $\gamma = |c_\uparrow|^2$ of α-spin and a fraction $|c_\downarrow|^2 = 1 - \gamma$ of β-spin. The exact energy of an isolated H atom is independent of γ, so we should have

$$E(\gamma) = \gamma E_\uparrow + (1 - \gamma)E_\downarrow = \text{const}, \qquad 0 \leq \gamma \leq 1. \tag{6.88}$$

Therefore, the plot of $E(\gamma)$ for an H atom should be a horizontal line segment. Weitao Yang and coworkers found that, instead of a horizontal line, all approximate density functionals predict a curve that is bent upward [58]. They also found that the maximum deviation from linearity, which occurs at the midpoint, coincides with the magnitude of the nondynamical correlation error in an infinitely stretched H_2 molecule. This discovery revealed an intimate connection between the fractional-spin error and nondynamical correlation.

The practical implications of the charge and spin delocalization errors are enormous. Binding energy curves for dissociating neutral molecules predicted with approximate DFT have massive positive errors at large internuclear distances. This occurs because neutral molecules dissociate into fractional-spin fragments for which approximate density functionals predict too high energies. In calculations of reaction barriers, theoretical energies of reactants are fairly accurate, but the energies of transition states are too low because transition states often consist of weakly interacting fractionally charged fragments for which approximate functionals predict too low energies. As a result, reaction barriers are severely underestimated. At the same time, molecular polarizability (a measure of the responsiveness of the electron density to an applied electric field) predicted by approximate density functionals is too high because fractional charges are artificially driven toward the edges of the molecule. In short, LDA, GGA, and meta-GGA fail to predict accurately many molecular properties because these approximations violate the important exact constraints of Equations 6.87 and 6.88.

The second fundamental result that follows from the analysis of Perdew et al. [60] is that the slope of the exact function $E(N)$, where N is a continuous electron number, changes discontinuously when N passes through an integer value. The significance of this fact will come to light once we reveal the physical meaning of the slope of $E(N)$.

Suppose first that N approaches the nearest integer J from above, that is, $N = J + q$, where q is a fractional electron number ($0 \leq q \leq 1$). From Equation 6.87 we obtain

$$\frac{\mathrm{d}E}{\mathrm{d}N} = \frac{\mathrm{d}E}{\mathrm{d}q} = E(J+1) - E(J), \qquad N = J + q \tag{6.89}$$

The quantity $E(J+1) - E(J)$ is the negative ionization potential of the $(J+1)$-electron system or, equivalently, the negative electron affinity of the J-electron system. Now let N approach J from below, that is, let us take $N = J - q$, where $q \geq 0$. We rewrite Equation 6.87 as

$$E(J-q) = qE(J-1) + (1-q)E(J), \qquad 0 \leq q \leq 1 \tag{6.90}$$

Differentiation of this equation with respect to N yields

$$\frac{\mathrm{d}E}{\mathrm{d}N} = -\frac{\mathrm{d}E}{\mathrm{d}q} = E(J) - E(J-1), \qquad N = J - q \tag{6.91}$$

This is the negative ionization potential of the J-electron system or, equivalently, the negative electron affinity of the $(J-1)$-electron system. It is instructive to rewrite these relations in terms of one-sided limits:

$$\left.\frac{\mathrm{d}E}{\mathrm{d}N}\right|_{N\to J+} = \lim_{\delta\to 0} \left.\frac{\mathrm{d}E}{\mathrm{d}N}\right|_{J+\delta} = -I_{J+1} = -A_J, \tag{6.92}$$

$$\left.\frac{\mathrm{d}E}{\mathrm{d}N}\right|_{N\to J-} = \lim_{\delta\to 0} \left.\frac{\mathrm{d}E}{\mathrm{d}N}\right|_{J-\delta} = -I_J = -A_{J-1}, \tag{6.93}$$

where I_J and A_J are, respectively, the ionization potential and electron affinity of the J-electron system. The last two equations mean that the exact derivative $\mathrm{d}E/\mathrm{d}N$ jumps by a constant when the number of electrons passes through an integer J. This constant is equal to

$$\left.\frac{\mathrm{d}E}{\mathrm{d}N}\right|_{N\to J+} - \left.\frac{\mathrm{d}E}{\mathrm{d}N}\right|_{N\to J-} = I_J - A_J. \tag{6.94}$$

Let us summarize. The exact ground-state energy of an N-electron system (i.e., a system with a continuous electron number N), plotted as a function of N, is a linkage of straight-line segments. The function $E(N)$ is itself continuous, but its first derivative, $\mathrm{d}E/\mathrm{d}N$, is discontinuous at all integer values of N. When N approaches an integer J from below, $\mathrm{d}E/\mathrm{d}N$ is the exact negative ionization potential of the J-electron system. When N approaches an integer J from above, $\mathrm{d}E/\mathrm{d}N$ is the exact negative electron affinity of the J-electron system.

Equations 6.87, 6.88, and 6.94 represent fundamental properties of the exact density functional. All semilocal density-functional approximations tend to violate these equations in many ways. The function $E(N)$ in approximate DFT no longer consists of straight-line segments but is a linkage of curves. Discontinuities of $\mathrm{d}E/\mathrm{d}N$ are observed only when a fraction of electron is added to a new orbital shell or subshell, whereas at other integer values of J, the curve $E(N)$ is smooth. Since the slopes of $E(N)$ are incorrect in approximate DFT, many physical properties including total energies, ionization potentials, electron affinities, band gaps, and polarizabilities are predicted with large errors.

Long-range-corrected hybrid density functionals and functionals that combine exact exchange with compatible nonlocal correlation violate the exact constraints of Equations 6.87, 6.88, and 6.94 to a lesser extent than the older (semilocal) approximations. For this reason, the newer functionals exhibit significantly better performance for a wider range of molecular properties than LDAs, GGAs, and meta-GGAs. Nevertheless, there is currently no approximate density functional that is entirely free from the fractional-charge and fractional-spin errors or which has correct derivative discontinuities at every integer electron number. Finding a way to construct functionals that respect the constraints of Equations 6.87, 6.88, and 6.94 would be a crucial step in overcoming the limitations of present-day DFT. If such a method is found, it will take DFT to the next level of predictive capability.

This brings us to an optimistic conclusion. If the history of DFT teaches us anything, it is that breakthroughs in density-functional development are usually preceded by years of scrutiny and introspection. This is why successful functionals tend to arrive in waves. The waves that have come ashore so far are LDAs, GGAs, global hybrids, and range-separated hybrids. The latest advances in our understanding of the limitations of existing density-functional approximations open exciting new opportunities for theorists and give us reasons to hope that the future of density-functional theory is secure.

REFERENCES

1. Kohn W. Rev Mod Phys 1999;71:1253.
2. Scuseria GE, Staroverov VN. In: Dykstra CE, Frenking G, Kim KS, Scuseria GE, editors. Theory and Applications of Computational Chemistry: The First Forty Years. Amsterdam: Elsevier; 2005. p 669–724.
3. Parr RG, Yang W. Density-Functional Theory of Atoms and Molecules. New York: Oxford University Press; 1989.
4. Szabo A, Ostlund NS. Modern Quantum Chemistry. New York: Macmillan; 1982.
5. Perdew JP, Kurth S. In: Fiolhais C, Nogueira F, Marques M, editors. A Primer in Density Functional Theory. Berlin: Springer; 2003. p 1–55.
6. Baerends EJ, Gritsenko OV. J Phys Chem A 1997;101:5383.
7. Kümmel S, Kronik L. Rev Mod Phys 2008;80:3.
8. Perdew JP, Schmidt K. In: Van Doren V, Van Alsenoy C, Geerlings P, editors. Density Functional Theory and Its Application to Materials. Melville (NY): AIP; 2001. p 1–20.
9. Dewar MJS. The Molecular Orbital Theory of Organic Chemistry. New York: McGraw-Hill; 1969.
10. Perdew JP, Wang Y. Phys Rev B 1992;45:13244.
11. Slater JC. Adv Quantum Chem 1972;6:1.
12. Brual G, Rothstein SM. J Chem Phys 1978;69:1177.
13. Becke AD. J Chem Phys 1986;84:4524.
14. Perdew JP, Burke K, Ernzerhof M. Phys Rev Lett 1996;77:3865.
15. Perdew JP, Ruzsinszky A, Tao J, Staroverov VN, Scuseria GE, Csonka GI. J Chem Phys 2005;123:062201.
16. Lieb EH, Oxford S. Int J Quantum Chem 1981;19:427.
17. Levy M. In: Gross EKU, Dreizler RM, editors. Density Functional Theory. New York: Plenum; 1995. p 11–31.
18. Becke AD. Can J Chem 1996;74:995.
19. Becke AD. J Chem Phys 1998;109:2092.
20. Becke AD. J Chem Phys 1988;88:1053.
21. Becke AD. J Chem Phys 1996;104:1040.
22. Schmider HL, Becke AD. J Chem Phys 1998;109:8188.
23. Boese AD, Handy NC. J Chem Phys 2002;116:9559.
24. Tao J, Perdew JP, Staroverov VN, Scuseria GE. Phys Rev Lett 2003;91:146401.
25. Van Voorhis T, Scuseria GE. J Chem Phys 1998;109:400.

26. Zhao Y, Truhlar DG. Theor Chem Acc 2008;120:215.
27. Perdew JP, Constantin LA, Sagvolden E, Burke K. Phys Rev Lett 2006;97:223002.
28. Perdew JP, Ruzsinszky A, Csonka GI, Vydrov OA, Scuseria GE, Constantin LA, Zhou X, Burke K. Phys Rev Lett 2008;100:136406.
29. Gill PMW. Aust J Chem 2001;54:661.
30. Becke AD. J Chem Phys 1993;98:5648.
31. Burke K, Perdew JP, Ernzerhof M. In: Dobson JF, Vignale G, Das MP, editors. Electronic Density Functional Theory: Recent Progress and New Directions. New York: Plenum Press; 1998. p 57–68.
32. Jaramillo J, Scuseria GE, Ernzerhof M. J Chem Phys 2003;118:1068.
33. Kaupp M, Arbuznikov AV, Bahmann H. Z Phys Chem 2010;224:545.
34. Becke AD, Roussel MR. Phys Rev A 1989;39:3761.
35. Savin A. In: Seminario JM, editor. Volume 4, Recent Developments and Applications of Modern Density Functional Theory, Theoretical and Computational Chemistry. Amsterdam: Elsevier; 1996. p 327–357.
36. Gill PMW, Adamson RD, Pople JA. Mol Phys 1996;88:1005.
37. Iikura H, Tsuneda T, Yanai T, Hirao K. J Chem Phys 2001;115:3540.
38. Nakano H, Nakajima T, Tsuneda T, Hirao K. In: Dykstra CE, Frenking G, Kim KS, Scuseria GE, editors. Theory and Applications of Computational Chemistry: The First Forty Years. Amsterdam: Elsevier; 2005. p 507–557.
39. Vydrov OA, Scuseria GE. J Chem Phys 2006;125:234109.
40. Heyd J, Scuseria GE, Ernzerhof M. J Chem Phys 2003;118:8207.
41. Heyd J, Scuseria GE. J Chem Phys 2004;121:1187.
42. Adamson RD, Gill PMW, Pople JA. Chem Phys Lett 1998;284:6.
43. Becke AD. J Chem Phys 1997;107:8554.
44. Becke AD. J Comput Chem 1999;20:63.
45. Schmider HL, Becke AD. J Chem Phys 1998;108:9624.
46. Hamprecht FA, Cohen AJ, Tozer DJ, Handy NC. J Chem Phys 1998;109:6264.
47. Zhao Y, Truhlar DG. Acc Chem Res 2008;41:157.
48. Becke AD. J Chem Phys 2005;122:064101.
49. Perdew JP, Staroverov VN, Tao J, Scuseria GE. Phys Rev A 2008;78:052513.
50. Becke AD. J Chem Phys 2003;119:2972.
51. Arbuznikov AV, Kaupp M. J Chem Phys 2009;131:084103.
52. Proynov E, Shao Y, Kong J. Chem Phys Lett 2010;493:381.
53. Jiménez-Hoyos CA, Janesko BG, Scuseria GE, Staroverov VN, Perdew JP. Mol Phys 2009;107:1077.
54. Perdew JP. In: Dreizler RM, da Providência J, editors. Density Functional Methods in Physics. New York: Plenum Press; 1985. p 265–308.
55. Perdew JP, Ruzsinszky A, Csonka GI, Vydrov OA, Scuseria GE, Staroverov VN, Tao J. Phys Rev A 2007;76:040501 (R).
56. Cohen AJ, Mori-Sánchez P, Yang W. Science 2008;321:792.
57. Mori-Sánchez P, Cohen AJ, Yang W. Phys Rev Lett 2008;100:146401.
58. Cohen AJ, Mori-Sánchez P, Yang W. J Chem Phys 2008;129:121104.
59. Mori-Sánchez P, Cohen AJ, Yang W. Phys Rev Lett 2009;102:066403.
60. Perdew JP, Parr RG, Levy M, Balduz JL Jr. Phys Rev Lett 1982;49:1691.

7

AN UNDERSTANDING OF THE ORIGIN OF CHEMICAL REACTIVITY FROM A CONCEPTUAL DFT APPROACH

ARINDAM CHAKRABORTY, SOMA DULEY, SANTANAB GIRI, AND PRATIM KUMAR CHATTARAJ

7.1 INTRODUCTION

One very basic question that has intrigued chemists is why a stable molecule, on coming close to a suitable counterpart under favorable conditions becomes excited, loses its own identity completely and thereby forms a new molecule! The energy criterion simply says that "... all spontaneous processes in the environment are associated with a lowering in the net free energy of the systems involved," which eventually leads to a stable state of the system in the form of favorable products. Scientists, experimentalists, as well as theoreticians, throughout the years, have tried hard to successfully rationalize the reactivity trends of molecular systems on chemical interaction. While the experimental chemists verified the above energy formalism through rigorous survey of a large set of chemical reactions, the theorists started developing intuitive mathematical paradigms toward setting a rationale behind such molecular reactivity. One very popular theoretical approach is conceptual density functional theory (CDFT) [1–4], which has proved to be an extremely successful method for describing the ground-state properties of molecules. The success of CDFT encompasses not only standard bulk materials but also complex materials such as proteins and carbon

A Matter of Density: Exploring the Electron Density Concept in the Chemical, Biological, and Materials Sciences, First Edition. Edited by N. Sukumar.

nanotubes. The main idea of DFT, stemming from the two theorems developed by Hohenberg and Kohn [5], is to describe an interacting system of fermions via its probability density, $\rho(\mathbf{r})$, and not via its many-body wave function, $\psi(\mathbf{r}_1, \mathbf{r}_2, \ldots, \mathbf{r}_N)$. Further, the many-body problem of N electrons with $3N$ spatial coordinates is reduced to just three coordinates. For an N-electron system, the corresponding Hamiltonian is completely determined by the number of electrons (N) and the external potential $\upsilon(\mathbf{r})$. Thus, all the properties of the system may be attained by appropriate variations of N and $\upsilon(\mathbf{r})$. This approach of analyzing chemical behavior has been termed *conceptual DFT* by Parr and Yang [1], who pioneered this DFT methodology. This method includes several global and local reactivity indices. These reactivity indices, also called *reactivity descriptors*, serve as mathematical response functions to describe chemical behavior of molecular motifs in the same manner as those observed from a variety of popular qualitative chemical concepts, such as electronegativity (χ) [6–8], hardness (η) [9–11], and electrophilicity (ω) [12–14]. Thus, the qualitative aspects of structure and bonding in chemical systems and their diverse reactivity patterns on chemical attack may be understood from a theoretical basis on a careful scrutiny of the different conceptual DFT-based global reactivity descriptors such as electronegativity (χ) [6–8], hardness (η) [9–11], and electrophilicity (ω) [12–14], as mentioned earlier, as well as local variants such ase atomic charges (Q_k) [15] and Fukui functions (FFs) (f_k) [16], which play a key role in ascertaining the local site selectivity in a molecule. Section 7.2 delineates an overview of the conceptual DFT-based global and local reactivity indices. An appraisal of the reactivity descriptors in determining molecular reactivity is presented in terms of several associated electronic structure principles in Section 7.3. Section 7.4 describes some applications of the conceptual DFT-based descriptors and the allied electronic structure principles in elucidating the stability, reactivity, and aromaticity of different organic and inorganic molecular systems. Section 7.5 delivers some concluding remarks.

7.2 REACTIVITY DESCRIPTORS

7.2.1 Global Reactivity Descriptors

The global reactivity descriptors determine the chemical behavior of a molecular species by considering it as a whole. The key global descriptors other than electronegativity (χ), hardness (η), and electrophilicity (ω) that determine chemical reactivity are chemical potential (μ), softness (S), polarizability (α), and magnetizability (ξ).

7.2.1.1 Electronegativity (χ) Electronegativity (χ), first proposed by Linus Pauling in 1932 as a development of valence bond theory [17, 18], is defined as the ability of an atom (or rarely a functional group) in a molecule to attract bonded electrons (or electron density) toward itself. Electronegativity, which has been shown to be correlated with a number of other chemical properties, is not strictly a property of an atom but rather that of an atom within a molecule [18]. Pauling's calculation of the electronegativity values for the elements is based on a thermodynamic consideration, which assumes that the covalent bond between two different atoms (A–B) is stronger than would be expected by taking the average of the strengths of the A–A and B–B bonds. According to valence bond theory, of which Pauling was a notable proponent, this "additional stabilization" of the heteronuclear bond is due to the contribution of ionic canonical forms to the bonding. The difference in electronegativity between atoms A and B is given by

$$\chi_{\mathrm{A}} - \chi_{\mathrm{B}} = (eV)^{-1/2}\sqrt{\frac{E_{\mathrm{d}}(\mathrm{AB}) - [E_{\mathrm{d}}(\mathrm{AA}) + E_{\mathrm{d}}(\mathrm{BB})]}{2}}, \tag{7.1}$$

where the dissociation energies, E_{d}, of the A–B, A–A, and B–B bonds are expressed in electron volts, the factor $(eV)^{-1/2}$ is included to ensure a dimensionless result. Hydrogen was chosen as the reference atom, and its electronegativity was arbitrarily assigned a value 2.2. An electronegativity scale ranging from 0.7 to 4.0 for the elements, as proposed by Pauling, was found to correlate well with common chemical intuition. Pauling's notion of the electronegativity of an element was supposed to vary with its chemical environment. Several other methods of calculating electronegativity have been proposed and, although there may be small differences in the numerical values, all methods show the same periodic trends between elements. Mulliken [19, 20] proposed that the arithmetic mean of two experimental observables viz., first ionization potential (IP) and the electron affinity (EA), should be a measure of the tendency of an atom to attract electrons. Thus, the electronegativity (χ) according to Mulliken's formalism can be mathematically expressed as

$$\chi = \frac{\mathrm{IP} + \mathrm{EA}}{2}. \tag{7.2}$$

As this definition is not dependent on an arbitrary relative scale, it is also invariant with the chemical environment and has been termed as *absolute electronegativity* by Pearson [21]. Allred and Rochow [22, 23] interpreted the electronegativity of an atom as an electrostatic force of attraction that exists between the nucleus and the valence electron(s), the electrons that are housed at a linear distance from the nucleus which is equivalent to the covalent radius (r_{cov}) of the atom. The electrostatic force experienced by the outermost electrons varies directly with the effective nuclear charge Z^*, while it is inversely proportional to the square of the covalent radius (r_{cov}). Thus, the Allred–Rochow electronegativity (χ_{AR}) can be formulated as

$$\chi_{\mathrm{AR}} = 0.359\frac{Z^*}{r_{\mathrm{cov}}^2} + 0.744. \tag{7.3}$$

Sanderson [24–30] noted the relationship between electronegativity and atomic size and proposed that the electronegativity (χ) should be proportional to the compactness of an atom. Specifically

$$\chi = \frac{D}{D_a}, \tag{7.4}$$

where D, the electron density of an atom, is expressed as $Z/r_{cov}{}^3$ (Z is the atomic number and $r_{cov}{}^3$ is its corresponding atomic volume (covalent radius cubed) and D_a is the expected electron density of an atom, calculated from extrapolation between the noble gas elements. This work underlies the concept of *electronegativity equalization*, which suggests that electrons distribute themselves around a molecule to minimize the energy or to equalize the electronegativity [31]. Allen [32] put forward a very simple definition of electronegativity that is related to the average energy of the valence electrons in a free atom and is expressed as

$$\chi = \frac{n_s \varepsilon_s + n_p \varepsilon_p}{n_s + n_p}, \tag{7.5}$$

where $\varepsilon_{s,p}$ are the one-electron energies of s- and p-electrons in the free atom and n_s and n_p are the number of s- and p-electrons in the valence shell. In other words, Allen's electronegativity can be correlated with the ionization potentials of the s- and p-orbitals of the atom. It is also considered as the first quantum mechanical realization of Pauling's electronegativity [32]. The one-electron energies can be determined directly from spectroscopic data, and so electronegativities calculated by this method are sometimes referred to as *spectroscopic electronegativities*.

7.2.1.2 Chemical Potential (μ) Gyftopoulos and Hatsopoulos [33], in an attempt to present a more physically meaningful explanation of electronegativity, considered a free atom or a free ion as a thermodynamic system. They provided a quantum thermodynamic definition of electronegativity of a system, which was identified as the negative of its electronic chemical potential (μ). The chemical potential (μ) of an atomic assembly may be evaluated by the theory of statistical ensembles. Thus if an atom or a molecule be considered as a member of a grand canonical ensemble where the energy (E) and the number of electrons (N) are continuous functions and vary independently, the chemical potential of the ensemble may be formulated as

$$\mu = \frac{\partial E}{\partial N} \text{ at constant entropy.} \tag{7.6}$$

The ground-state electron density is determined by a *constrained* variational optimization of the electronic energy. The Lagrange multiplier enforcing the density normalization constraint is also called *the chemical potential* (μ). Thus

μ can be simply written as the partial derivative of the system's energy (E) with respect to the number of electrons at fixed external potential $\upsilon(\mathbf{r})$

$$\mu = \left(\frac{\partial E}{\partial N}\right)_{\upsilon(\mathbf{r})}. \tag{7.7}$$

Furthermore, as the chemical potential (μ) of a system from the perspective of density functional theory (DFT) [1, 8] is considered as "the escaping tendency of an electronic cloud," electronegativity (χ) is the negative thereof. Thus,

$$\chi = -\mu = -\frac{\partial E}{\partial N} \text{ at constant entropy,} \tag{7.8}$$

which varies continuously with N and temperature (θ).

The process of chemical potential equalization is sometimes referred to as *the process of electronegativity equalization*. This connection comes from Mulliken's definition of electronegativity. By inserting the energetic definitions of the ionization potential and electron affinity into the Mulliken electronegativity, it is possible to show that the Mulliken chemical potential is a finite difference approximation to the derivative of the electronic energy with respect to the number of electrons, that is, the ionization potential, or ionization energy, of an atom or molecule is the energy required to strip it of an electron. The EA, E_{ea}, of an atom or molecule is the energy required to detach an electron from a singly charged negative ion, that is,

$$\mu_{\text{Mulliken}} = -\chi_{\text{Mulliken}} = -\frac{\text{IP} + \text{EA}}{2} = \left[\frac{\delta E[N]}{\delta N}\right]_{N=N_0}, \tag{7.9}$$

where IP and EA are the ionization potential and electron affinity of the atom, respectively. Furthermore, according to the Koopmans' theorem [34], the ionization potential (IP) and electron affinity (EA) of a molecular system can be expressed in terms of the energies of the frontier molecular orbitals (FMOs) as

$$\text{IP} = -\varepsilon_{\text{HOMO}} \text{ and } \text{EA} = -\varepsilon_{\text{LUMO}}$$

$$\text{Thus, } \mu = \frac{1}{2}(\varepsilon_{\text{HOMO}} + \varepsilon_{\text{LUMO}}). \tag{7.10}$$

7.2.1.3 Chemical Hardness (η) and Softness (S) The unique idea of chemical hardness for a molecular species initially stems from some conclusions drawn by inorganic chemists in explaining the interactions of several metal ions with suitable anions. The classification of chemical systems as "hard" or "soft" was successfully implemented by Pearson [35–39] in the early 1960s in connection with the study of generalized Lewis acid–base reactions

$$\text{A} + :\text{B} \rightarrow \text{A:B}$$

where A, the electron acceptor, is the acid and the electron-pair donor B is the base.

The acids and bases were classified as hard or soft by Pearson [35–39] on a purely qualitative basis where hardness simply meant the degree of compactness of the electron cloud surrounding the nucleus in a molecular system and softness was the extent to which the electronic environment around the nucleus/nuclei of an atomic/molecular species loosens itself. The general characteristics of hard and soft acids and bases, according to Pearson, may be illustrated as below:

	Hard	Soft
Acids	High positive charge Low polarizability Small size	Low positive charge High polarizability Larger size
Bases	High electronegativity difficult to oxidize low polarizability	Low electronegativity easily oxidized higher polarizability

Further experimental investigations revealed that hard acids preferred to bind with hard bases, while soft acids preferentially bind to soft bases. This preference for a "hard–hard" and "soft–soft" combination among two classes of acids and bases became more or less a rule of thumb, known as the hard–soft acid–base (HSAB) principle [40]. The HSAB principle sheds light on how a pair of interacting systems possessing a hard–hard or soft–soft combination attains an extra thermodynamic stabilization, but it does not give a quantitative measure of hardness and softness. Later, Parr and Pearson [10, 41, 42] mathematically defined the chemical hardness (η) for a molecular system, identified as the first derivative of the chemical potential (μ) or the second derivative of the energy (E) as a function of the number of electrons N at a fixed external potential $\upsilon(\mathbf{r})$

$$\eta = \left(\frac{\partial \mu}{\partial N}\right)_{\upsilon(\mathbf{r})} = \left(\frac{\partial^2 E}{\partial N^2}\right)_{\upsilon(\mathbf{r})}. \tag{7.11}$$

The convexity of the E versus N curve renders the value of η positive, and from the method of finite differences, the curvature equals IP − EA, which signifies hardness. Therefore,

$$\eta = \left(\frac{\partial \mu}{\partial N}\right)_{\upsilon(\mathbf{r})} = \left(\frac{\partial^2 E}{\partial N^2}\right)_{\upsilon(\mathbf{r})} = \mathrm{IP} - \mathrm{EA}. \tag{7.12}$$

Thus for a normal charge-transfer (CT) process such as that in a Lewis acid–base pair, the chemical hardness (η) of a species is a measure of its resistance to further shift of electrons to the other species and corresponds qualitatively to its compactness.

The reciprocal of the global hardness (η) for a molecular species is defined as its global softness (S) [43], which is expressed as

$$S = \frac{1}{2\eta} = \frac{1}{2}\left(\frac{\partial N}{\partial \mu}\right)_{\upsilon(\mathbf{r})}. \tag{7.13}$$

Thus, the global softness (S) of a system quantitatively measures the relative diffuseness of the electron density between two interacting pairs. Chemical hardness (η) and softness (S) can therefore be quantitatively well correlated with molecular polarizability (α), following the trend predicted qualitatively by the HSAB principle.

7.2.1.4 Electrophilicity Index (ω) In terms of conceptual chemistry, an electrophile is designated as a species that has a special affection for electrons. It is generally an electron-deficient system, might possess some positive charge, or even be a free radical with a tendency to attract electron-rich components called *nucleophiles*. As chemical reaction involves the shifting and rearrangement of electrons between reactant moieties to create new products by the rupture of electron–electron linkages (bonds) and the formation of new ones, the electrophiles and nucleophiles play a sheet anchor role in devising the mechanistic pathways of almost all types of organic and inorganic chemical reactions, such as acid–base, redox, addition, substitution, elimination, and molecular rearrangement. Thus, an electrophile on interaction with an electron-rich species (nucleophile) will strongly attract the electron density of the latter and get stabilized with the gradual lowering of energy and formation of a stable covalent bond. Initially, there was no quantitative designation of this simple qualitative idea as to the extent to which the electrophilic system is energetically stabilized on gradual transfer of electron density from the adjacent nucleophile. Maynard et al. [44], on experimenting with the human immunodeficiency virus type 1 (HIV-1) nucleocapsid protein p7 (NCp7) with a variety of electrophilic agents, showed that the fluorescence decay rates vary almost linearly with the ratio of the square of electronegativity (χ) to hardness (η), χ^2/η. This quantity is the capacity of an electrophile to attract electrons from a nucleophile to give a covalent bond. This innovative qualitative idea proposed by Maynard et al. [44] soon caught the eyes of Parr and coworkers [12] and, in an attempt to quantify the electron-attracting power of a species, they coined a new descriptor called *electrophilicity index* (ω). Parr's postulate of electrophilicity index (ω) unlike Maynard et al. (which was developed from kinetic considerations by studying reaction rates) is based on thermodynamics. Thus, ω is a measure of the favorable change in energy on saturation of a system with electrons. The electrophilicity index (ω) as defined by Parr et al. [12] is

$$\omega = \frac{\mu^2}{2\eta} = \frac{\chi^2}{2\eta}. \tag{7.14}$$

The above expression for electrophilicity (ω) is comparable with the equation of electrical power (W) in classical physics, where $W = V^2/R$, V and R being the voltage and electrical resistance, respectively. Thus, ω represents the "electrophilic power" of a species. Comprehensive reviews on electrophilicity index (ω) [13, 14, 45] have dealt with its genesis and rigorous applications toward an understanding of chemical reactivity.

Electrodonating (ω^-) and Electroaccepting (ω^+) Powers In order to correlate the energy changes associated between the corresponding acceptors and donors in a CT process, Gazquez et al. [46] utilized the second-order Taylor series energy expansion formula as a function of the number of electrons (N) in the intervals between $N-1$ and N, and N and $N+1$, to show that the electrodonating (ω^-) and the electroaccepting (ω^+) powers may be defined as

$$\omega^- = \frac{(\mu^-)^2}{2\eta^-}; \; \omega^+ = \frac{(\mu^+)^2}{2\eta^+}, \tag{7.15}$$

where μ^- and μ^+ are the chemical potentials for electron donation and electron acceptance, respectively, and η^- and η^+ signify the hardness for electron donation and electron acceptance, respectively. Furthermore, μ^- and μ^+ can in fact be equated to μ [46], so that $\mu^- = \mu^+ = \mu$ and likewise $\eta^- = \eta^+ = \eta$. This eventually equates the electrodonating (ω^-) and the electroaccepting (ω^+) powers with the original concept of electrophilicity (ω) owing to which ω, ω^-, or ω^+ may be expressed in terms of chemical potential (μ) and hardness (η) as $\omega^- = \omega^+ = \omega = \mu^2/2\eta$. Gazquez et al. [46] proposed two sets of definitions for ω^- or ω^+ based on two different approaches, one exploiting the original formula above, expressed as

$$\omega^+ = \frac{\mathrm{EA}^2}{2\,(\mathrm{IP} - \mathrm{EA})}, \tag{7.16}$$

$$\omega^- = \frac{\mathrm{IP}^2}{2\,(\mathrm{IP} - \mathrm{EA})}, \tag{7.17}$$

and the other utilizing an alternative expression for energy

$$\omega^+ = \frac{(\mathrm{IP} + 3\mathrm{EA})^2}{16\,(\mathrm{IP} - \mathrm{EA})}, \tag{7.18}$$

$$\omega^- = \frac{(3\mathrm{IP} + \mathrm{EA})^2}{16\,(\mathrm{IP} - \mathrm{EA})}, \tag{7.19}$$

where IP and EA are the first ionization energy and electron affinity, respectively, of the system. It was further shown that this alternative approach of expressing ω^- or ω^+ yields better correlations [46].

Net Electrophilicity ($\Delta\omega^{\pm}$) The concept of net electrophilicity ($\Delta\omega^{\pm}$) for a system recently proposed by Chattaraj et al. [47] is an attempt to assess the electron-accepting power of a molecule on chemical reaction due to the combined attractive and repulsive effects arising out of the presence of both electrons and nuclei. In other words, this new dual descriptor ($\Delta\omega^{\pm}$) is an appraisal of the electrophilicity of a system relative to its own nucleophilicity and definitely serves to provide a physically more meaningful understanding of the electrophilic power of a system. From energy considerations, it is apparent that a larger value of ω^{+} for a system corresponds to an enhanced capability to accept charge, whereas a smaller value of ω^{-} implies it acts as a better donor. Therefore, the mathematical foundation of $\Delta\omega^{\pm}$ arises out of a parity between ω^{+} and ω^{-} where the negative (or reciprocal) of ω^{-} is compared with ω^{+}. Thus, net electrophilicity ($\Delta\omega^{\pm}$) in terms of ω^{+} and ω^{-} is formulated as

$$\Delta\omega^{\pm} = \{\omega^{+} - (-\omega^{-})\} = (\omega^{+} + \omega^{-}) \tag{7.20}$$

or

$$\Delta\omega^{\pm} = \left\{\omega^{+} - \left(\frac{1}{\omega^{-}}\right)\right\}.$$

7.2.1.5 ***Polarizability*** **(α)** ***and Magnetizability*** **(ξ)** The polarizability (α) of an atom or molecule is described as the lowest order response of its electron cloud to an external weak electric field [48, 49]. The static dipole polarizability (α) is a linear response property and is defined as *the second derivative of the total electronic energy* (E) with respect to the external homogeneous electric field as

$$\alpha_{\alpha\beta} = -\left(\frac{\partial^2 E}{\partial F_\alpha \partial F_\beta}\right)_{F=0}, \tag{7.21}$$

where F_α and F_β are the electric field components for a fixed coordinate system with $\alpha, \beta, \gamma = x, y, z$. The polarizability ($\alpha$) is very sensitive to basis set, electron correlation, and relativistic effects and to the vibrational structure in case of a molecule. Qualitatively, polarizability (α) has been found to vary inversely with global hardness (η) [9,50–56] with increasing softness, a molecule becomes more polarizable, and an interesting linear correlation is established between the static dipole polarizability (α) and the third power of molecular softness (S) [57–59].

The magnetizability (ξ) of a chemical system is a measure of the linear response of its electron cloud to an externally applied magnetic field. It is expressed as

$$\left|\xi = -\left(\frac{\partial^2 \varepsilon(B)}{\partial B^2}\right)\right|_{B=0}, \tag{7.22}$$

where B is the external magnetic field.

The magnetizability (ξ) of a chemical species is found to vary with its softness (S) and polarizability (α). Thus, a softer species is found to be more polarizable and hence more magnetizable [60].

7.2.2 Local Reactivity Descriptors

The various CDFT-based global reactivity parameters discussed above show how to predict the structural changes and reactivity patterns of simple molecular systems. However, a deeper insight into the activity of a particular atomic site in a molecule during chemical interactions or excitations can be gained from a careful survey of different local reactivity descriptors such as electron density ($\rho(\mathbf{r})$), FF ($f(\mathbf{r})$), local softness ($s(\mathbf{r})$), local hardness ($\eta(\mathbf{r})$), philicity ($\omega(\mathbf{r})$), and nucleophilicity excess ($\Delta\omega_{\mathrm{g}}^{\pm}$). The local descriptors thus help to determine the site selectivity in a chemical reaction.

7.2.2.1 Electron Density ($\rho(\mathbf{r})$) The electron density ($\rho(\mathbf{r})$), a local reactivity descriptor by itself, is a measure of the first-order variation of the energy as a function of the external potential ($\upsilon(\mathbf{r})$) [1, 61] It is expressed as

$$\rho(\mathbf{r}) = \left(\frac{\delta E}{\delta \upsilon(\mathbf{r})}\right)_N . \tag{7.23}$$

The electron density provides useful site reactivity information for a molecular species.

7.2.2.2 Fukui Function ($f(\mathbf{r})$) The FF ($f(\mathbf{r})$) [16, 62] is a very popular local reactivity descriptor that is usually applied to rationalize the site selectivity of chemical systems. It is actually defined as *the differential change* in electron density ($\rho(\mathbf{r})$) because of an infinitesimal change in the number of electrons (N) and is expressed as

$$f(\mathbf{r}) = \left(\frac{\partial \rho(\mathbf{r})}{\partial N}\right)_{\upsilon(\mathbf{r})} = \left(\frac{\delta \mu}{\delta \upsilon(\mathbf{r})}\right)_N . \tag{7.24}$$

Since there is a discontinuity in the derivative of Equation 7.24 for integral values of N, three different types of FFs can be defined by applying the finite difference and frozen core approximations as follows [16, 62]:

$$f^{+}(\mathbf{r}) = \left(\frac{\partial \rho(\mathbf{r})}{\partial N}\right)^{+}_{\upsilon(\mathbf{r})} \approx \rho_{N+1}(\mathbf{r}) - \rho_N(\mathbf{r}) \approx \rho_{\mathrm{LUMO}}(\mathbf{r}) \text{ for nucleophilic attack,} \tag{7.24a}$$

$$f^{-}(\mathbf{r}) = \left(\frac{\partial \rho(\mathbf{r})}{\partial N}\right)^{-}_{\upsilon(\mathbf{r})} \approx \rho_N(\mathbf{r}) - \rho_{N-1}(\mathbf{r}) \approx \rho_{\mathrm{HOMO}}(\mathbf{r}) \text{ for electrophilic attack,} \tag{7.24b}$$

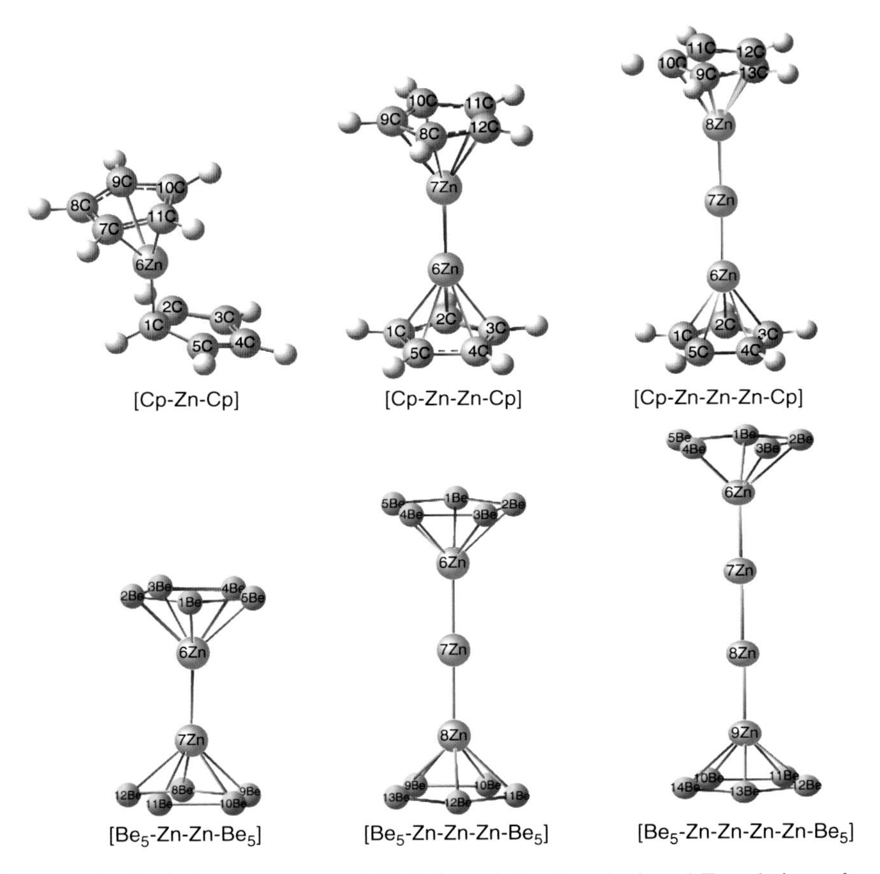

Figure 7.5 Optimized structures of $[C_5H_5]^-$ and $[Be_5]^-$ substituted Zn_n chain analogs.

A Matter of Density: Exploring the Electron Density Concept in the Chemical, Biological, and Materials Sciences, First Edition. Edited by N. Sukumar.

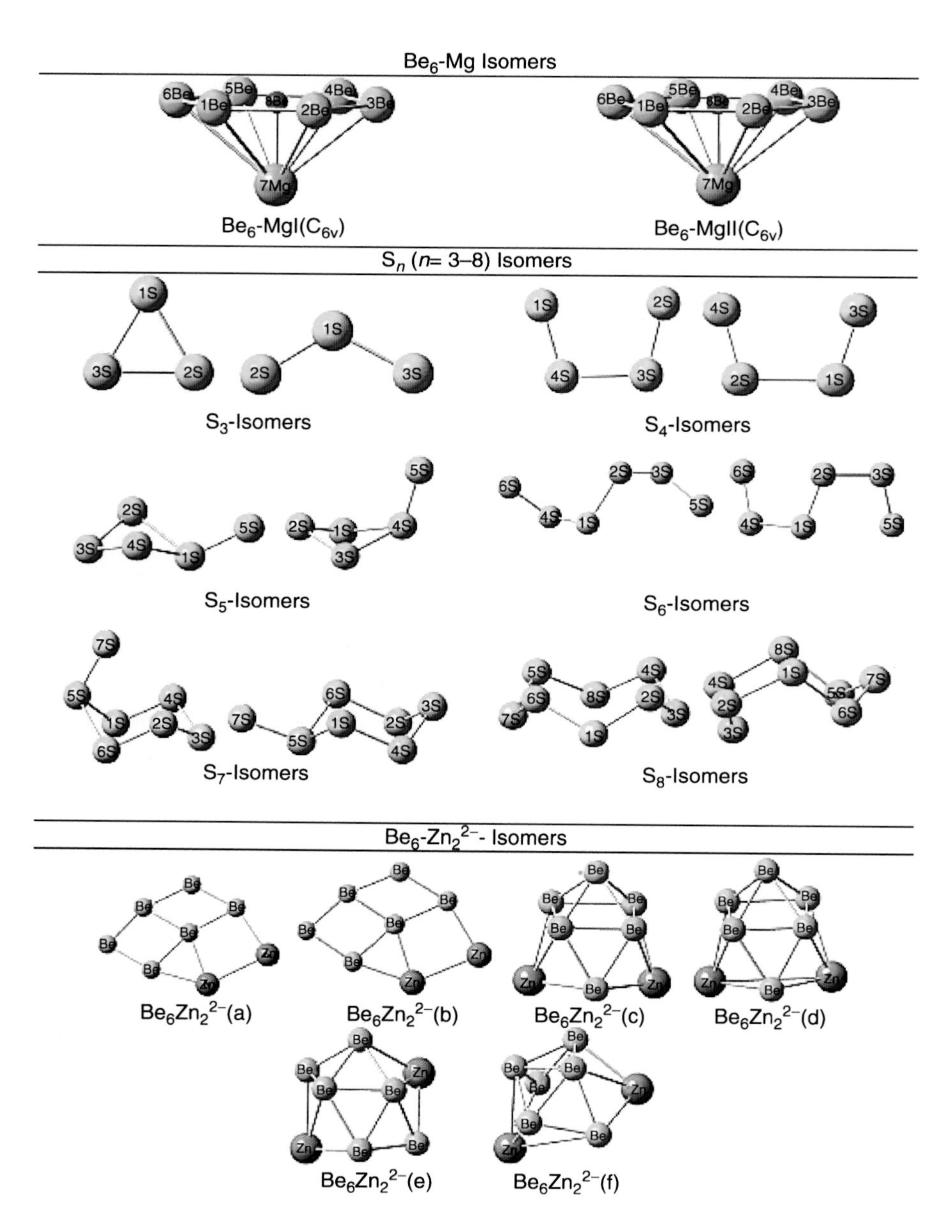

Figure 7.8 Some dominant bond—stretched conformers of all-metal and nonmetal systems.

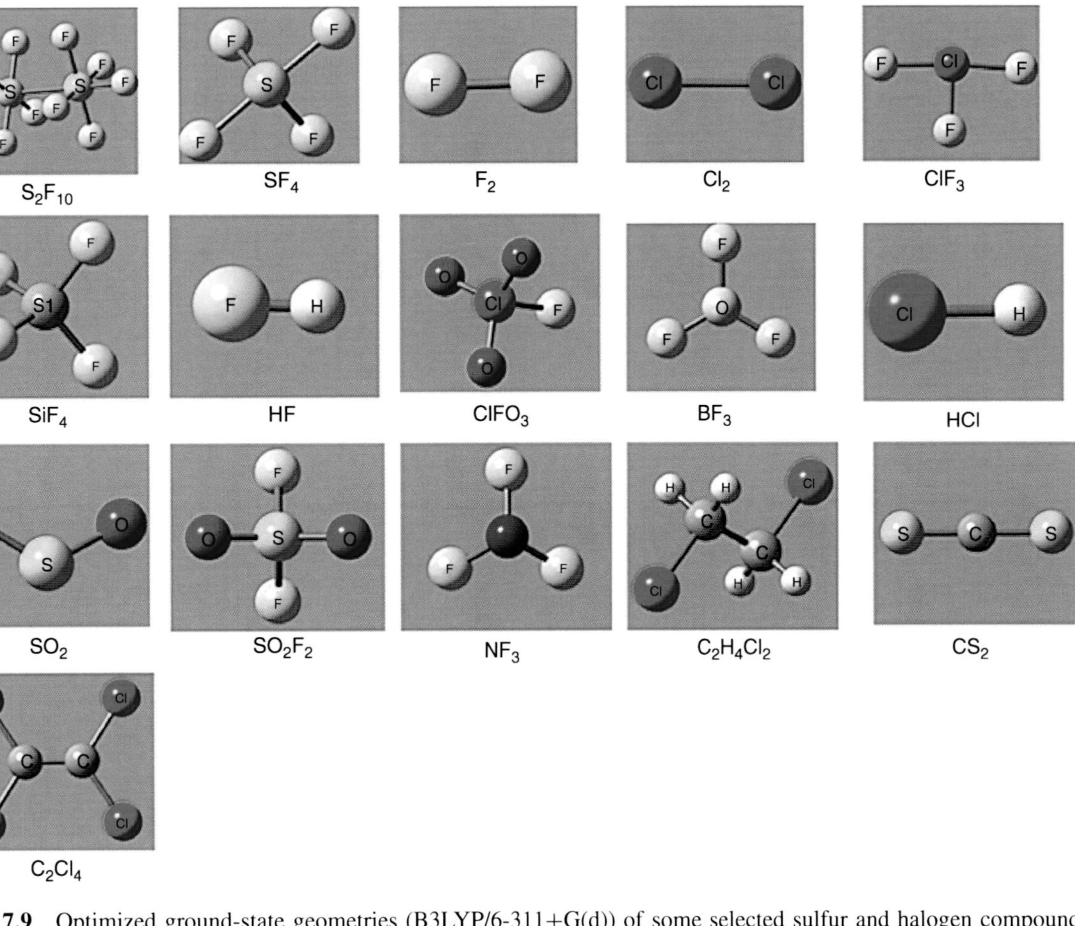

Figure 7.9 Optimized ground-state geometries (B3LYP/6-311+G(d)) of some selected sulfur and halogen compounds.

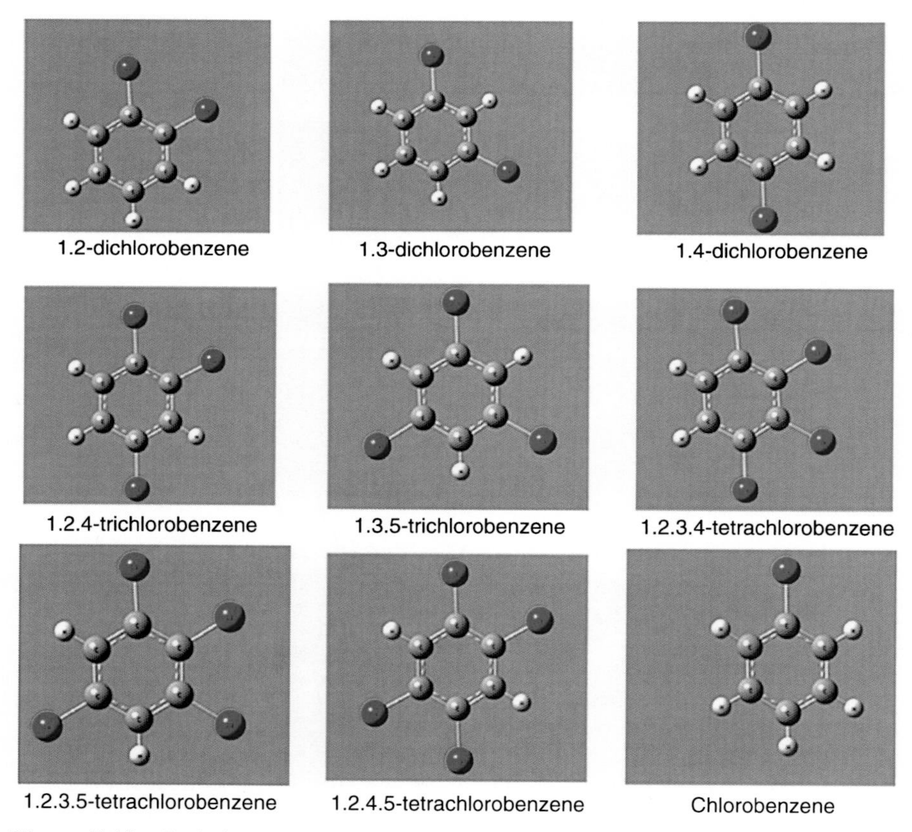

Figure 7.10 Optimized ground-state geometries (B3LYP/6-311+G(d)) of chlorinated aromatic compounds.

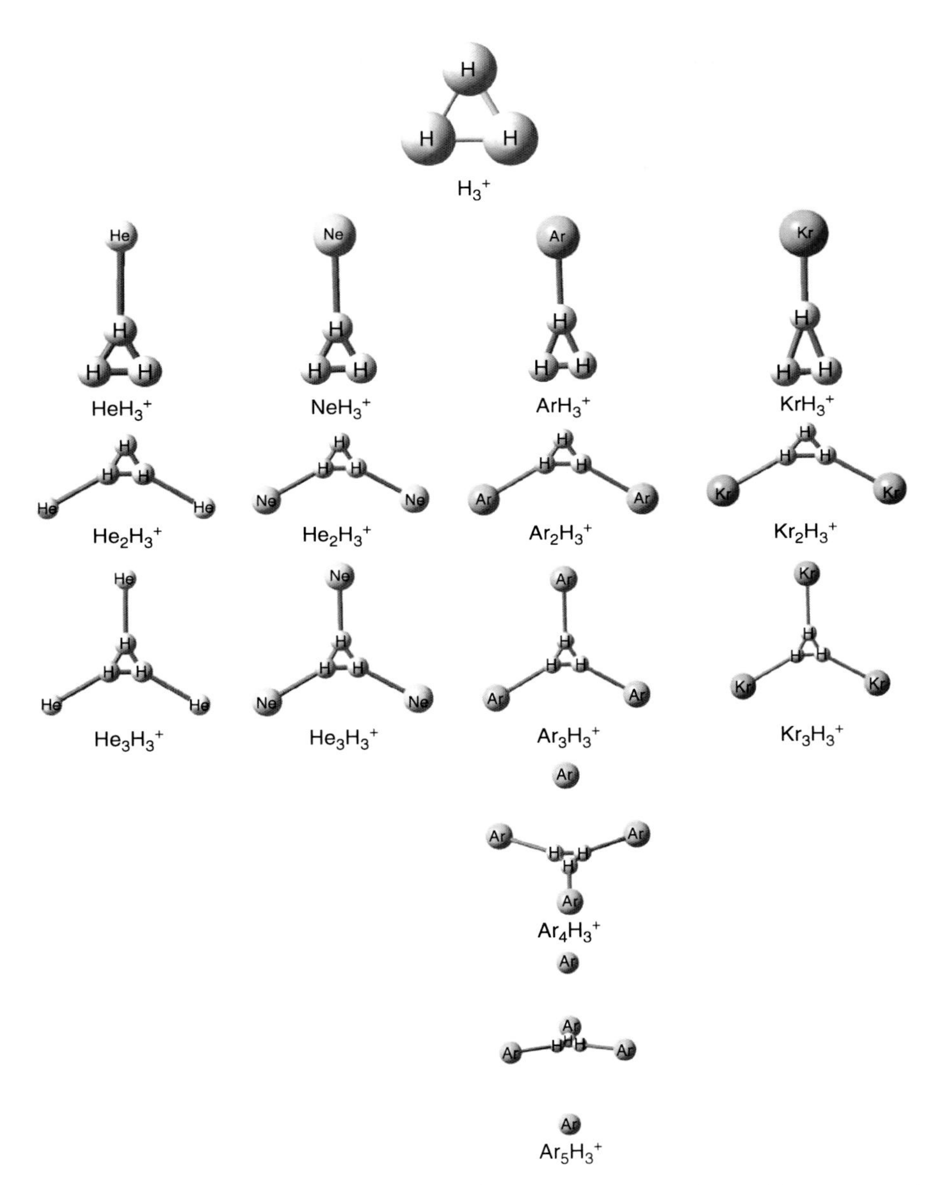

Figure 7.13 Geometrical structure of H_3^+ and its corresponding noble-gas-trapped clusters optimized at the B3LYP/6-311+G(d) level of theory.

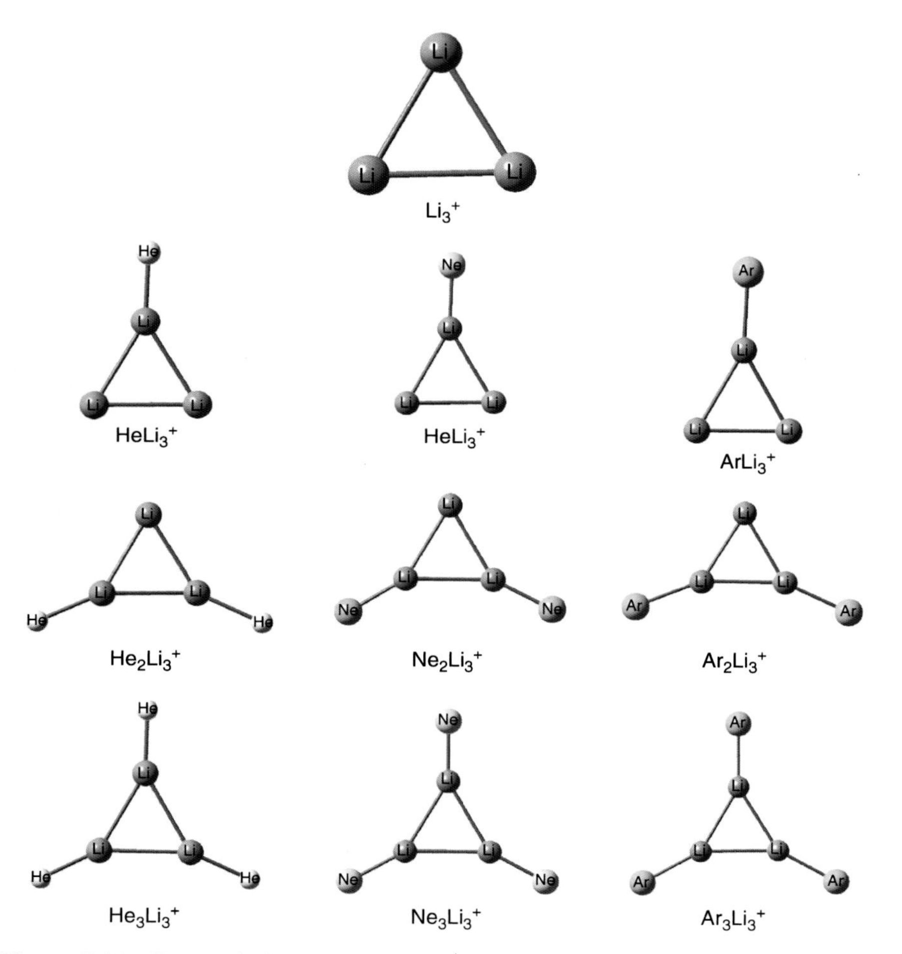

Figure 7.14 Geometrical structure of Li_3^+ and its corresponding noble-gas-trapped clusters optimized at the B3LYP/6-311+G(d) level of theory.

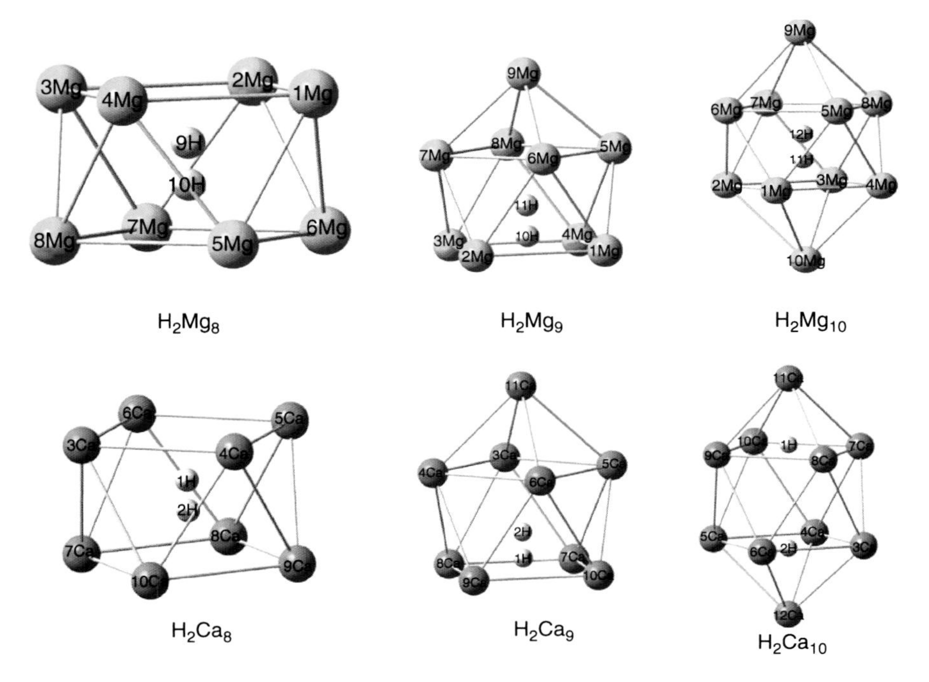

Figure 7.15 Optimized geometries (at B3LYP/6-311+G(d) level) of H_2M_n (where M = Mg, Ca; n = 8, 9, 10) clusters.

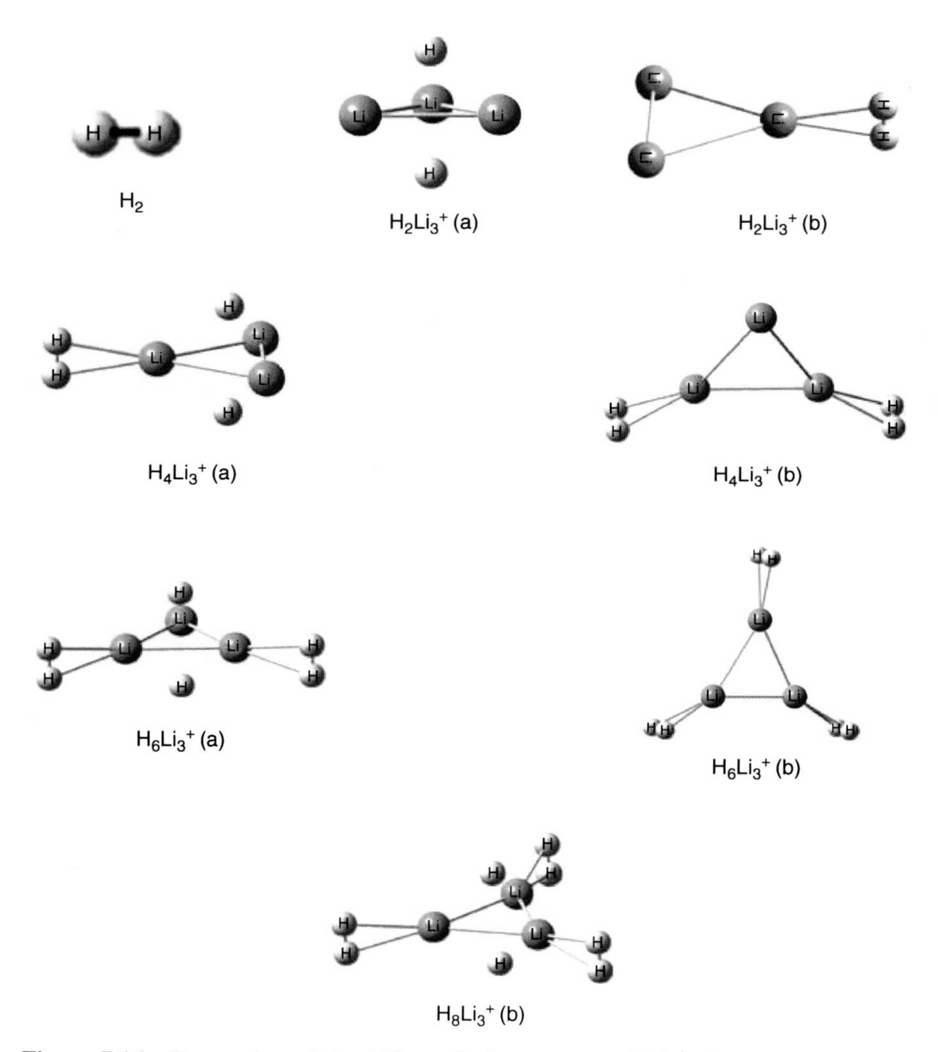

Figure 7.16 Geometries of the different hydrogen-trapped Li_3^+ clusters optimized at B3LYP/6-311+G(d) level of theory.

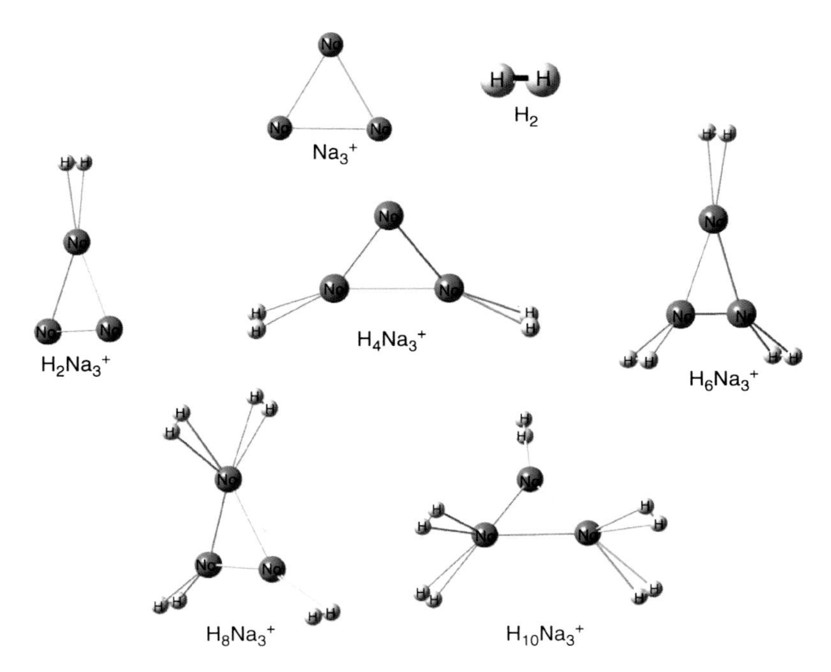

Figure 7.17 Geometries of different H_2 trapped Na_3^+ clusters (B3LYP/6-311+G(d)).

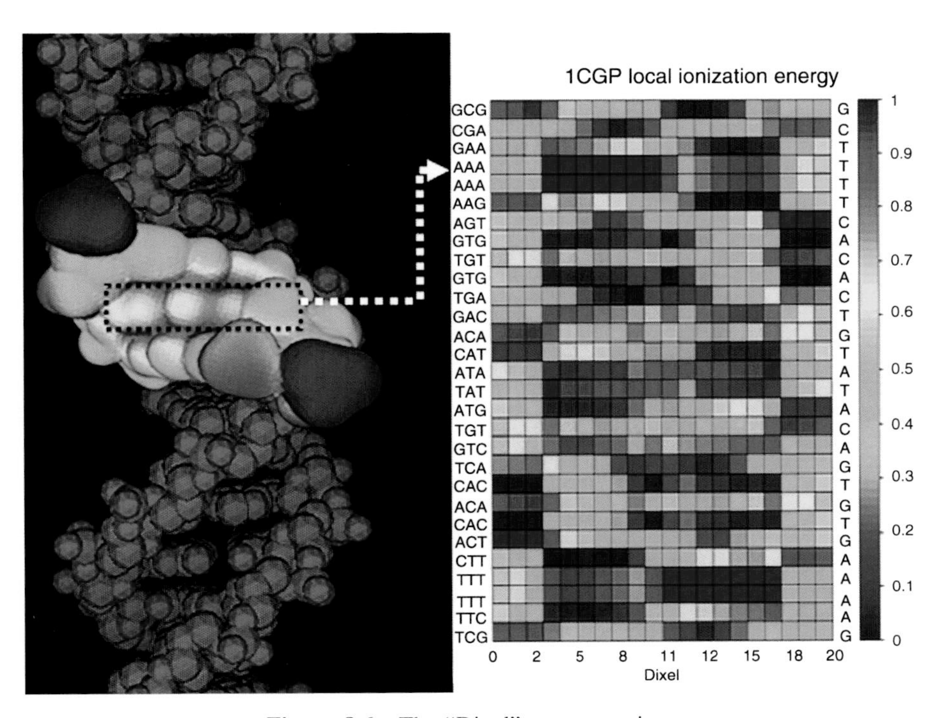

Figure 8.6 The "Dixel" representation.

Figure 8.7 PEST ray trace showing ray length and angle-of-reflection information recorded at each point of intersection of the rays with the molecular surface [111]. (Reproduced with permission from Breneman CM, Sundling CM, Sukumar N, Shen L, Katt WP, Embrechts MJ. J Comput Aided Mol Des 2003;17:231–240, Copyright 2003 Springer).

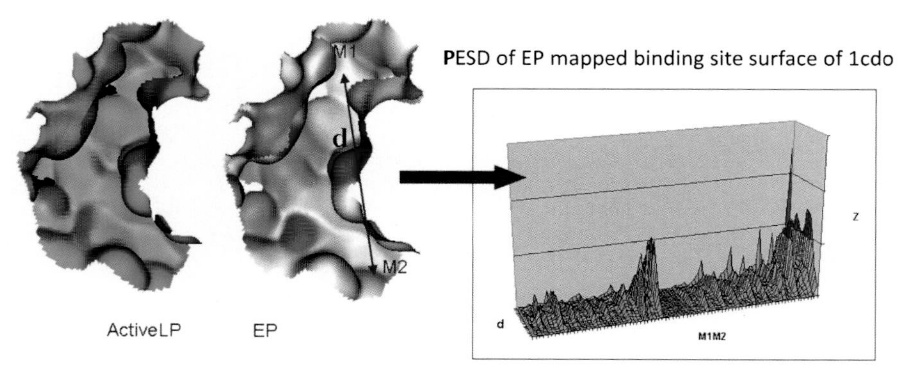

Figure 8.8 Construction of PESD signatures employing MEP and MLP [114]. (Reproduced with permission from Das S, Krein MP, Breneman CM. J Chem Inf Model 2010; 50 (2): 298–308, Copyright 2010 American Chemical Society).

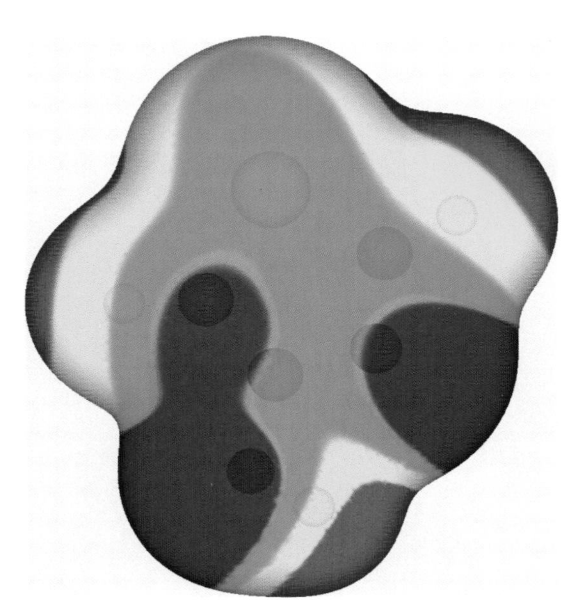

Figure 9.1 Color-coded diagram of the electrostatic potential on the molecular surface of 4-hydroxy-1,3-thiazole computed at the B3PW91/6-31G(d,p) level. The locations of the atomic nuclei are visible through the surface; sulfur is at the top, and nitrogen is at the lower right. Color ranges, in kilocalories per mole, are red > 15.0 > yellow > 0.0 > green > −8.0 > blue.

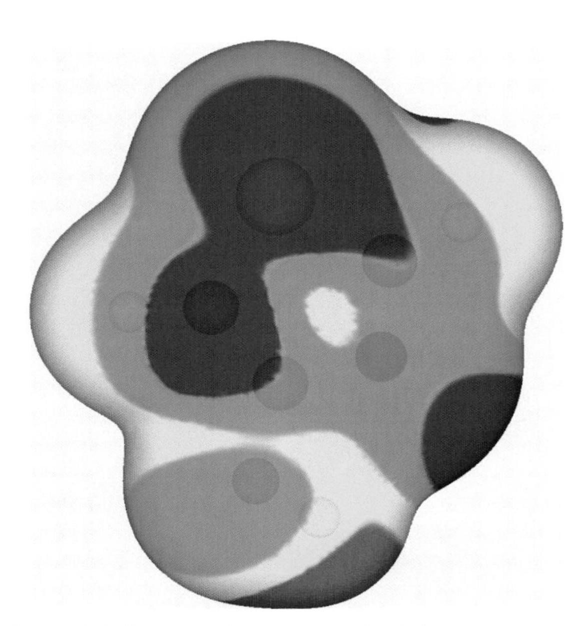

Figure 9.2 Color-coded diagram of the average local ionization energy on the molecular surface of 4-hydroxy-1,3-thiazole computed at the B3PW91/6-31G(d,p) level. The locations of the atomic nuclei are visible through the surface; sulfur is at the top, and nitrogen is at the lower right. Color ranges, in electro volts, are red > 14.0 > yellow > 12.0 > green > 10.0 > blue.

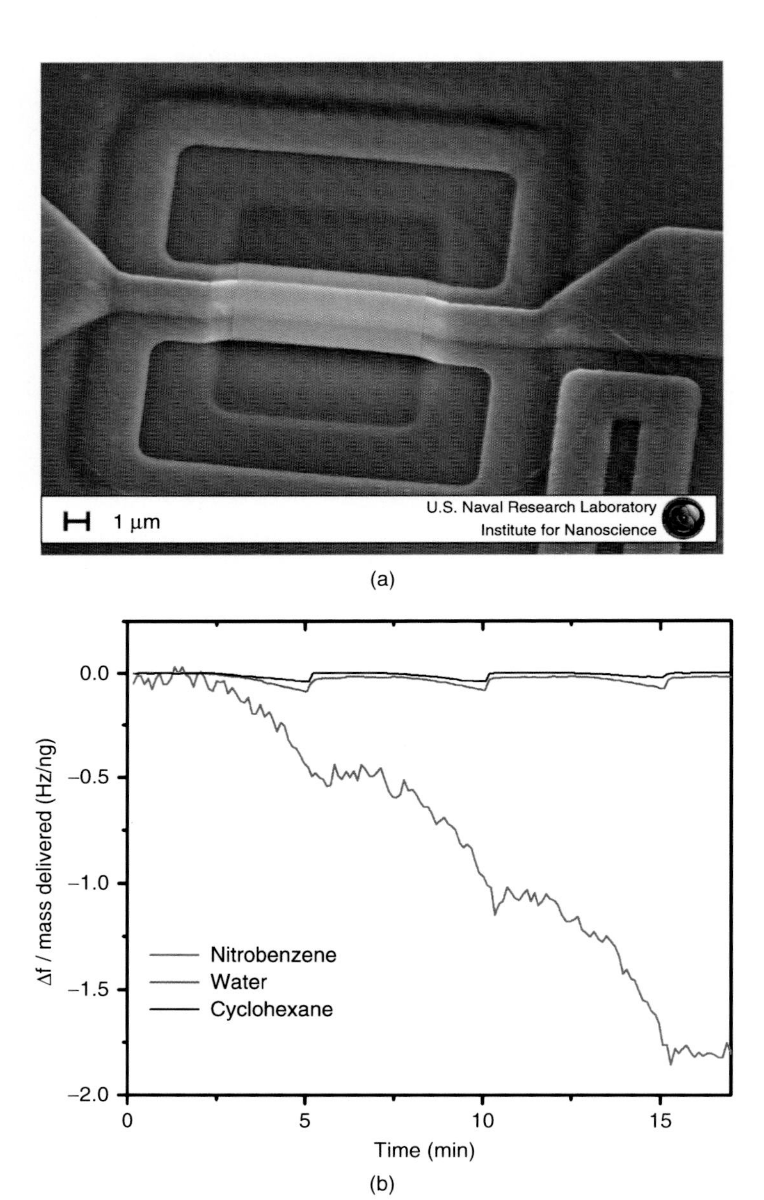

Figure 9.3 (a) Chemically functionalized CMOS nanomechanical resonator (shaded bridge) with electrical actuation and transduction. (b) Exposure of CMOS nanomechanical resonator to nitrobenzene (analyte), water (interferent), and cyclohexane (interferent) shows the selectivity of sensor. On the abscissa, f = frequency.

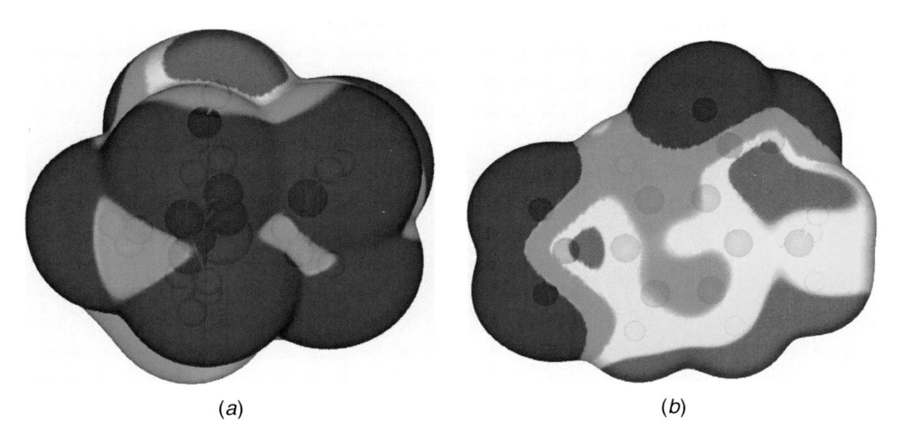

Figure 9.5 Calculated molecular electrostatic potentials on the 0.001 au molecular surfaces of HFIPA (a) and DNT (b), from B3LYP/6-311G(d,p) wavefunctions. In HFIPA, the hydroxyl group is at the top; in DNT, the methyl group is at the right. Color ranges, in kilocalories per mole, are (a, HFIPA) red $>$ 30 $>$ yellow $>$ 17.0 $>$ green $>$ 0.0 $>$ blue, and (b, DNT) red $>$ 20 $>$ yellow $>$ 13.0 $>$ green $>$ 0.0 $>$ blue.

Figure 9.6 Various stable complexes formed between HFIPA and DNT, with binding energies ranging from 7.7 kcal/mol to 1.8 kcal/mol at the B3LYP/6-311G(d,p) level. Some key structural elements (distances between closest contacts) are shown.

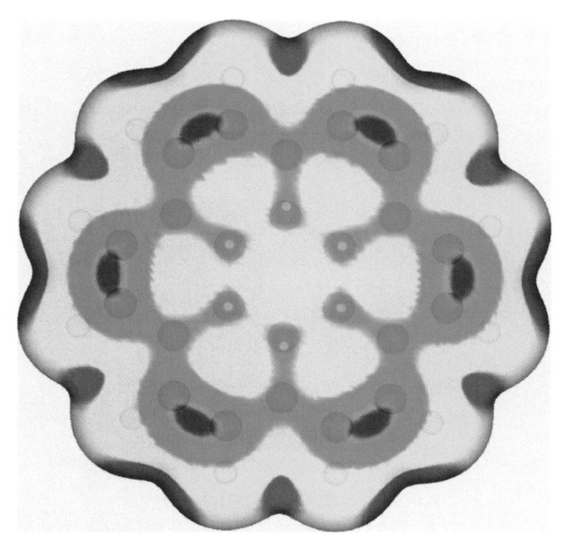

Figure 9.7 Diagram of the average local ionization energy on the molecular surface of coronene, computed at the B3PW91/6-311 G(d,p) level. Color ranges, in electron volts, are red >12 > yellow > 10 > green > 9 > blue. The light blue circles indicate the positions of the local minima found in the central ring.

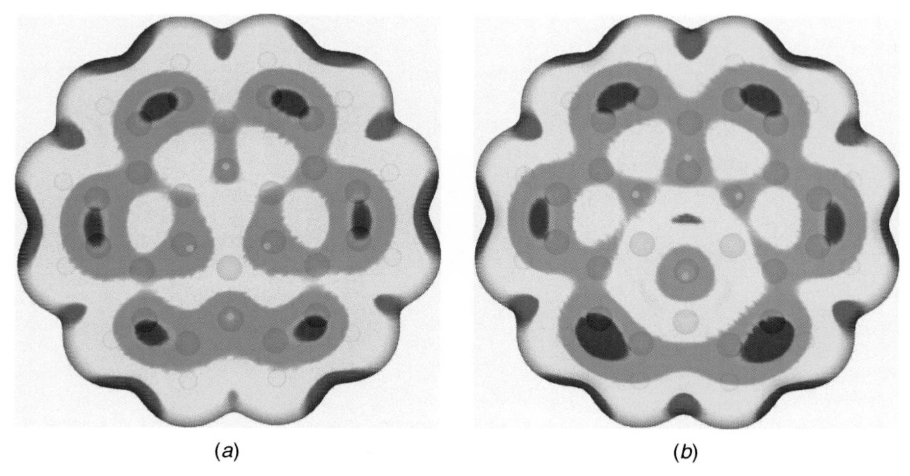

Figure 9.8 Diagram of the average local ionization energy on the 0.001 au molecular surface of the 1-H-coronene radical computed at the B3PW91/6-31G(d,p) level. The added hydrogen atom is in (b), coming out of the plane of the figure, in the lower portion of it. Color ranges, in electron volts, are red > 12 > yellow > 10 > green > 9 > blue. The light blue circles indicate the positions of the local minima found in the central ring and that associated with C9 (a).

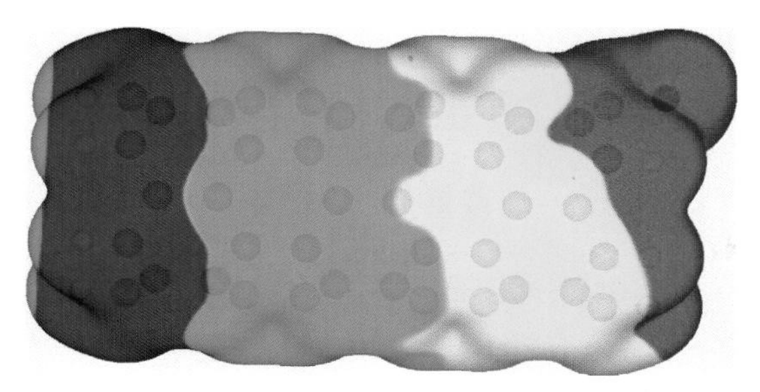

Figure 9.9 Diagram of the electrostatic potential on the 0.001 au surface of an NH_2-end-substituted (6,0) carbon nanotube, at the HF/STO-4G level. The NH_2 group is at the right. The tube is otherwise terminated with hydrogens. Color ranges, in kilocalories per mole, are red $>20 >$ yellow $>0 >$ green $> -20 >$ blue.

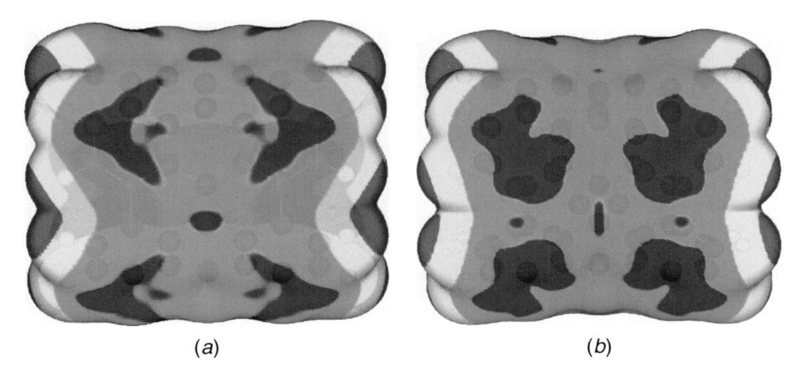

(*a*) (*b*)

Figure 9.10 Diagram of the electrostatic potential on the 0.001 au surface of a pristine (a) and a Stone–Wales defective (b) (5,5) carbon nanotube, at the HF/STO-4G level. Both tubes are terminated with hydrogens. The defect is in the central portion of the tube facing the reader (b). Color ranges, in kilocalories per mole, are red $> 9 >$ yellow $> 0 >$ green $> -5 >$ blue.

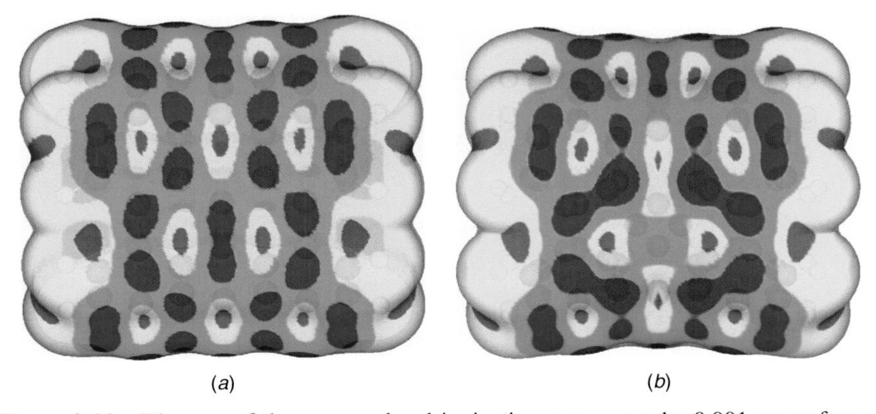

(*a*) (*b*)

Figure 9.11 Diagram of the average local ionization energy on the 0.001 au surface of a pristine (a) and a Stone–Wales defective (b) (5,5) carbon nanotube at the HF/STO-4G level. Both tubes are terminated with hydrogens. The defect is in the central portion of the tube facing the reader (b). Color ranges, in electron volts, are red $> 18.0 >$ yellow $> 16.0 >$ green $> 14.5 >$ blue.

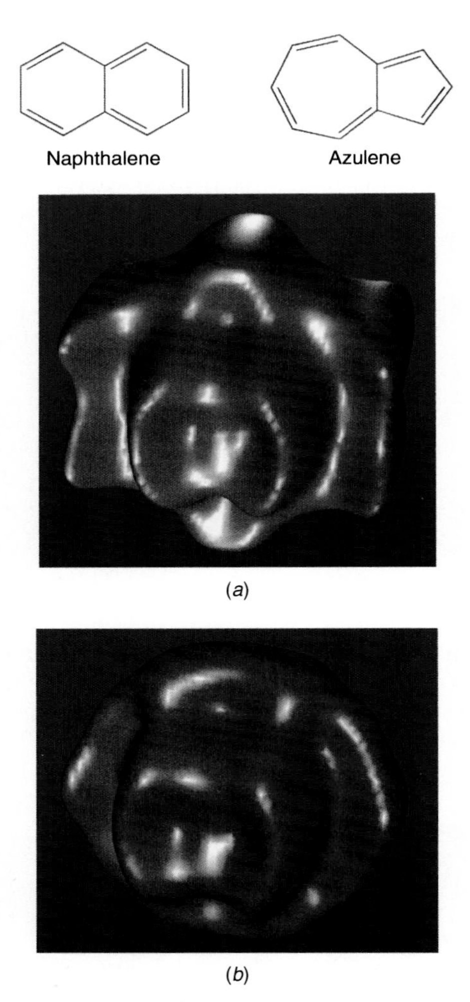

Figure 10.2 Isovalue surfaces for $\nabla^2\Pi = -0.015$ au for naphthalene (a) and azulene (b). All momenta coordinates inside the envelopes are momenta for which the electron dynamics are locally laminar ($\nabla^2\Pi < 0$). The vertical axis in both images corresponds to the component of electron momentum parallel to the direction of the C–C bond that is shared by both rings. The axis that appears to be coming toward the viewer is the component of electron momentum that is perpendicular to the plane containing the nuclei in both molecules. The momentum coordinates in both images extend out to $\pm$ 1.0 au.

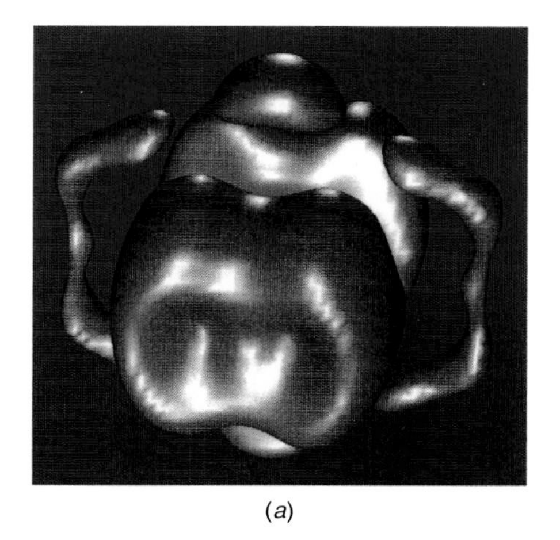

(a)

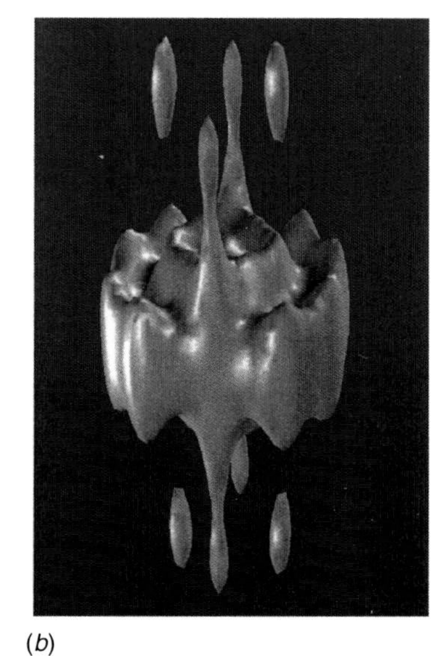

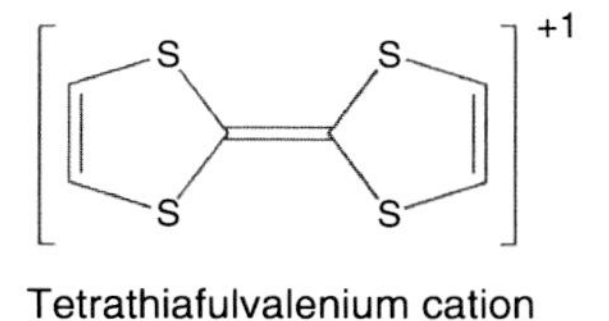

(b)

Figure 10.3 (a) An isovalue surface for $\nabla^2\Pi = -0.030$ au for naphthalene, with the same axis system shown in Figure 10.2. All momenta inside the envelope, therefore, have locally more laminar electron dynamics than points on the surface shown in Figure 10.2. (b) Isovalue surface for which $\nabla^2\Pi$ is marginally less than zero, computed for the TTF radical cation. The vertical axis corresponds to the component of electron momentum that is perpendicular to the plane containing the nuclei. The axis that appears to be coming toward the viewer corresponds to the component of electron momentum that is parallel to the central C–C bond. The momentum coordinates in both images extend out to $\pm$ 1.0 au.

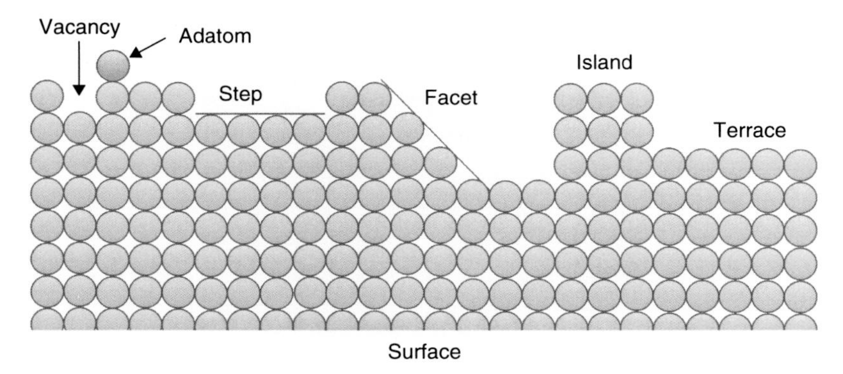

Figure 11.3 A schematic illustration depicting a microscopic view of a surface. In "real-life" situations, a surface of any material may significantly deviate from its idealistic bulk-terminated geometry and often contains zero-, one-, and two-dimensional defects; surface adsorbates; etc.

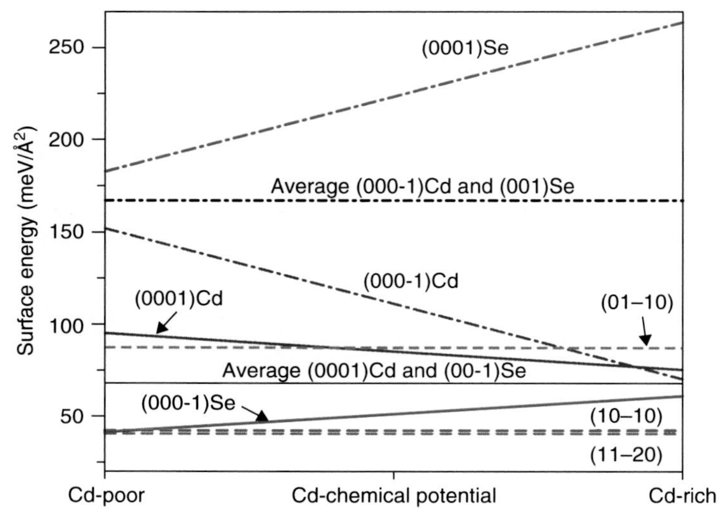

Figure 11.6 Surface energies for polar and nonpolar wurtzite CdSe facets as a function of Cd chemical potential (μ_{Cd}^{CdSe}).

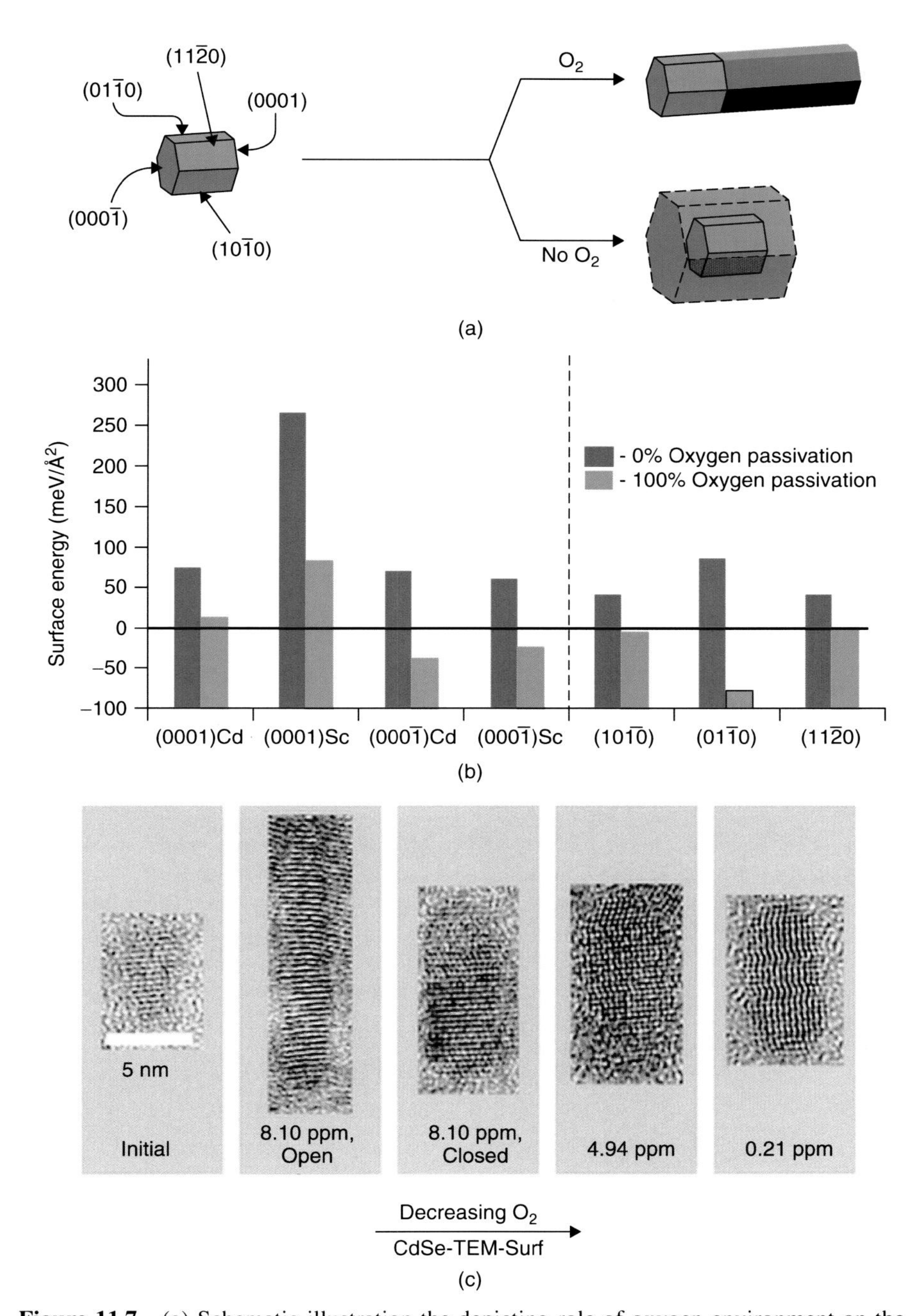

Figure 11.7 (a) Schematic illustration the depicting role of oxygen environment on the growth of wurtzite CdSe nanocrystals. (b) DFT-calculated surface energies for the polar and nonpolar wurtzite CdSe facets before and after oxygen passivation. (c) High resolution transmission electron microscopy images of representative CdSe nanocrystals for various oxygen concentrations. The 5 nm scale bar shown in the leftmost panel applies to all images [55]. (Reprinted with permission from J. D. Doll, G. Pilania, R. Ramprasad, and F. Papadimitrakopoulos, Nano Lett., 2010, 10, 680. Copyright 2010 American Chemical Society.)

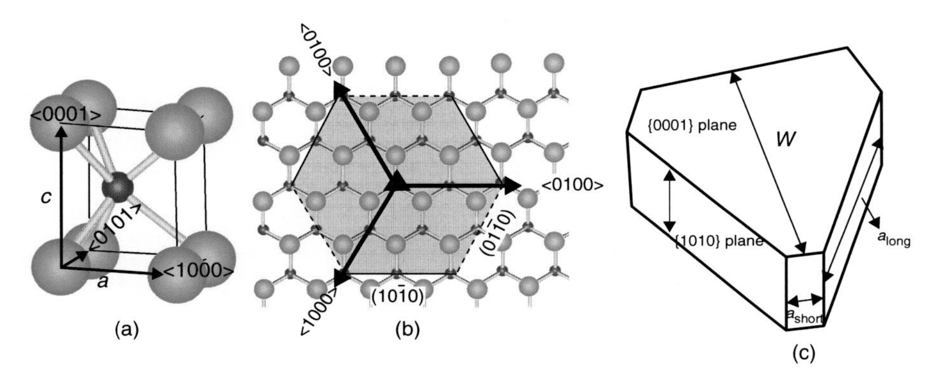

Figure 11.8 (a) The unit cell of WC. (b) Projection along the 0001 direction with the (1010) and (0110) facets represented by solid and dashed lines, respectively. Stands for the threefold rotational symmetry about 0001 axis. (1010) surface could be terminated by C with four dangling bonds or W with two dangling bonds. Similarly, (1010) could be terminated by C with two dangling bonds and W with four dangling bonds. (c) Schematic representation of the equilibrium shape for a WC particle. The shape can be characterized by two shape factors, that is, $r = \mathrm{a_{short}}/\mathrm{a_{long}}$ and $k = t/w$. Green (large) and red (small) spheres represent W and C atoms, respectively.

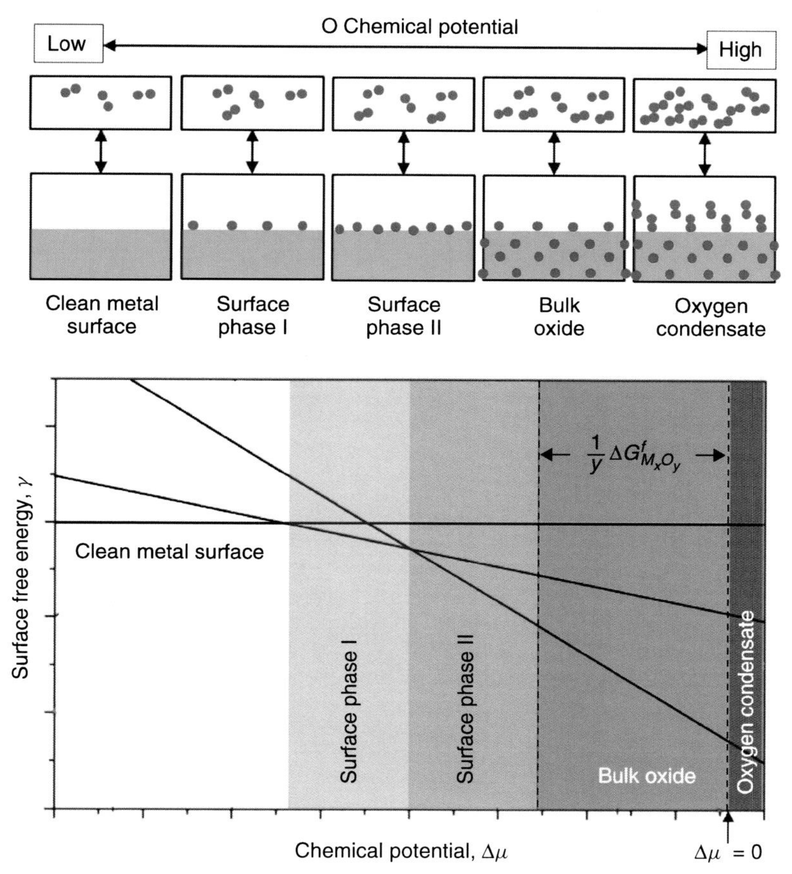

Figure 11.14 Example plot of Gibbs surface free energy (γ) versus oxygen chemical potential ($\mu O = \mu O(T, P) - 1/2E_{\mathrm{DFT}}\ \mathrm{O_2}$) for a surface in equilibrium with a surrounding gas-phase. The surface energy of the clean metal surface serves as a zero reference in the plot. Adsorption of an O adatom will become favorable on the metal surface only if the Gibbs free energy of the resultant oxygen adatom-covered surface is lower than that of the clean surface. Also note that as the O content of the surface increases, the slope of the γ versus μO plot becomes increasingly negative. With increasing gas-phase chemical potential, first the clean metal phase is stable, then the first and second adsorbate phases become stable followed by an oxide phase, and finally, oxygen gas phase condensation takes place on the surface.

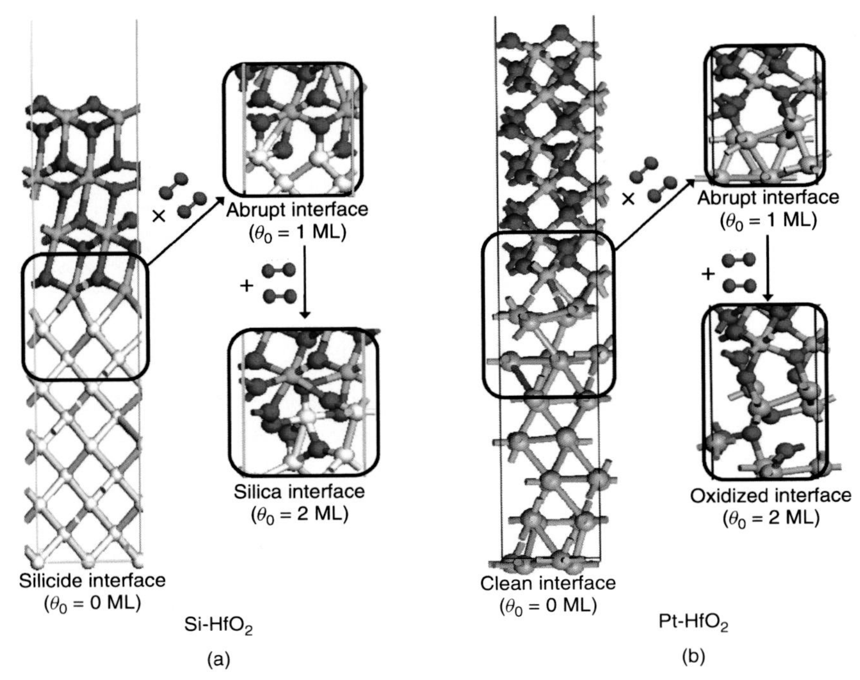

Figure 11.15 The representative atomic structures of the X-HfO_2 heterostructures with X at the bottom: (a) Si-HfO_2 interface and (b) Pt-HfO_2 interface. The Hf and O atoms of the top HfO_2 parts in both (a) and (b) are represented by light blue (gray) and red (dark) spheres, respectively [99]. (Reprinted with permission from H. Zhu, C. Tang, and R. Ramprasad, Phys Rev, 2010, B 82, 235413 and H. Zhu and R. Ramprasad, Phys. Rev, 2011, B 83, 081416(R). Copyright 2010 by the American Physical Society.)

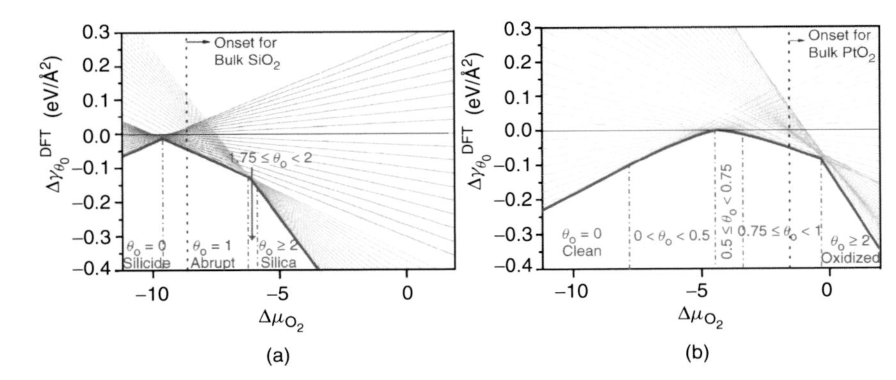

Figure 11.16 The relative interface free energy as a function of $\Delta\mu_{O_2}$ for the (a) Si-HfO_2 and (b) Pt-HfO_2 interfaces. The inner boundaries represent the lowest interface energy at each $\Delta\mu_{O_2}$ value. The corresponding θ_O value is indicated in the figure. The vertical dash-dotted lines are boundaries between two different stable interface structures. The onset $\Delta\mu_{O_2}$ to form bulk SiO_2 and α-PtO_2 is indicated by the vertical dotted line in (a) and (b), respectively [99]. (Reprinted with permission from H. Zhu, C. Tang, and R. Ramprasad, Phys Rev, 2010, B 82, 235413 and H. Zhu and R. Ramprasad, Phys. Rev, 2011, B 83, 081416(R). Copyright 2010 by the American Physical Society.)

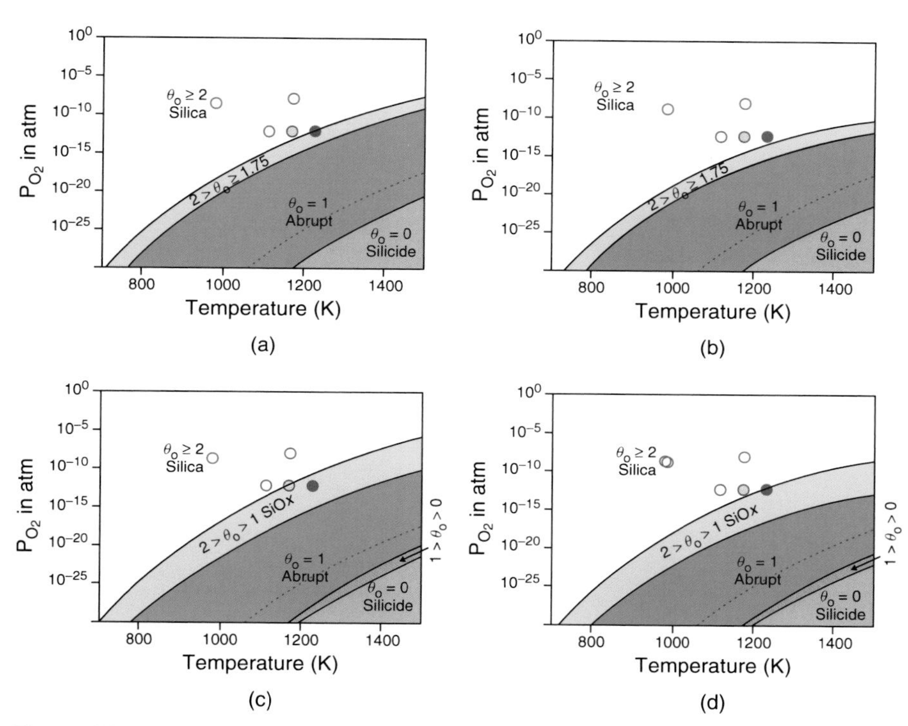

Figure 11.17 (a) The interface phase diagrams for the Si-HfO_2 interface, determined using Equation 27), *i.e.*, after neglecting the vibrational and configurational entropic contributions of the condensed phases to the relative interface free energy. (b) Same as (a), but with the vibrational entropic contribution ($\Delta F^{v}_{\theta_O}$) included. (c) Same as (a), but with the configurational entropic contribution ($\Delta F^{c}_{\theta_O}$) included. (d) The interface phase diagrams determined using Equation 11.26; this is the same as (a), but with both the vibrational and configurational entropic contributions included. The solid curves indicate the interface phase boundaries, and the dotted curves represent the onset of formation of bulk SiO_2. The open and solid circles stand for the condition to form SiO_2 and SiO at the Si-HfO_2 interfaces in experiments, respectively [92, 93]. The gray filled circle stands for the critical point for interfacial SiO_2 to decompose to SiO [93, 99]. (Reprinted with permission from H. Zhu, C. Tang, and R. Ramprasad, Phys Rev, 2010, B 82, 235413 and H. Zhu and R. Ramprasad, Phys. Rev, 2011, B 83, 081416(R). Copyright 2010 by the American Physical Society.)

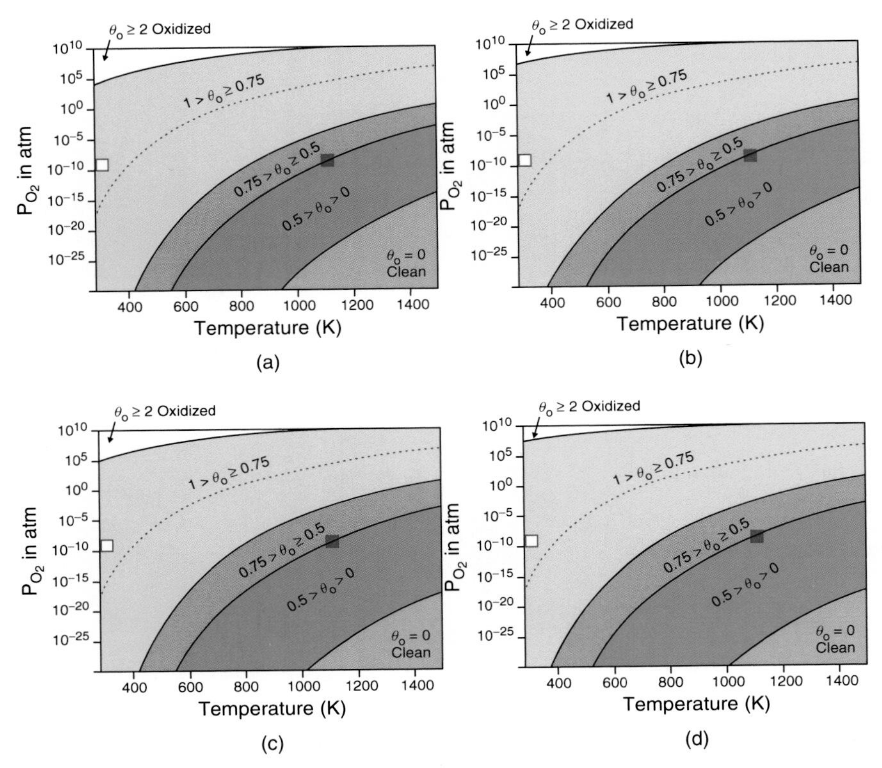

Figure 11.18 (a) The interface phase diagrams for the Pt-HfO_2 interface, determined using Equation 11.27, that is, after neglecting the vibrational and configurational entropic contributions of the condensed phases to the relative interface free energy. (b) Same as (a), but with the vibrational entropic contribution ($\Delta F^{v}_{\theta_O}$) included. (c) Same as (a), but with the configurational entropic contribution ($\Delta F^{c}_{\theta_O}$) included. (d) The interface phase diagrams determined using Equation 11.26; this is the same as (a), but with both the vibrational and configurational entropic contributions included. The solid curves indicate the interface phase boundaries, and the dotted curves represent the onset of formation of bulk PtO_2. The open and solid squares are the T and P_{O_2} when a 0.25 ML O-adsorbed (111) Pt surface and a clean (111) Pt surface were observed, respectively [99, 104]. (Reprinted with permission from H. Zhu, C. Tang, and R. Ramprasad, Phys Rev., 2010, B 82, 235413 and H. Zhu and R. Ramprasad, Phys. Rev., 2011, B 83, 081416(R). Copyright 2010 by the American Physical Society.)

and

$$f^0(\mathbf{r}) = \frac{1}{2}[f^+(\mathbf{r}) + f^-(\mathbf{r})] \text{ for radical attack.} \tag{7.24c}$$

Thus, $f(\mathbf{r})$ physically signifies the propensity of a particular atomic site in a molecule toward chemical reactivity by virtue of the different local parameters, in the sense of the frontier orbital theory as proposed by Fukui et al. [63–66]. The FFs are therefore referred to as *the DFT analog* of the frontier orbitals [67]. Equation 7.24 also implies that a larger variation of the chemical potential ($\delta\mu$) for a particular reactive site in a molecule corresponds to an increase in the value of $f(\mathbf{r})$. A higher $f(\mathbf{r})$ value for an active site in a molecule thus signifies an increase in its local reactivity [43] as well, and so, the most reactive site of a chemical species is associated with the largest $f(\mathbf{r})$ value, which originates from the following rule of chemical reactivity: " large $|d\mu|$ is good." Thus, at a point r in a system, a higher value of $f^+(\mathbf{r})$ presupposes a greater reactivity toward a nucleophilic attack that results in an electron increase in the system [63–66, 68, 69]. Similarly, a higher $f^-(\mathbf{r})$ value implies an increasing tendency toward electrophilic attack, resulting in an electron decrease in the system [63–66, 68, 69].

7.2.2.3 Condensed Fukui Functions Computational chemists generally prefer to consider an atom rather than any "point" in a molecule as the plausible site of electrophilic/nucleophilic attack. Consequently, a coarse-grained atom by atom representation of the FF, called *condensed-to-atom Fukui function*, was proposed by Yang and Mortier [70], based on a finite difference approach in terms of the Mulliken population analysis (MPA) scheme that may be represented as

$$f_k^+ = q_k(N+1) - q_k(N) \text{ for nucleophilic attack,} \tag{7.25a}$$

$$f_k^- = q_k(N) - q_k(N-1) \text{ for electrophilic attack,} \tag{7.25b}$$

$$f_k^o = [q_k(N+1) - q_k(N-1)]/2 \text{ for radical attack,} \tag{7.25c}$$

where q_k refers to the electron population at a particular atomic site k in a molecule. Equation 7.25(a–c) can be easily evaluated from population analysis data.

The calculation of the FF using the variational technique for a nondegenerate ground state of a system has been proposed [71], which determines $f(\mathbf{r})$ and η. Assuming the ground-state electron density ρ to be known, the hardness kernel may be expressed as [72]

$$\eta(\mathbf{r}, \mathbf{r}') = \frac{\delta^2 F[\rho]}{\delta\rho(\mathbf{r})\delta\rho(r')}, \tag{7.26}$$

where $F[\rho]$ is the Hohenberg–Kohn universal functional comprising the electronic kinetic energy and the electron–electron repulsion energy. Subsequently, a hardness functional (of an arbitrary function g) may be defined as

$$\eta[g] = \int\int g(\mathbf{r})g(\mathbf{r}')\eta(\mathbf{r}, \mathbf{r}')\mathrm{d}\mathbf{r}\mathrm{d}\mathbf{r}'. \quad (7.27)$$

If $\eta[g]$ with respect to $g(\mathbf{r})$ is minimized subject to the normalization condition

$$\int g(\mathbf{r})\mathrm{d}\mathbf{r} = 1, \quad (7.28)$$

the exact FF for the system is obtained, and the extremum value of $\eta[f]$ can be considered to be the exact hardness of the system.

For an inhomogeneous gas, the necessary gradient correction in the FF has been obtained [73]. For a homogeneous system of N electrons constrained to move in a cavity of volume V, the FF within the local density approximation (LDA) is given by

$$f(\mathbf{r}) = \left(\frac{\partial\rho(\mathbf{r})}{\partial N}\right)_{\upsilon(\mathbf{r})} = \frac{\rho(\mathbf{r})}{N}, \quad (7.29)$$

where $\rho(\mathbf{r})$ and $f(\mathbf{r})$ are constants equal to N/V and $1/V$, respectively.

For an inhomogeneous gas, the necessary corrections are incorporated as [73]

$$f(\mathbf{r}) = \frac{\rho(\mathbf{r})}{N}[1 + \alpha\phi(\mathbf{r}; \rho(\mathbf{r}), \nabla\rho, \nabla^2\rho, \ldots)], \quad (7.30)$$

where α is a parameter.

Using the normalization condition, the FF can be approximated as a dimensionless quantity, ϕ [73]

$$\phi \approx \frac{1}{\rho}\nabla.\left[\frac{\nabla\rho}{\rho^{2/3}}\right] = \frac{\nabla^2\rho}{\rho^{5/3}} - \frac{2}{3}\frac{\nabla\rho.\nabla\rho}{\rho^{8/3}}. \quad (7.31)$$

7.2.2.4 Local Softness ($s(\mathbf{r})$) and Local Hardness ($\eta(\mathbf{r})$) The earlier discussions on global softness (S) and FFs $f(\mathbf{r})$ have clearly settled the fact that a "softer" molecule is supposed to be more polarizable and hence more reactive through a covalent interaction because of the greater delocalization of its electron cloud and an increase in the chemical potential (μ). To explain such reactivity behavior for the local sites in a molecule, Parr and Yang [1] postulated that the preferred sites for attacking groups in a molecule are described by the maxima of the FF ($f(\mathbf{r})$) [16]. Thus, for a case of covalent bonding where the main interaction is between the FMOs, the FF, ($f(\mathbf{r})$), and local softness ($s(\mathbf{r})$) may be considered equivalent from the viewpoint of local reactivity, as the FFs quantify the tendency of a molecular species to accept (or donate) electrons from (to) another chemical system [16, 62]. So for a soft–soft, covalent, frontier-controlled interaction [74], an increase in the $f(\mathbf{r})$ value for a particular reactive site of a molecular system results in a corresponding increase in the corresponding local softness ($s(\mathbf{r})$). The local softness [43] ($s(\mathbf{r})$) is expressed as

$$s(\mathbf{r}) = \left(\frac{\partial \rho(\mathbf{r})}{\partial \mu}\right)_{\upsilon(\mathbf{r})}. \tag{7.32}$$

The global softness (S) is a summation of the individual local softness values for all the reactive sites over the entire molecule. Thus S and $s(\mathbf{r})$ are related through a normalization condition

$$S = \int s(\mathbf{r})\mathrm{d}\mathbf{r}, \tag{7.33}$$

which stems from the fact that $f(\mathbf{r})$ is normalized [1] to unity.

The local softness $s(\mathbf{r})$ is related to the FFs $f(\mathbf{r})$ through a chain rule as

$$s(\mathbf{r}) = \left(\frac{\partial \rho(\mathbf{r})}{\partial \mu}\right)_{\upsilon(\mathbf{r})} = \left(\frac{\partial \rho(\mathbf{r})}{\partial N}\right)_{\upsilon(\mathbf{r})} \cdot \left(\frac{\partial N}{\partial \mu}\right)_{\upsilon(\mathbf{r})} = f(\mathbf{r}) \cdot S. \tag{7.34}$$

It should, however, be mentioned that while the FF ($f(\mathbf{r})$) seems to assess the reactivity pattern of the different local active sites of the same molecule, thereby behaving as an intramolecular reactivity descriptor, the local softness ($s(\mathbf{r})$) compares and correlates the propensity of a pair of interacting molecular neighbors toward chemical attack and hence, unlike $f(\mathbf{r})$, serves as an intermolecular reactivity descriptor.

The idea of local hardness ($\eta(\mathbf{r})$) was introduced by Berkowitz et al. [75, 76] in the same spirit as that of local softness to describe the lack of propensity toward chemical reactivity of an atom or group in a molecule. However, the definition of $\eta(r)$, unlike $s(\mathbf{r})$, is not so simple and neither can $\eta(\mathbf{r})$ be normalized to η. $\eta(\mathbf{r})$ is expressed in a different and a slightly more complex manner as

$$\eta(\mathbf{r}) = \frac{1}{N}\int \frac{\delta^2 F[\rho(\mathbf{r})]}{\delta\rho(\mathbf{r})\delta\rho(\mathbf{r}')}\rho(\mathbf{r}')\mathrm{d}\mathbf{r}', \tag{7.35}$$

where $F[\rho(\mathbf{r})]$ is the Hohenberg–Kohn universal functional. Local hardness is also defined [75, 76] in another way at par with global hardness (η) by replacing the number of electrons (N) with electron density ($\rho(\mathbf{r})$). Thus $\eta(\mathbf{r})$ can also be expressed as

$$\eta(\mathbf{r}) = \left(\frac{\delta \mu}{\delta \rho(\mathbf{r})}\right)_{\upsilon(\mathbf{r})}. \tag{7.36}$$

This definition for $\eta(\mathbf{r})$ as expressed in Equation 7.36 is ambiguous, because of the interdependence of $\rho(\mathbf{r})$ and $\upsilon(\mathbf{r})$, and therefore, it cannot be considered as a basic definition for $\eta(\mathbf{r})$. Thus, unlike η and S, $\eta(\mathbf{r})$ and $s(\mathbf{r})$ do not even have a reciprocal relationship and are interconnected by

$$\int \eta(\mathbf{r})s(\mathbf{r})d\mathbf{r} = 1. \tag{7.37}$$

A clear definition of local hardness and subsequent identification of the hard reactive sites in a molecule is wanting [74,77–82]. The "minimum FF rule" prescribed by Li and Evans [79] asserts that, hard reactions unlike softer ones, prefer sites with minimum FF values. Wide applications [83–85] of this rule and subsequent criticisms owing to its over simplicity [81, 86] have been reported. The fact that hard–hard interactions are electrostatic and charge-controlled, where the role of frontier orbitals is quite irrelevant, cannot be corroborated from the "minimum FF rule." The basic problem of this rule is that although it justifies the insignificance of the frontier orbitals, it misses the role of electrostatic interactions, which is a trademark of hard–hard interactions [81, 82, 86].

In a recent article, it has been shown that, for polyatomic systems, the FF is the best option to obtain reliable local hardness profiles [87].

A mathematical correlation between local hardness ($\eta(\mathbf{r})$) and FF ($f(\mathbf{r})$) may be drawn on the basis of Equations 7.34 and 7.37, which is

$$\eta = \int \eta(\mathbf{r})\, f(\mathbf{r})dr. \tag{7.38}$$

Furthermore, in an attempt to draw a physical rationale between local softness ($s(\mathbf{r})$) and local hardness ($\eta(\mathbf{r})$), it has been suggested that $s(\mathbf{r})$ being a measure of electronic fluctuations [43] should be considered as an electronic reactivity index, $\eta(\mathbf{r})$, on the other hand should be considered as a nuclear reactivity index [88]. Torrent-Sucarrat et al. [89] recently interpreted local softness ($s(\mathbf{r})$) and local hardness ($\eta(\mathbf{r})$) as mathematical functions that are pointwise measures of the "local abundance" or "concentration" of their corresponding global quantities. Further vital applications of this new approach toward understanding $\eta(\mathbf{r})$ and $s(\mathbf{r})$ and its implications for the HSAB theory have recently been reported [90].

7.2.2.5 Philicity ($\omega^{\alpha}(\mathbf{r})$) and Group Philicity ($\omega_g^{\alpha}$) The unique concept of philicity ($\omega^{\alpha}(\mathbf{r})$) was introduced [91, 92] via resolution of the identity, which presumes that the local electro- (or nucleo-) philicity at a particular reactive site in a molecule may be increased (or decreased) on electronic interactions, without rendering any change in the overall philicity values. The local electrophilicity ($\omega(\mathbf{r})$), along the lines of local softness ($s(\mathbf{r})$) and FF ($f(\mathbf{r})$), may be normalized to global electrophilicity (ω) as

$$\omega = \omega \int f(\mathbf{r})d\mathbf{r} \quad \text{or,} \quad \omega = \int \omega f(\mathbf{r})d\mathbf{r} = \int \omega(\mathbf{r})d\mathbf{r} \quad \text{thus,} \quad \omega(\mathbf{r}) = \omega f(\mathbf{r}). \tag{7.39}$$

It is interesting to note that $\omega(\mathbf{r})$ contains information about both ω and $f(\mathbf{r})$ but $f(\mathbf{r})$ alone cannot provide any information regarding $\omega(\mathbf{r})$ without the knowledge of ω. Further calculations invoked three different types of $\omega(\mathbf{r})$ similar to that of the three different condensed-to-atom FF ($f_k^\alpha(\mathbf{r})$) variants. Thus $\omega(\mathbf{r})$, better symbolized as $\omega^\alpha(\mathbf{r})$, was termed "philicity" as it takes care of all types of reactions depending on the sign of α, that is, for $\alpha = +, -,$ and 0 represent nucleophilic, electrophilic, and radical attacks, respectively. Thus

$$\omega^\alpha(\mathbf{r}) = \omega \cdot f^\alpha(\mathbf{r}) \quad \text{or,} \quad \omega_k^\alpha = \omega \cdot f_k^\alpha, \tag{7.40}$$

where ${\omega_k}^\alpha$ refers to the condensed-to-atom local philicity variants for the kth atomic site in a molecule. Local philicity (${\omega_k}^\alpha$) has proved to be a more powerful quantity than its global counterpart because ${\omega_k}^\alpha$, besides providing information regarding electrophilicity (ω), also decides the site selectivity of a molecule toward electrophilic, nucleophilic, or radical attacks. The propensity of a particular atomic site toward electrophilic, nucleophilic, or radical attack may be enhanced with a proportionate decrease in the other surrounding sites by keeping the global electrophilicity (ω) conserved. Furthermore, for a pair of different reactants having different electrophilicity values, the appropriate local atomic sites for electronic interactions can be fruitfully determined in terms of philicity rather than the FF, as the former has been shown to be more informative than the latter. Local philicity (${\omega_k}^\alpha$) with the aid of electronegativity (χ) is able to provide information on $s(\mathbf{r})$, S, and η as well.

The group philicity concept [93] (${\omega_g}^\alpha$) is applicable in correlating the reactivity of an atomic assembly or a group, which is generally expressed as a sum of the individual condensed-to-atom philicities (${\omega_k}^\alpha$) over all the relevant atoms. So

$$\omega_g^\alpha = \sum_{k=1}^{n} \omega_k^\alpha, \tag{7.41}$$

where n denotes the number of atoms present in the reacting group and $\alpha = +, -,$ and 0 refers to nucleophilic, electrophilic, and radical attacks, respectively. The additive behavior of local philicity as described above for a reactive group, goes hand-in-hand with similar trends exhibited by $f(\mathbf{r})$ and $s(\mathbf{r})$, in a conceptual scheme.

7.2.2.6 Nucleophilicity Excess ($\Delta\omega_g^\mp$) and Electrophilicity Excess ($\Delta\omega_g^\pm$) On the basis of the group concept, a nucleophile in a molecule should possess greater group philicity to electrophilic attack than to nucleophilic attack. This difference, named nucleophilicity excess ($\Delta\omega_g^\mp$) [94], was expressed at par with the dual descriptor [95] as

$$\Delta\omega_g^\mp = \omega_g^- - \omega_g^+ = \omega(f_g^- - f_g^+), \tag{7.42}$$

where $\omega_g^-(\equiv \sum_{k=1}^{n} \omega_k^-, k = 1 \text{ to } n)$ and $\omega_g^+(\equiv \sum_{k=1}^{n} \omega_k^+, \text{k} = 1 \text{ to } n)$ refer to the group philicities of the nucleophile in the molecule toward electrophilic and nucleophilic attacks, respectively. It is quite obvious to expect a positive value of the nucleophilicity excess ($\Delta\omega_g^{\mp}$) for a nucleophile while the same shall turn negative for an electrophile in a molecule.

Correspondingly, the electrophilicity excess ($\Delta\omega_g^{\pm}$) for an electrophile in a molecule reflects greater group philicity to nucleophilic attack than to electrophilic attack. The electrophilicity excess ($\Delta\omega_g^{\pm}$) for an electrophile can be expressed as the negative of nucleophilicity excess ($\Delta\omega_g^{\mp}$)

$$\Delta\omega_g^{\mp} = -\Delta\omega_g^{\mp} = -(\omega_g^- - \omega_g^+) = \omega_g^+ - \omega_g^- = \omega(f_g^+ - f_g^-), \quad (7.43)$$

where ${\omega_g}^+$ and ${\omega_g}^-$ as usual define the group philicities of the electrophile in the molecule due to nucleophilic and electrophilic attacks, respectively. The sign of the electrophilicity excess ($\Delta\omega_g^{\pm}$) for an electrophile is expected to be positive, whereas it is negative for a nucleophile in a molecule.

Furthermore, for a molecular system with only two distinct units, the nucleophilicity excess ($\Delta\omega_g^{\mp}$) of the nucleophile should be equal to the electrophilicity excess ($\Delta{\omega_g}^{\pm}$) of the electrophile, as expected from the conservation of FF and philicity, that is

$$\Delta\omega_g^{\mp} \text{ (Nucleophile)} = \Delta\omega_g^{\pm} \text{ (Electrophile)}. \quad (7.44)$$

7.2.2.7 Quantum Dissimilarity ($\Delta\omega^{ij}$) During the interaction of two different molecular systems, the quantum dissimilarity ($\Delta\omega^{ij}$) refers to the minimum squared difference in philicity and may be defined [96] as

$$\Delta\omega^{ij} = [{\omega_{\max(i)}}^+(\text{electrophile}) - {\omega_{\max(j)}}^-(\text{nucleophile})]^2, \quad (7.45)$$

where ${\omega_{\max}}^+$ and ${\omega_{\max}}^-$ refer to the maximum philicity values at any atomic site due to nucleophilic and electrophilic attack, respectively.

7.3 MOLECULAR ELECTRONIC STRUCTURE PRINCIPLES

It is now quite obvious that the various CDFT-based global and local reactivity descriptors play a significant role in describing a reaction mechanism. The manner in which these reactivity parameters influence the reaction behavior of molecules can be understood through associated electronic structure principles, which are discussed in the following sections.

7.3.1 Electronegativity Equalization Principle

Electronegativity (χ) of an atom or molecule plays a vital role in determining its tendency to react with other atoms or molecules during a chemical combination. As chemical reactions are nothing but electronic interactions, it is obvious that between two systems having different χ values, the electrons will face a spontaneous drift from the system with lower electronegativity to the one with higher electronegativity, that is, electron drift between two systems occurs only when there exists an electronegativity gradient between them, a condition quite analogous to that of heat flow (temperature gradient) or of matter flow (chemical potential gradient) in macroscopic thermodynamics. Therefore, electronegativity (χ) in comparison with the temperature (T) in thermodynamics can be attributed to the electronic condition of an atom or molecule that dictates the direction of electron flow during chemical interactions, just as T being assigned as the thermal condition of a body in macroscopic thermodynamics determines the direction of heat flow between two systems at different temperatures. The well-known "principle of calorimetry" for macrosystems might have therefore influenced Sanderson [24, 25] to postulate an "electronegativity equalization principle," which seemed to dictate quite successfully the direction and extent of electron flow in the microdomain. The principle of electronegativity equalization states that "*When two or more atoms initially different in electronegativity combine chemically, they adjust to have the same intermediate electronegativity within the compound*." This intermediate electronegativity (χ_{GM}) is given by the geometric mean of the individual electronegativities of the isolated component atoms [97–99] and is expressed as

$$\chi_{GM} \approx \left(\prod_{k=1}^{P} \chi_k \right)^{1/P}, \tag{7.46}$$

where the molecule contains P atoms (same or different) and $\{\chi_k, k = 1, 2, \ldots, P\}$ signify their isolated atom electronegativities.

In other words, the electron density will flow from the more electropositive atom to the more electronegative atom, creating a partial positive charge on the former and a partial negative charge on the latter. As the positive charge on the electropositive atom increases, its effective nuclear charge increases, hence its electronegativity increases. The same trend happens in the opposite direction for the more electronegative atom, until the two have the same electronegativity. Thus, the equalization of electronegativity occurs through the adjustment of the bond polarities.

7.3.2 Electrophilicity Equalization Principle

On the basis of the fact [100] that the ratio between hardness (η) and electronegativity (χ) is approximately a constant value for atoms that belong to the same periodic group [101] and similar molecules, and by using Equation 7.46, an equalization principle for hardness similar to that for electronegativity may be derived as [102–104]

$$\eta_{GM} \approx \left(\prod_{k=1}^{P} \eta_k \right)^{1/P}, \tag{7.47}$$

where $\{\eta_k, k = 1, 2, \ldots, P\}$ correspond to the associated isolated atom hardness values. The idea of electrophilicity equalization proposed recently [105] finds its basis in Equations 7.46 and 7.47; the equalized electrophilicity (ω_{GM}) is the consequence of the electronegativity and hardness equalization phenomena. Thus, ω_{GM} is expressed as [105]

$$\omega_{GM} = \frac{{\chi_{GM}}^2}{2\eta_{GM}} = \left(\prod_{k=1}^{P} \omega_k \right)^{1/P}, \tag{7.48}$$

where $\omega_k = \frac{\chi_k^2}{\eta_k}$, $(k = 1, 2, \ldots, P)$ are the electrophilicities of the isolated atoms.

The final equalized electrophilicity (conceptually in the same spirit of electronegativity) may be conceived to be the geometric mean of the corresponding isolated atom values.

The model of equalized electrophilicity (ω_{GM}) further requires that its local variant [91, 106, 107] be constant everywhere throughout the system and equal to the global electrophilicity.

7.3.3 Hard–Soft Acid–Base (HSAB) Principle

The inception of the idea of assigning molecules as "hard" and "soft" by Pearson [33–39] and the subsequent classification of acids and bases as hard or soft has already been discussed briefly in Section 7.2.1.3, in connection with chemical hardness and softness. That a hard species prefers a hard one and a soft one desires another soft moiety for binding, popularly known as *the HSAB principle* [40], is considered one of the most useful concepts in chemical bonding theory for acids and bases. This principle has been employed to study a large number of acid–base reactions [41]. The essence of the HSAB theory is based on a CT process between a pair of donor (Lewis base) and acceptor (Lewis acid) molecules. Although the theory justifies the formation of a coordination linkage between a pair of donor and acceptor moieties and even determines the direction of electron transfer, its basis was nevertheless merely qualitative, lacking a rigorous mathematical proof. The CDFT-based reactivity descriptors were employed by Parr and Pearson [10] to quantify the so-called CT phenomenon between donors and acceptors. They set up a simple model of electron transfer between a Lewis base B (donor) and a Lewis acid A (acceptor), exploiting the concept of absolute hardness. The driving force behind the transfer of electrons from B to A is essentially due to an electronic chemical potential gradient or electronegativity gradient that exists between the two interacting systems. Thus, the CT process in the HSAB theory is directly linked with the principles of electronegativity equalization or chemical potential equalization. The total amount of electronic charge (ΔN) transferred and subsequent energy change (ΔE) between A and B on the formation of the adduct A:B was expressed as [10]

$$\Delta N = \frac{\chi_{\mathrm{A}}^{0} - \chi_{\mathrm{B}}^{0}}{2(\eta_{\mathrm{A}} + \eta_{\mathrm{B}})}, \tag{7.49}$$

$$\Delta E = -\frac{(\chi_{\mathrm{A}}^{0} - \chi_{\mathrm{B}}^{0})^2}{4(\eta_{\mathrm{A}} + \eta_{\mathrm{B}})}. \tag{7.50}$$

Thus, the gradual lowering of energy on electron transfer may be explained from the corresponding electronegativity (χ) and hardness (η) values of A and B. It is also quite relevant that the electronegativity difference drives the electron transfer, while the sum of the absolute hardness values hinders the same.

In another approach, Pearson [108] showed that the inherent acid–base strengths (σ) can be correlated with the softness (S) and stability constant (K) of a reaction through the four-parameter relation

$$\log K = S_{\mathrm{A}} S_{\mathrm{B}} + \sigma_{\mathrm{A}} \sigma_{\mathrm{B}}, \tag{7.51}$$

where σ_A, σ_B and S_A, S_B signify the inherent strengths and softness factors of acid (A) and base (B), respectively. The σ_A and σ_B are expected to correlate with the corresponding χ_A, χ_B values.

Among other significant articles [78, 79, 109–113] that accentuate a theoretical understanding of the HSAB principle, the article by Chattaraj et al. [110] deserves special mention. On the basis of the "maximum hardness principle" or "minimum softness principle" [40, 114] (as described later), these authors established that "among potential partners of a given electronegativity, hard likes hard and soft likes soft." Thus, they showed that the HSAB principle in chemistry mimics the well-known Pareto principle in economics, which states that "Efficiency is the highest when partners are both well satisfied" [115]. On the basis of an approximate expression of the total interaction energy (E) between two reacting acid–base pairs A and B, Gázquez [116] provided additional quantitative support for a better understanding of the HSAB principle. The total energy (E) in terms of chemical potential (μ) and hardness (η) is given by

$$E[\rho] = N_{\mathrm{e}}\mu - \frac{1}{2}N_{\mathrm{e}}^2\eta + E_{\mathrm{core}}[\rho], \tag{7.52}$$

where N_{e} represents an effective number of valence electrons and $E_{\mathrm{core}}[\rho]$ signifies the core contribution to the total energy. It was further revealed that interaction between species having closely comparable softness values is energetically favorable relative to those whose softness differs significantly. This alternative approach was quite at par with the earlier theoretical justification [110].

7.3.4 Principle of Maximum Hardness

From the preceding discussions on the concepts of local hardness ($\eta(\mathbf{r})$), local softness ($s(\mathbf{r})$), and the HSAB principle, it can now be concluded with certainty that the favorable "hard–hard" interactions are electrostatic in nature and hence charge-controlled, whereas "soft–soft" interactions are covalent and frontier-controlled in nature. The above-mentioned principles and directives can be more lucidly explained when one considers the molecular orbitals (MOs) of the interacting molecules. A plot of the MOs of different molecules as a function of their energies demonstrates that, for a hard species, the energy gap between the corresponding frontier orbitals, the highest occupied molecular orbital (HOMO), and the lowest unoccupied molecular orbital (LUMO) is quite high as compared to a softer molecule. A large HOMO–LUMO gap for a hard species thus corresponds to a large energy gap between the ground state and the corresponding manifold of excited states of same multiplicity, thereby minimizing the mixing of electron density ρ between the two frontier levels, which results in a minimal response to perturbation of the electron cloud on chemical attack. Thus, harder systems tend to show a natural reluctance toward changes in ρ unlike softer moieties. So, harder systems tend to show lower reactivity and greater stability, unlike the softer ones that readily undergo changes in electron density ρ during chemical response. This qualitative trend of reactivity patterns of harder and softer molecules was dictated by Pearson [40, 117, 118] as a statement popularly known as the principle of maximum hardness (PMH), which states that "there seems to be a rule of nature that molecules arrange themselves so as to be as hard as possible." The use of statistical mechanics and DFT provides a rigorous and general proof of this maximum hardness principle [114]. The validity of the proof, however, depends on the constancy of temperature and chemical potential [119–121]. Thus, the PMH implies that chemical systems at equilibrium have a propensity to remain "as hard as possible." Therefore, the PMH sometimes serves as a fair guideline to interpret the correct wave function for a molecular system.

7.3.5 Minimum Polarizability Principle

The inverse relationship between polarizability (α) and chemical hardness (η) (see Section 7.2.1.5 for details) has already established the fact that a stable atomic system or an atomic assembly is characterized by a high value of η and a low value of α. Furthermore, the PMH [40, 114, 117, 118] provides a corollary statement for a minimization of the polarizability [122–124] from a dynamical perspective. The minimum polarizability principle (MPP) states that "the natural direction of evolution of any system is toward a state of minimum polarizability." Thus, the criteria of maximum hardness and minimum polarizability describe the ground state of a chemical system and also complement the minimum energy condition for stability. In a recent article highlighting the changes in polarizability (α) and chemical hardness (η) of molecules during the course of chemical reactions of various types, viz., dissociation, exchange, or isomerization, Ghanty and Ghosh [52] also observed that a condition of minimum polarizability is achieved with maximization of the hardness, which eventually accounts for an energetically more stable situation.

7.3.6 Minimum Electrophilicity Principle

The term electrophilicity (ω), also envisioned as the electron-attracting power for a molecular system, plays a pivotal role in determining reaction pathways. A system with a high ω value will naturally be more prone to accepting electrons and thus be highly reactive. But in the course of a chemical interaction as the reactants gradually transform themselves to form stable products through bond dissociation and bond formation, their reactivity decreases, with a consequential drop in the ω values. Chamorro et al. [107] studied the variation of the electrophilicity (ω) as a function of reaction coordinates for some isomerization and rearrangement reactions and showed that a maximum or a minimum in the ω value can be obtained during the course of a reaction. While the transition state (TS) in a reaction profile is characterized by the existence of the interacting systems in their most excited states, ω turns out to be the maximum. Similarly, the formation of products characterizes a stable state, a condition where ω shows its minimum value. Thus, a stable condition for a system is described with a minimization of the energy as well as electrophilicity (ω) values. So in accordance with the PMH [40, 114, 117, 118] for a stable state, there will also be a minimum electrophilicity principle (MEP).

7.3.7 Minimum Magnetizability Principle

The magnetizability of a molecular system can be segregated into its diamagnetic component (ξ_{dm}) and paramagnetic component (ξ_{pm}), of which the former is negative.

$$\xi_{total} = \xi_{dm} + \xi_{pm}. \tag{7.53}$$

On the basis of the already established fact that an energetically stable system attains a state of maximum hardness and minimum polarizability and that the polarizability (α) of the system varies linearly with its magnetizability, Tanwar et al. [125] proposed an electronic structure principle that correlates the stability of a molecule with its magnetizability and asserted that "a stable configuration/conformation of a molecule or a favorable chemical process is associated with a minimum value of the magnetizability," better known as the minimum magnetizability principle (MMP). Thus the total magnetizability (ξ_{total}) of a chemical system reaches its minimum value at the equilibrium geometry.

7.4 CONCEPTUAL DFT AS A USEFUL TOOL TOWARDS ANALYZING CHEMICAL REACTIVITY

DFT falls within the broad domain of quantum chemistry and, with the gradual evolution of the latter, has turned out to become a very constructive theoretical paradigm toward providing innovative ideas in determining molecular reactivity. CDFT [1–4] derives its basic mathematical formalism from DFT and is attributed primarily to Robert G Parr, who once said that "to calculate a molecule is not to understand it." The various global and local reactivity descriptors and the different molecular electronic structure principles associated with chemical reactivity that have been discussed in the preceding sections basically build the "road map" for molecular interactions and provide a clear idea to ascertain reaction pathways from a theorist's standpoint.

Applications of CDFT in molecular reaction mechanism are so widespread that it is a daunting task to summarize them all in one chapter. Here, we thus opt for a survey of the applications of conceptual DFT toward a realistic understanding of the structure, bonding, stability, and aromaticity of a selection of all-metal and nonmetal cluster assemblies. Furthermore, we also discuss the toxic effects of different hazardous compounds through suitable quantitative structure toxicity relationship (QSTR)-based models with the aid of conceptual DFT. Last but not the least, the trapping of hydrogen (atomic as well as molecular) onto suitable storage materials and their plausible use in industry as an alternative future fuel reserve to combat the probable energy crisis is also discussed within the CDFT framework.

7.4.1 Computational Details

All computations presented here have been performed at several higher levels of theory (B3LYP, MP2) at various basis sets (cc-pvdz, 6-31G, 6-31+G(d), 6-311+G(d), and 6-311+G(d,p)) by utilizing the GAUSSIAN 03 [126] program package. The choice for the level of theory and basis set, however, varies with the system under study. All the structures reported were fully optimized and their number of imaginary frequencies (NIMAG) was zero, thereby confirming their existence at positions of minima (global or local) on the potential energy surface (PES). The ionization potential (IP) and electron affinity (EA) values were either calculated with the aid of Koopmans theorem [34] or by using a ΔSCF technique. The atomic charges [15] (Q_k) and FFs [16] ($f(\mathbf{r})$) were computed from the MPA [70] scheme.

7.4.2 Structure, Bonding, Reactivity, and Aromaticity in Cluster Assemblies

Metal clusters [127–129] serve as a bridge between coordination chemistry and solid-state chemistry. The applications of metal clusters are so enormous that it has spawned a new branch of chemistry known as *cluster chemistry*. The bonding and stability aspects of such cluster moieties are generally judged from the perspective of a strong metal–metal bond formation along with the coordination behavior of the metals with different ligands.

CDFT armed with its powerful global and local reactivity descriptors serves as a promising approach to achieve a realistic understanding of the structure, bonding, and stability of various metal clusters from a theoretical standpoint. On the basis of the discovery of ferrocene [130–133], a unique sandwich complex, the further design, and the study of several multidecker sandwich compounds or metallocenes have been performed both experimentally [134–136] and theoretically [137, 138]. Great progress in the study of such cluster assemblies was achieved when Boldyrev et al. [139] extended the novel concept of aromaticity [140–142], used hitherto only by organic chemists to account for an extra stability of cyclic, planar, and conjugated $(4n+2)\pi$ (Hückel's rule) electron systems, to an "all-metallic" system, $[Al_4]^{2-}$ (Fig. 7.1).

Subsequently, they introduced a new term—"all-metal aromaticity." This unique idea of all-metal aromaticity, veeringly away as it does from the traditional ideas of bonding, helped in elucidating the stability of various metal clusters. The $Al_4{}^{2-}$ ring possesses a perfect square-planar geometry with two π electrons delocalized through the entire all-metal σ-framework. Widespread applications [143–146] of Boldyrev's concept of all-metal aromaticity are partly due to the fact that the square-planar $Al_4{}^{2-}$ system is considered to be both π-aromatic and doubly σ-aromatic [139] owing to the conspicuous σ- and π-electron delocalization. The stabilities of these diverse cluster moieties in terms of the aromaticity criterion have been quantitatively assessed from the nucleus-independent chemical shift (NICS) values, which are in turn computed through the explicit procedure of Schleyer et al. [147]. The idea of all-metal aromaticity introduced by the Boldyrev group [139] and its subsequent quantitative evaluation by exploiting Schleyer's [147] procedure prompted many enthusiasts to devise a variety of all-metal and nonmetal clusters [148–167].

Attempts have been made [150–174] toward analyzing the bonding and associated structural changes of varied cluster molecules during chemical interactions. It was also established quite unequivocally that the aromaticity criterion for these all-metal clusters has a direct bearing on their reactivity patterns. The existence and reactivity of ferrocene-type sandwich complexes consisting of many of those aromatic $[Al_4]^{2-}$ units, for example, MAl_4TiMAl_4 (M = Li, Na, K) [150, 171] have been widely studied. The stability, reactivity, and aromaticity of many such all-metal systems along with the organic [151] and inorganic [155, 171–173] moieties have been assessed under the paradigm of CDFT [1–4]. Effective

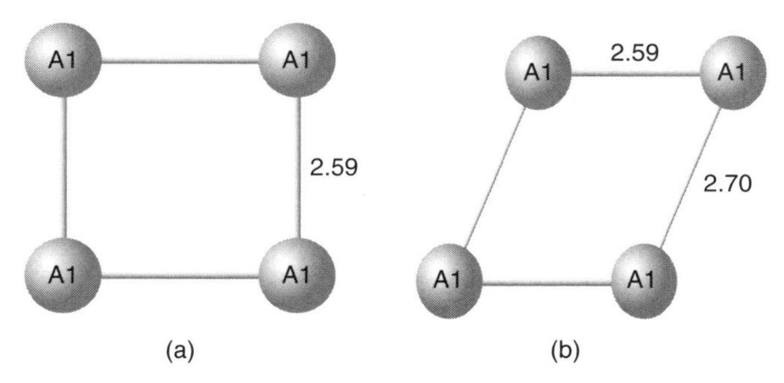

Figure 7.1 Optimized geometries of $Al_4{}^{2-}$ (a) delocalized (singlet) and (b) localized (triplet) molecules.

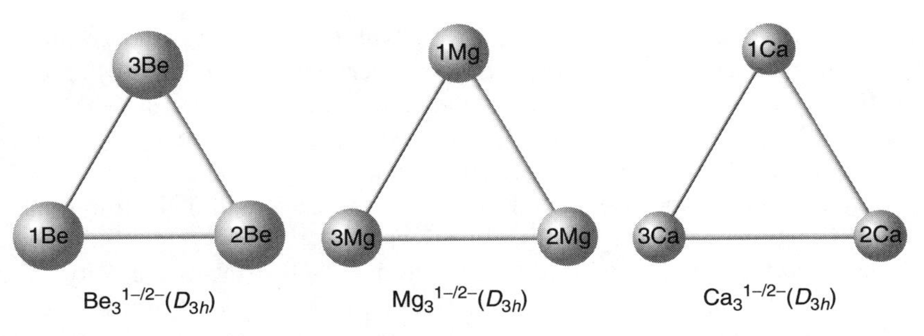

Figure 7.2 Optimized geometries of $[Be_3]^{q-}$, $[Mg_3]^{q-}$, and $[Ca_3]^{q-}$ ($q = 1, 2$).

inroads have been made toward predicting the reactivity and establishing the existence of exclusive π-aromaticity in trigonal $[X_3]^{n-}$ (X = Be, Mg, Ca; $n = 1$, 2) systems [159, 160, 166, 167] (Fig. 7.2).

The results are found to corroborate nicely with other parallel theoretical and experimental findings [174–178]. As the anionic all-metal trigonal clusters contain an excess of electrons, the metal centers in spite of being electropositive in nature, bear some negative charge that requires to be stabilized by coupling with suitable counterions. Thus, the trigonal anionic $[X_3]^{n-}$ (X = Be, Mg, Ca; $n = 1$, 2) systems on binding with another suitable metal counter-cation serves as the building blocks to produce double [161, 162] and multidecker [163] metal clusters.

The formation of a stable M–M (M = metal) bond is actually the essence of cluster chemistry. Resa et al. [179] in their landmark article proved the very existence of a direct Zn–Zn linkage stabilized in a sandwiched form on coordination with a $[C_5Me_5]^-$ group. Inspired by the work of Resa et al. [179] the aromatic ($[Be_3]^{2-}$) [159] and ($[Zn_3]^{2-}$) [180] units have been adopted [161, 165] to stabilize the Zn–Zn linkage (Fig. 7.3).

Subsequent substitution of all the Zn atoms by Be atoms yielded the double-decker $[Be_3\text{-Be-Be-}Be_3]^{2-}$ $[Be_8]^{2-}$, which showed that the aromatic $[Be_3]^{2-}$ is also capable of stabilizing a direct Be–Be linkage [161]. The $[Zn_3]^{2-}$ unit [180] has already been established to exhibit π-delocalization throughout the trigonal ring. Using the $[Zn_3]^{2-}$ unit as the base and coupling it with another Zn^{2+} counter-cation, a number of anionic "all-zinc" chain clusters have been theoretically conceived [165] (Fig. 7.4).

It was found that the aromatic $[Zn_3]^{2-}$ trigonal units are capable of holding a Zn chain of up to four Zn atoms. The stability of such large all-zinc clusters is mainly attributed to the existence of an aromaticity criterion in the $[Zn_3]^{2-}$ units that hold the metal chain. Further stabilization of the Zn–Zn linkage with the aromatic, nonmetallic $[C_5H_5]^-$, and another all-metal pentagonal $[Be_5]^-$ rings has also been performed [165] (Fig. 7.5).

It was observed that unlike the $[Zn_3]^{2-}$ and $[C_5H_5]^-$ units, the slightly aromatic $[Be_5]^-$ ring on complexation with Zn turns antiaromatic and hence confers

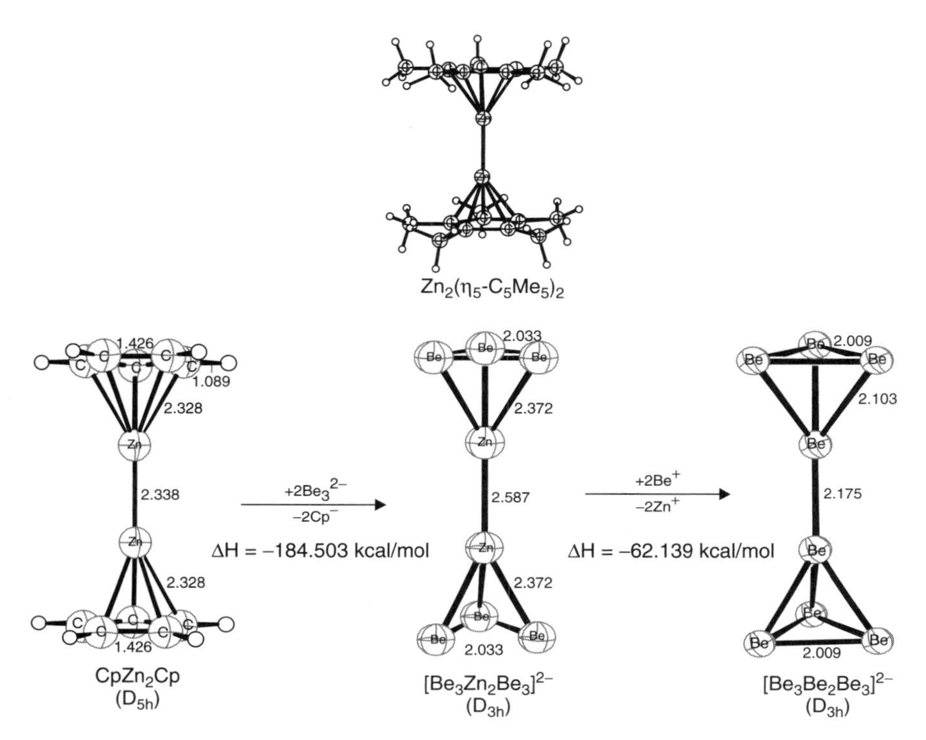

Figure 7.3 Structure of $Zn_2(\eta^5\text{-}C_5Me_5)_2$ (reported by Resa et al. [179]) and two-step substitution reaction of CpZnZnCp to produce $[Be_3BeBeBe_3]^{2-}$.

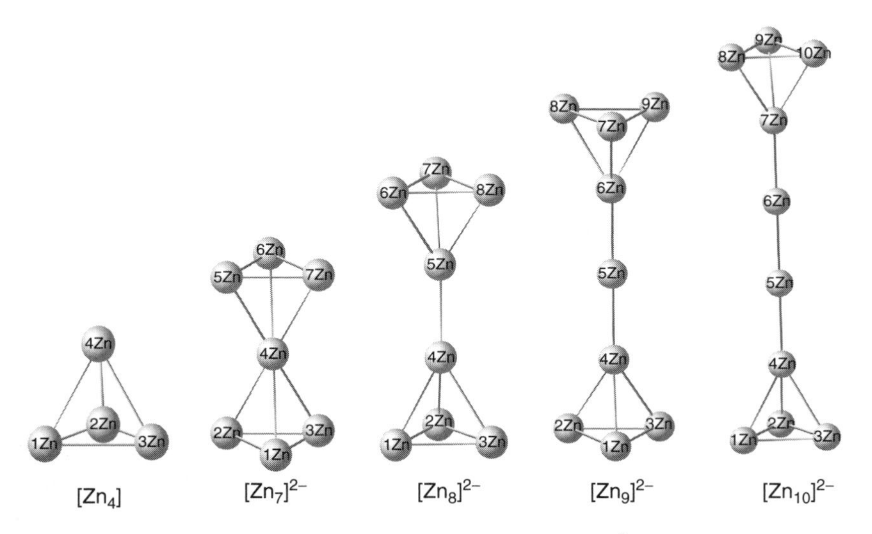

Figure 7.4 Optimized geometries of the $[Zn_n]^{2-}$ clusters.

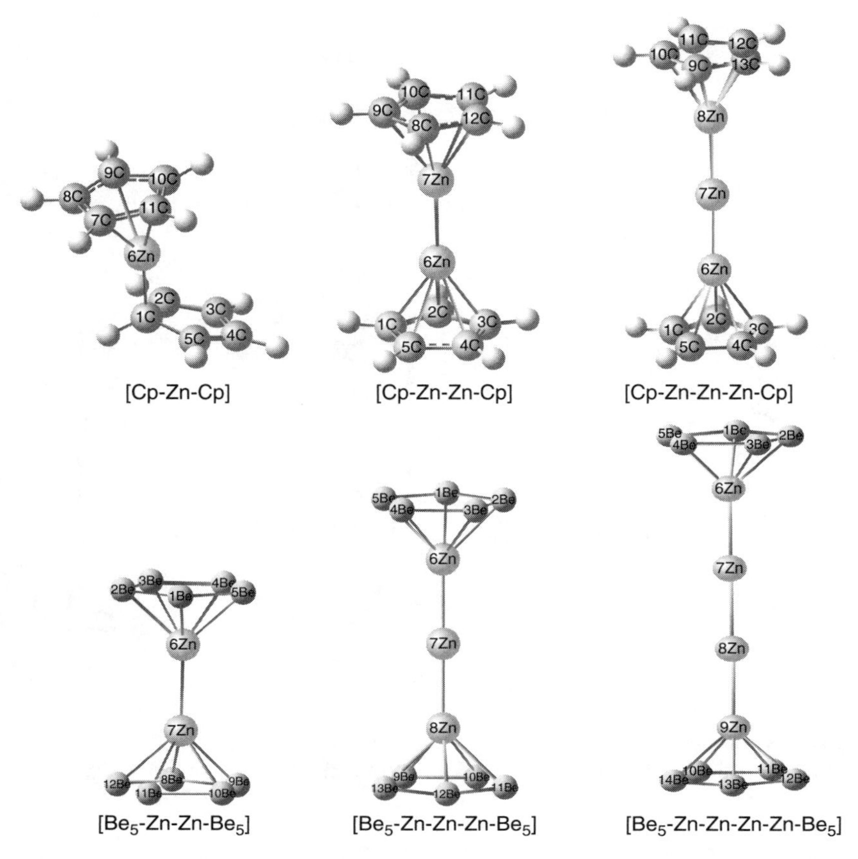

Figure 7.5 Optimized structures of $[C_5H_5]^-$ and $[Be_5]^-$ substituted Zn_n chain analogs. (*See insert for color representation of the figure.*)

some instability to the $[Be_5]^-$ bound chain clusters. This fact is also quite apparent from the corresponding reaction enthalpy (ΔH) and reaction electrophilicity ($\Delta\omega$) values of some probable substitution reactions.

Therefore, the existence of an aromaticity criterion in the $[C_5H_5]^-$ and all-metal $[Zn_3]^{2-}$ systems is supposed to be the driving force behind the above-mentioned thermodynamically spontaneous reactions. The design of unique multidecker [163] sandwich complexes utilizing the aromatic $[Be_3]^{2-}$ and $[Mg_3]^{2-}$ units as the base and their subsequent plausible substitution reactions has also been studied [163] (Fig. 7.6).

The stability of these large all-metal clusters is also attributed to the existence of an all-metal aromaticity in the basal, trigonal rings. The corresponding reactions are also thermodynamically feasible. A detailed documentation of a spectacular range of some all-metal as well as nonmetallic sandwiched, chairlike as well as flanked clusters has already been reported [164, 169].

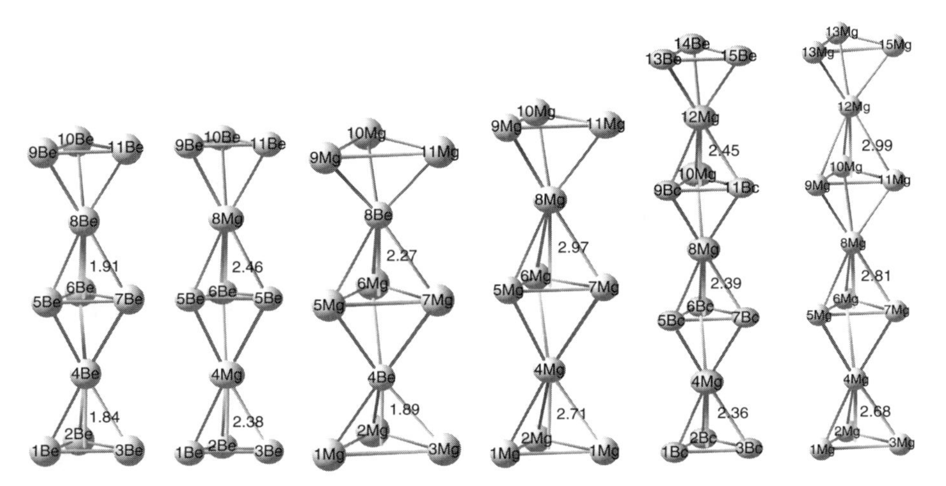

Figure 7.6 Optimized geometries of various "all-metal" multidecker complexes.

7.4.3 Bond-Stretch Isomerism

Bond-stretch isomerism [181–190] may be defined as the *phenomenon* whereby molecules of similar spins lying on the same PES, differ only in the length of one or several bonds. In contrast to exhibiting a single minimum on the PES for the stretching of a bond, the existence of bond-stretch isomers requires the presence of a double minimum, with a barrier between the two minima (Fig. 7.7).

The phenomenon of "bond-stretch isomerism" has been observed for a variety of metal cluster assemblies [164, 168, 191] (Fig. 7.8).

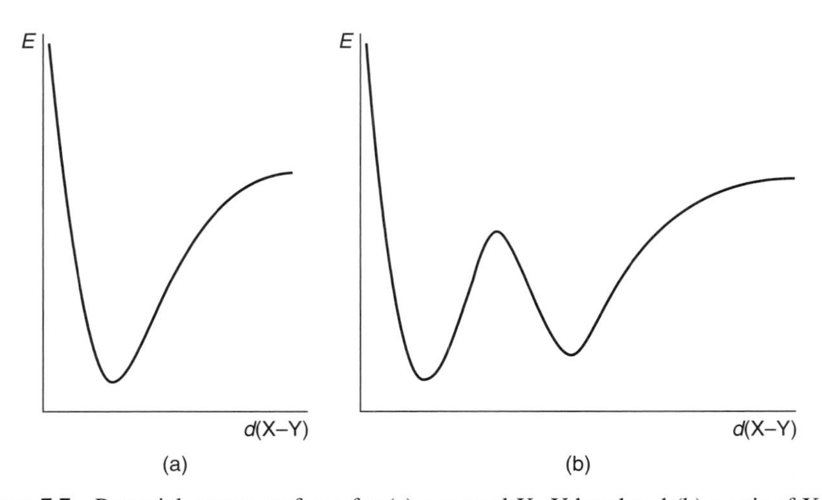

Figure 7.7 Potential energy surfaces for (a) a normal X–Y bond and (b) a pair of X–Y bond-stretch isomers.

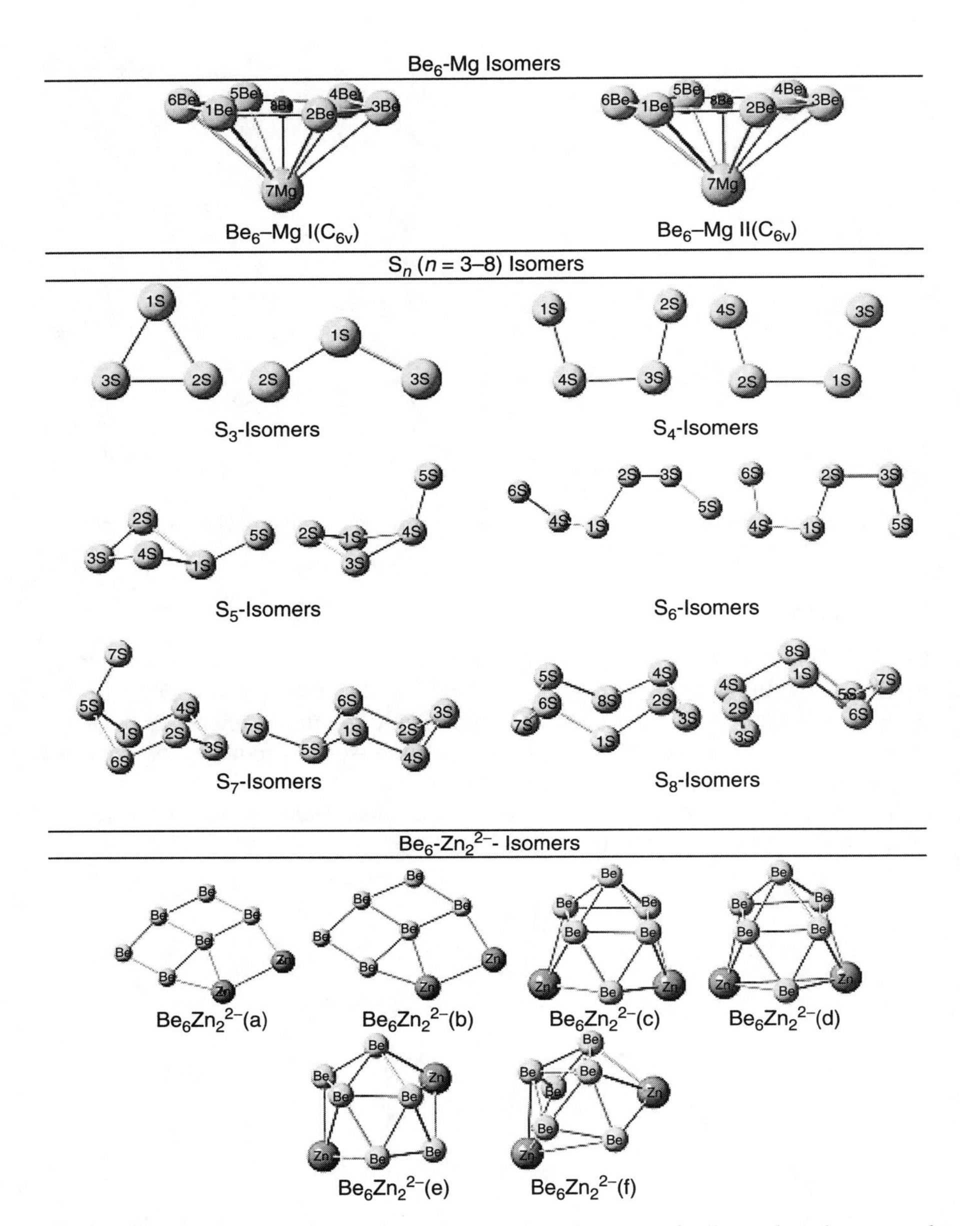

Figure 7.8 Some dominant bond—stretched conformers of all-metal and nonmetal systems. (*See insert for color representation of the figure.*)

A comparative study of the molecular conformations, ground-state energy (E, au), and the allied global reactivity parameters of the bond-stretch isomers shows that for any unique pair, the ground-state energy differences are very low, which indicates that the corresponding molecular structures are easily interconvertible and one geometric form can be converted to another by simple stretching or flipping. The similar magnitudes of the ground-state energy (E, au) for each isomeric pair lends evidence in favor of a low barrier height for conversion from one conformation to another. Thus a relatively fleeting existence [192] of the bond-stretched isomeric pairs at the local minima on the PES seems to be quite feasible. The PES profiles of bond-stretch isomeric pairs therefore roughly resemble a double well, the structures being kinetically stable.

7.4.4 Analysis of the Toxic Effects of Different Molecules through QSAR/QSTR-Based Model

Quantitative structure–activity relationship (QSAR) is a method that mathematically correlates the electronic structures of several compounds with their chemical reactivity. Suitable one-parameter and multi-parameter regression models are developed by correlating the important chemical properties such as biological activity and toxicity as a function of several DFT-based reactivity descriptors. A recent study [193] assessed the toxic effects of some halogen, sulfur, and chlorinated aromatic compounds using quantitative structure–toxicity relationship (QSTR)-based models. In this QSTR study, the toxicities of the various halogen, sulfur, and chlorinated aromatic compounds were correlated with the different CDFT-based global reactivity descriptors viz., electrophilicity index (ω), chemical potential (μ), and the newly proposed net electrophilicity [47] ($\Delta\omega^{\pm}$). Two sets of compounds, containing mainly halogen and sulfur inorganic compounds in the first set and chlorinated aromatic compounds in the second set, were considered for investigation (Figs. 7.9 and 7.10).

The corresponding R^2, R^2_{CV}, and R^2_{adj} values as listed in Table 7.1 (Table 7.2) and the experimental versus calculated log (LC_{50}) plots utilizing the one-parameter regression equations for sets 7.1 and 7.2 in Figures 7.11 and 7.12, respectively, reveal that in the case of the first set, the newly proposed net electrophilicity descriptor [47] ($\Delta\omega^{\pm}$) provides the best result, whereas, for the second set, both electrophilicity index (ω) as well as net electrophilicity index ($\Delta\omega^{\pm}$) show comparable results. Thus, the newly proposed descriptor, net electrophilicity index ($\Delta\omega^{\pm}$) shows potential for constructing effective regression models suitable for QSAR/QSTR studies.

In a recent communication [194], it has been explained that suitable QSAR-based regression models are quite capable of explaining the toxicity trends of several alkali metal ions and various arsenic compounds. A suitable three-parameter-based regression model consisting of electrophilicity (ω),

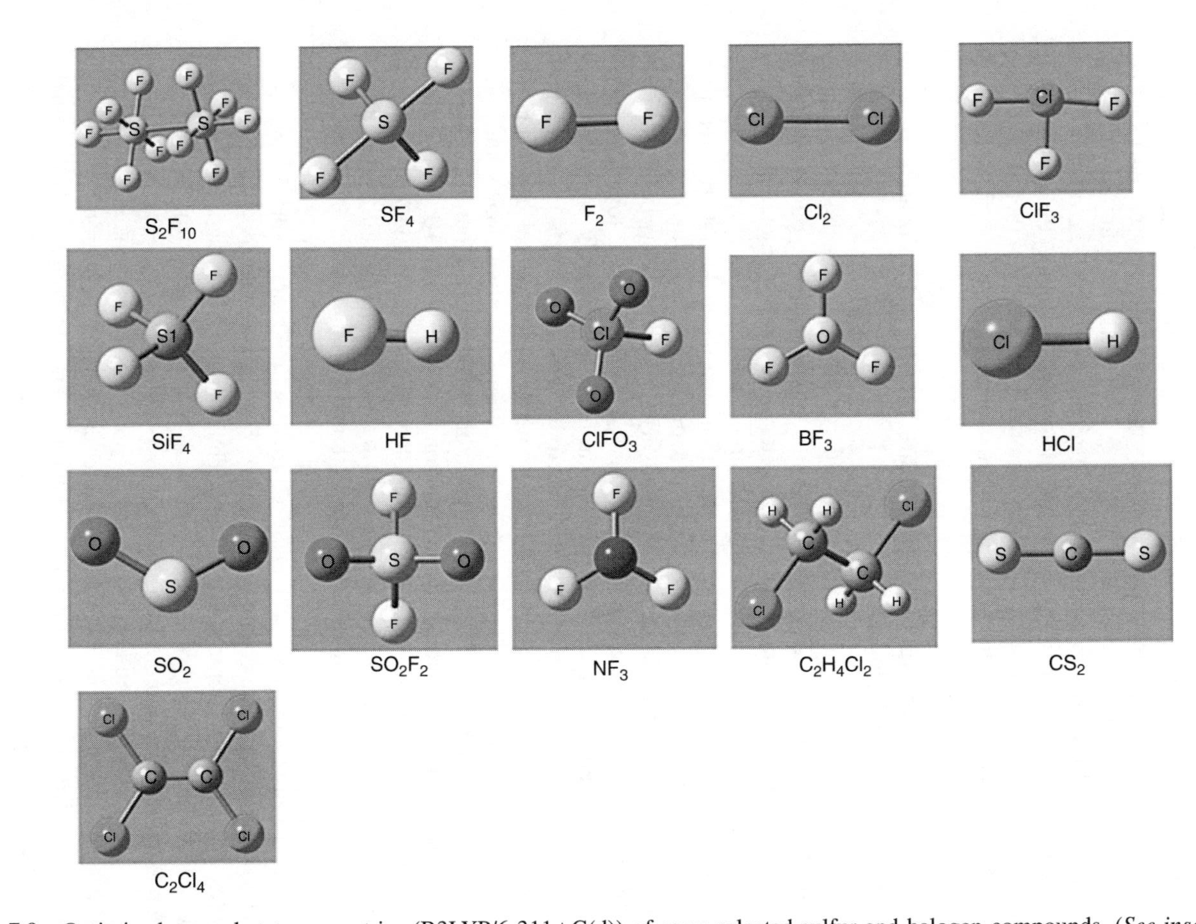

Figure 7.9 Optimized ground-state geometries (B3LYP/6-311+G(d)) of some selected sulfur and halogen compounds. (*See insert for color representation of the figure.*)

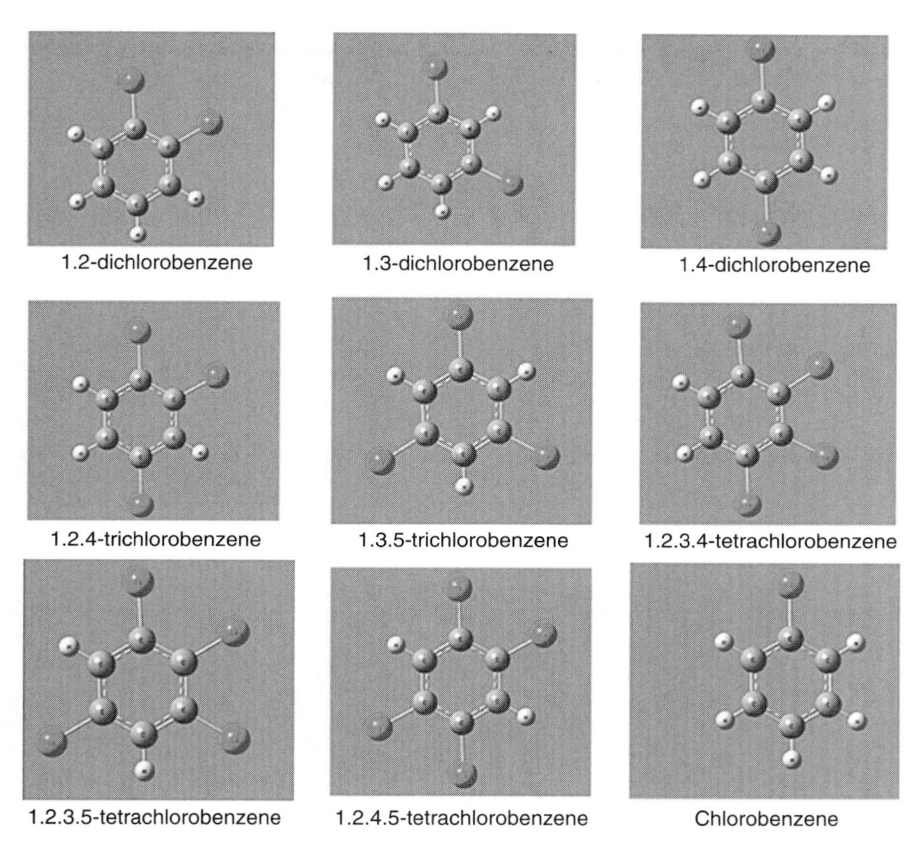

Figure 7.10 Optimized ground-state geometries (B3LYP/6-311+G(d)) of chlorinated aromatic compounds. (*See insert for color representation of the figure.*)

TABLE 7.1 Some Plausible Substitution Reactions Involving the Replacement of the Marginally Aromatic $[Be_5]^-$ or Fairly Aromatic $[C_5H_5]^-$ with the Corresponding All-Metal Systems

No.	Reactions	ΔH (kcal/mole)	$\Delta\omega$ (eV)
1.	Be_5-Zn-Zn-Be_5 + $2Zn_3^{2-}$ = $[Zn_3$-Zn-Zn-$Zn_3]^{2-}$ + $2Be_5^-$	−149.28	−15.925
2.	Be_5-Zn-Zn-Zn-Be_5 + $2Zn_3^{2-}$ = $[Zn_3$-Zn-Zn-Zn-$Zn_3]^{2-}$ + $2Be_5^-$	−154.33	−16.545
3.	Be_5-Zn-Zn-Zn-Zn-Be_5 + $2Zn_3^{2-}$ = $[Zn_3$-Zn-Zn-Zn-Zn-$Zn_3]^{2-}$ + $2Be_5^-$	−159.53	−16.994
4.	C_5H_5-Zn-C_5H_5 + $2Zn_3^{2-}$ = $[Zn_3$-Zn-$Zn_3]^{2-}$ + $2C_5H_5^-$	−144.51	−11.308
5.	C_5H_5-Zn-Zn-C_5H_5 + $2Zn_3^{2-}$ = $[Zn_3$-Zn-Zn-$Zn_3]^{2-}$ + $2C_5H_5^-$	−138.85	−12.007
6.	C_5H_5-Zn-Zn-Zn-C_5H_5 + $2Zn_3^{2-}$ = $[Zn_3$-Zn-Zn-Zn-$Zn_3]^{2-}$ + $2C_5H_5^-$	−144.30	−12.808
7.	Be_5-Zn-Zn-Be_5 + $2C_5H_5^-$ = C_5H_5-Zn-Zn-C_5H_5 + $2Be_5^-$	−10.43	-3.917
8.	Be_5-Zn-Zn-Zn-Be_5 + $2C_5H_5^-$ = C_5H_5-Zn-Zn-Zn-C_5H_5 + $2Be_5^-$	−10.03	-3.737

TABLE 7.2 Regression Models and Various Coefficients of Determination with the Various Combinations of μ, ω, $\Delta\omega^{\pm a}$, and $\Delta\omega^{\pm b}$ for the Set 7.1 and Set 7.2

Regression Model	R^2	R^2_{CV}	R^2_{adj}
Set 7.1 (halogen and sulfur compounds)			
$\text{Log LC}_{50} = (0.521 \times \mu) + 6.60$	0.694	0.618	0.673
$\text{Log LC}_{50} = (-0.417 \times \omega) + 4.64$	0.686	0.618	0.662
$\text{Log LC}_{50} = (-0.2341 \times \Delta\omega^{\pm a}) + 6.33$	0.745	0.713	0.742
$\text{Log LC}_{50} = (-0.233 \times \Delta\omega^{\pm b}) + 6.33$	0.728	0.696	0.727
Set 7.2 (chlorinated aromatic compounds)			
$\text{Log LC}_{50} = (-2.95 \times \mu) - 7.89$	0.729	0.549	0.691
$\text{Log LC}_{50} = (2.66 \times \omega) - 0.4$	0.872	0.798	0.854
$\text{Log LC}_{50} = (1.74 \times \Delta\omega^{\pm\ a*}) - 10.3$	0.686	0.449	0.641
$\text{Log LC}_{50} = (1.43 \times \Delta\omega^{\pm b\dagger}) - 2.50$	0.844	0.751	0.822

[a]Calculated as per Equations 7.16 and 7.17.
[b]Calculated as per Equations 7.18 and 7.19.

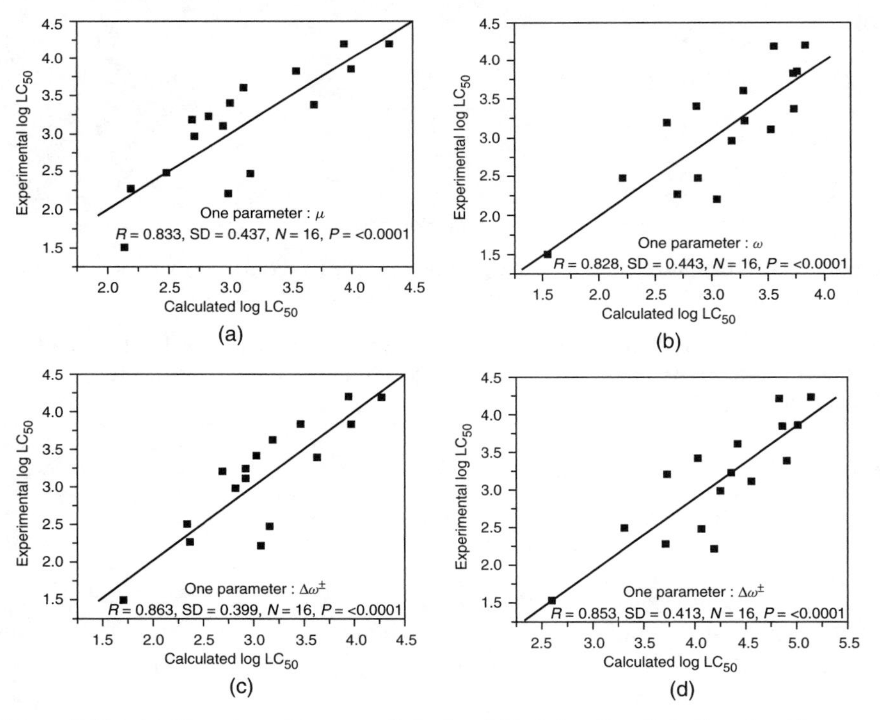

Figure 7.11 (a) Experimental versus Calculated Toxicity (Log(LC_{50})) values for compounds containing *F* and *S*, using (a) chemical potential, (μ); (b) electrophilicity index, (ω); (c) net electrophilicity ($\Delta\omega^{\pm}$) from Equations 7.16, 7.17, and 7.20; (d) $\Delta\omega^{\pm}$ from Equations 7.18–7.20.

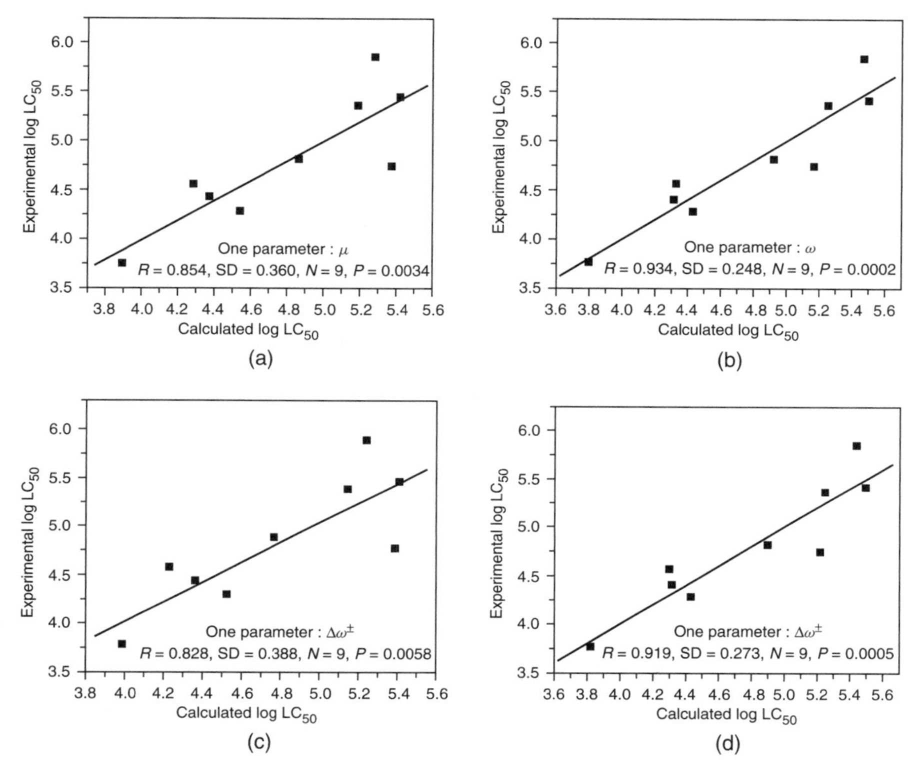

Figure 7.12 Experimental versus Calculated Toxicity (Log(LC_{50})) values for chlorinated aromatic compounds, using (a) chemical potential (μ); (b) electrophilicity index; (ω); (c) net electrophilicity ($\Delta\omega^{\pm}$) from equations (7.16), (7.17), and (7.20); (d) $\Delta\omega^{\pm}$ from equations (7.18–7.20).

philicity (${\omega^{+}}_{As}$), and atomic charge (Q_{As}) has been found to be quite effective in predicting the toxicity of the arsenic compounds.

7.4.5 CDFT as a Novel Tool toward Designing Suitable Storage Materials for Trapping Hydrogen

Hydrogen, the third most abundant and ubiquitous element on the earth's surface [195], is found everywhere, in rocks, soil, air, and particularly in water. It is conceived as an important energy source in this century with the potential to combat the energy crisis. Hydrogen as a future fuel in transportation engines shows the promise of substantially cleaner emissions as compared to the existing oil-powered combustion engines. This offers the advantage of hydrogen being an environmentally benign fuel and an efficient energy carrier. Hydrogen has a high gravimetric energy content of 120.7 MJ/kg, which is the highest for any known

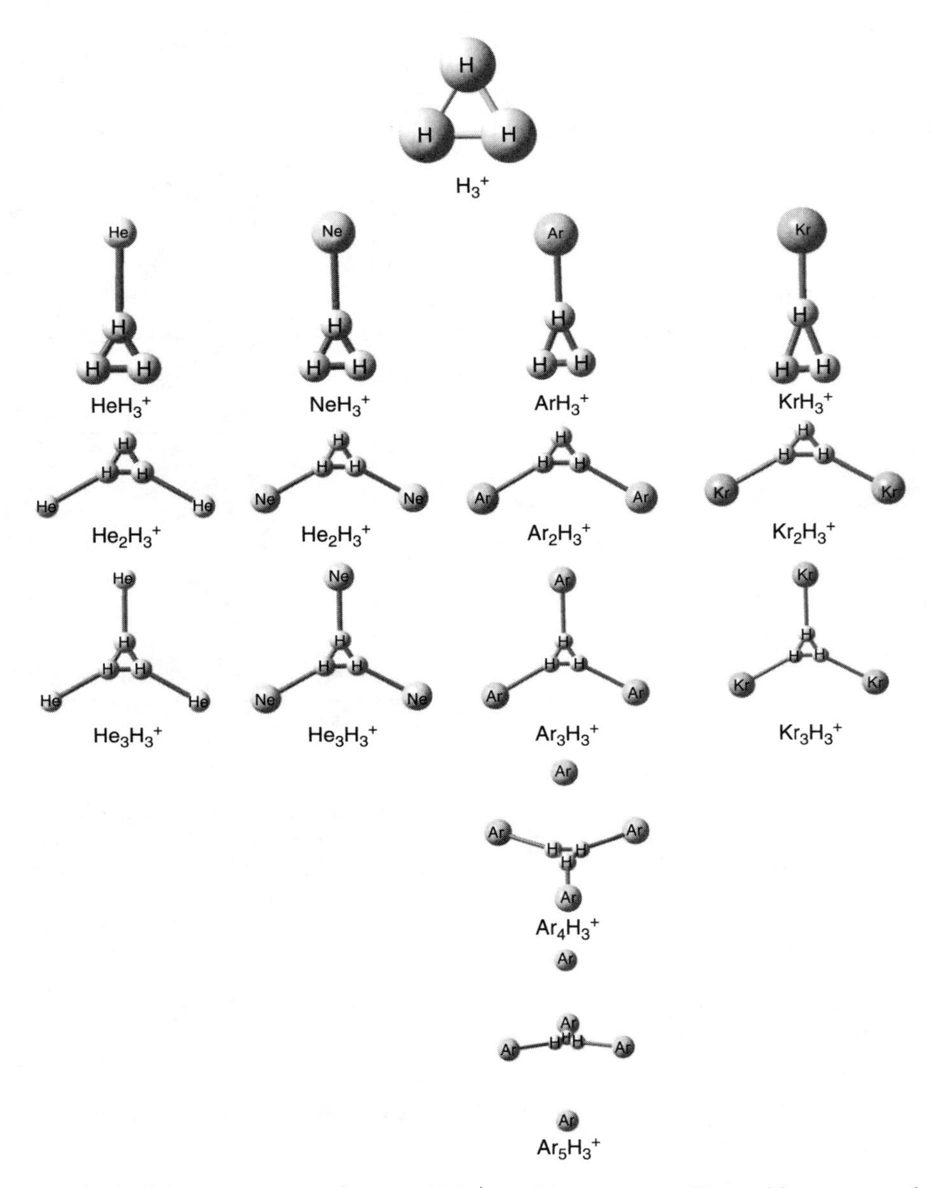

Figure 7.13 Geometrical structure of H_3^+ and its corresponding noble-gas-trapped clusters optimized at the B3LYP/6-311+G(d) level of theory. (*See insert for color representation of the figure.*)

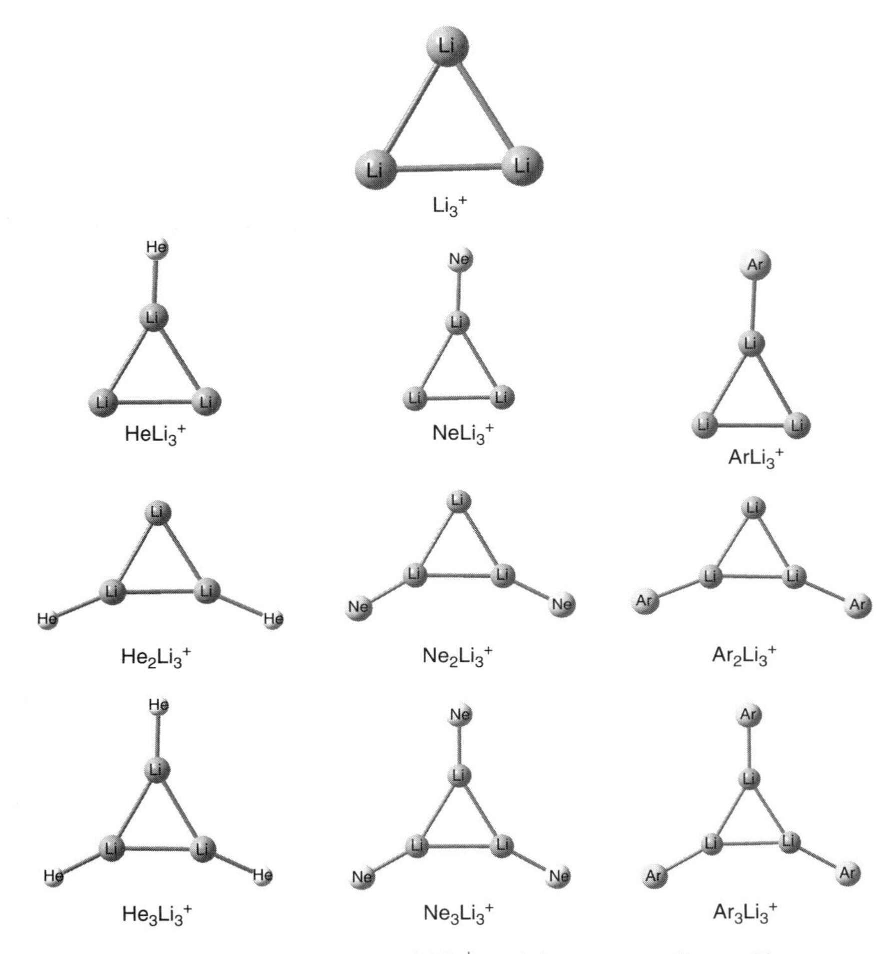

Figure 7.14 Geometrical structure of Li_3^+ and its corresponding noble-gas-trapped clusters optimized at the B3LYP/6-311+G(d) level of theory. (*See insert for color representation of the figure.*)

fuel. However, its volumetric energy content is rather low. This poses challenges for developing safe and effective capacities for storage and transportation of hydrogen, thus limiting its extensive use in practice, as compared to fossil fuels. Thus there is a need for effective trapping materials that can bind hydrogen either in its atomic or molecular form, thereby functioning as latent storage for fuel. A recent article [196] has demonstrated with the aid of CDFT [1–4] and its various global and local reactivity variants that the trapping of noble gases (He–Kr) onto trigonal and aromatic H_3^+ and Li_3^+ moieties is a favorable process (Figs. 7.13 and 7.14).

The stability of the noble-gas-trapped cationic clusters has been attributed to the presence of an "aromaticity criterion," which was theoretically justified

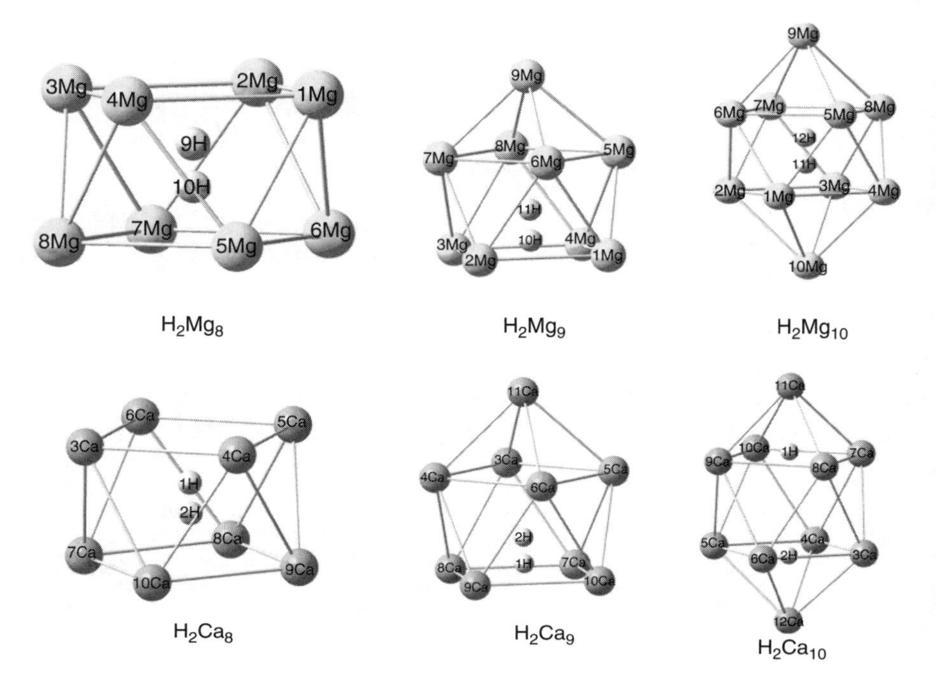

Figure 7.15 Optimized geometries (at B3LYP/6-311+G(d) level) of H_2M_n (where M = Mg, Ca; $n = 8, 9, 10$) clusters. (*See insert for color representation of the figure.*)

from NICS [147] calculations. Several materials such as AlN nanostructures [197], transition-metal-doped BN systems [198], alkali-metal-doped benzenoid [199] and fullerene clusters [200], bare as well as light metal- and transition-metal-coated boron buckyballs, B_{80} [201], and magnesium clusters [202] have been effectively implemented experimentally and theoretically shown to be capable of hydrogen storage. Likewise, MgH_2 has proved to be very effective as a hydrogen-storage material. Mg clusters doped with H_2 molecule have been theoretically investigated and found to be weakly stable or metastable depending on the cluster size [203]. By exploiting the useful tools [204] of CDFT, an attempt has been made to utilize the cagelike all-metal Mg clusters predicted by McNelles and Naumkin [203] and similar Ca analogs of those Mg cages and some novel trigonal, aromatic $Li_3{}^+$ and $Na_3{}^+$ units as potential traps for hydrogen (Figs. 7.15–7.17).

An in-depth study of the different hydrogen-trapped metal cluster units reveals that on gradual increase in the amount of hydrogen being trapped accompanied by a cluster growth, the energy (E), hardness (η), and electrophilicity (ω) of the poly-hydrogen-bound metal complexes suggest a gradual increment in stability. The NICS(0,1) for the upper and lower rings of the H_2-bound Mg_n and $Ca_n (n = 8 - 10)$ cages as well as the $NICS_{zz}(0)$ values of the free and hydrogen-trapped $Li_3{}^+$ and $Na_3{}^+$ rings are negative. Thus the presence of an all-metal aromaticity

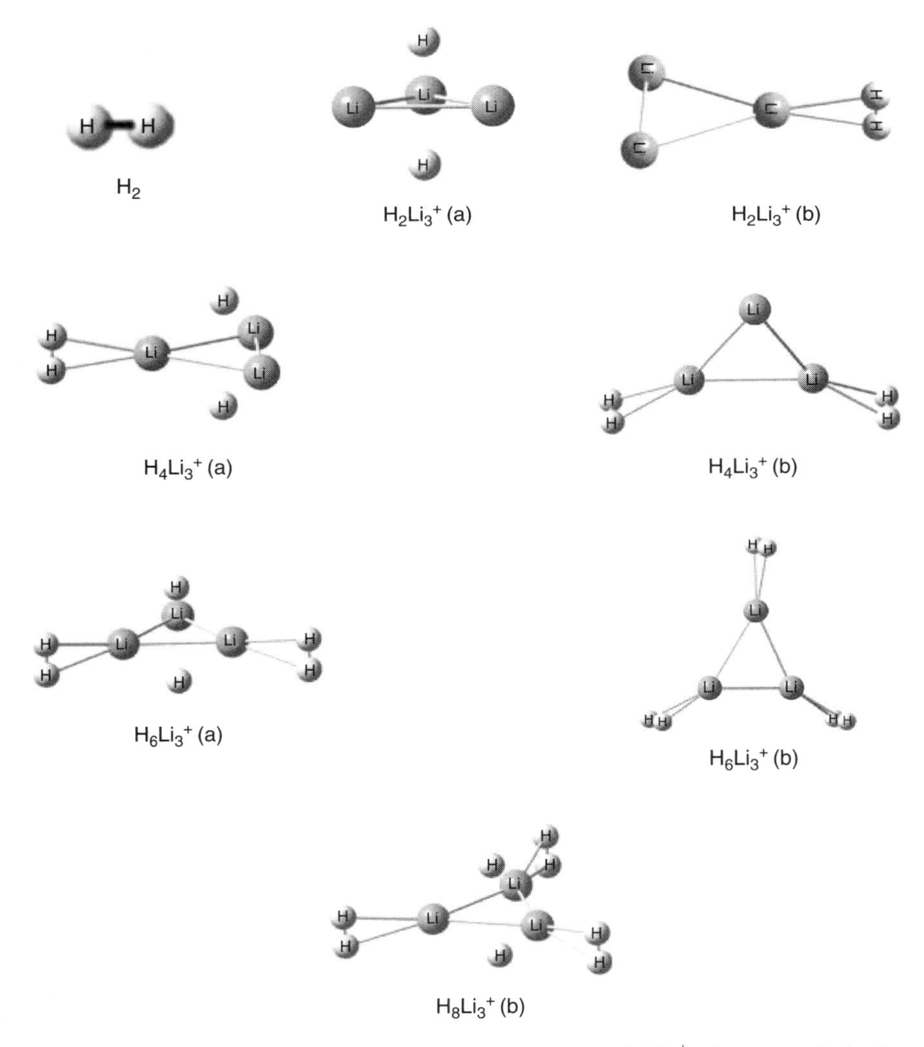

Figure 7.16 Geometries of the different hydrogen-trapped Li_3^+ clusters optimized at B3LYP/6-311+G(d) level of theory. (*See insert for color representation of the figure.*)

in the different cages and rings is verified. The stability of the hydrogen-trapped complexes also receives some support with regard to the thermodynamic aspects from the negative reaction energy (ΔE) values for the gradual stepwise binding of H_2 by the aromatic Li_3^+ and Na_3^+ clusters. These all-metal cages and rings can therefore be fruitfully employed as trapping materials for hydrogen, a future fuel reserve. Further work employing various molecular assemblies as potential traps for hydrogen is in progress.

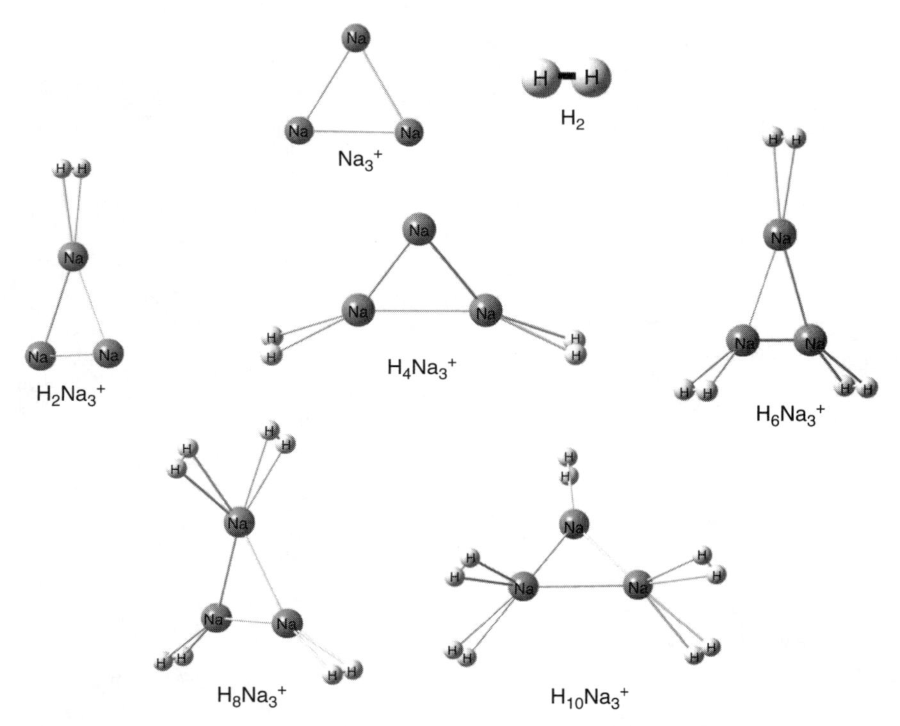

Figure 7.17 Geometries of the different hydrogen-trapped Na_3^+ clusters optimized at B3LYP/6-311+G(d) level of theory. (*See insert for color representation of the figure.*)

7.5 CONCLUDING REMARKS

CDFT, born from the ideas of Hohenberg and Kohn and later developed by Parr et al. has established itself as a unique and dominant theory for predicting chemical reactivity. CDFT along with its various allied global and local reactivity descriptors, in conjunction with the associated important molecular electronic structure principles, has indeed revolutionized chemical thinking and has now become a very useful approach to rationalize the structure, bonding, and reactivity patterns of molecular systems. The preceding discussions have demonstrated that DFT is a very effective mathematical tool to study the chemical reactivity patterns of several different metal clusters. The stability of such cluster assemblies using the concept of all-metal aromaticity has been given a firm footing through the paradigm of CDFT. The role of CDFT in building useful structure–activity relationship-based regression models for analyzing the toxicity trends of different alkali metals—arsenic as well as that of some sulfur and halogenated inorganic and benzenoid compounds—further affirms the effectiveness of theory in predicting chemical behavior. The usefulness of DFT in designing appropriate molecular materials for the storage of hydrogen, conceived as a future alternative fuel source, adds further support to the potential of DFT in modern chemical reactivity theory.

ACKNOWLEDGMENTS

Financial assistance from CSIR, New Delhi, and the Indo-EU (HYPOMAP) is gratefully acknowledged. PKC and AC would like to specially thank DST, New Delhi, for the J. C. Bose National Fellowship and CTS, IIT Kharagpur for a Visitors' Fellowship, respectively. We would also like to thank Prof. V. Subramanian, Prof. P. Bultinck, Dr. G. Merino, and Dr. D. R. Roy for helpful discussions.

REFERENCES

1. Parr RG, Yang W. *Density Functional Theory of Atoms and Molecules*. New York: Oxford University Press; 1989.
2. Chattaraj PK, editor. *Chemical Reactivity Theory: A Density Functional View*. Boca Raton (FL): Taylor & Francis/CRC Press; 2009.
3. Geerlings P, De Proft F, Langenaeker W. Chem Rev 2003;103:1793.
4. Chattaraj PK, Giri S. Annu Rep Prog Chem C: Phys Chem 2009;105:13.
5. Hohenberg P, Kohn W. Phys Rev 1964;136:B864.
6. Sen KD, Jorgenson CK, editors. Volume 66, *Structure and Bonding: Electronegativity*. Berlin: Springer; 1987.
7. Chattaraj PK. J Indian Chem Soc 1992;69:173.
8. Parr RG, Donnelly RA, Levy M, Palke WE. J Chem Phys 1978;68:3801.
9. Sen KD, Mingos DMP, editors.Volume 80, *Structure and Bonding: Chemical Hardness*. Berlin: Springer; 1993.
10. Parr RG, Pearson RG. J Am Chem Soc 1983;105:7512.
11. Pearson RG. *Chemical Hardness: Applications from Molecules to Solids*. Weinheim: Wiley-VCH; 1997.
12. Parr RG, Szentpaly LV, Liu S. J Am Chem Soc 1999;121: 1922.
13. Chattaraj PK, Sarkar U, Roy DR. Chem Rev 2006;106:2065.
14. Chattaraj PK, Roy DR. Chem Rev 2007;107:PR46.
15. Mulliken RS. J Chem Phys 1955;23: 1833.
16. Parr RG, Yang W. J Am Chem Soc 1984;106:4049.
17. Pauling L. J Am Chem Soc 1932;54:3570.
18. Pauling L. *The Nature of the Chemical Bond*. Ithaca: Cornell University Press; 1960.
19. Mulliken RS. J Chem Phys 1934;2:782.
20. Mulliken RS. J Chem Phys 1935;3:573.
21. Pearson RG. J Am Chem Soc 1985;107:6801.
22. Allred AL, Rochow EG. J Inorg Nucl Chem 1958;5:264.

23. Allred AL. J Inorg Nucl Chem 1961;17:215.
24. Sanderson RT. Science 1951;114:670.
25. Sanderson RT. Science 1955;116: 41; 1952;121:2078.
26. Sanderson RT. J Chem Educ 1952;29:539.
27. Sanderson RT. J Am Chem Soc 1983;105:2259.
28. Sanderson RT. *Inorganic Chemistry*. New York: Van Nostrand-Reinhold; 1967.
29. Sanderson RT. *Chemical Bonds and Bond Energy*. New York: Academic Press; 1976.
30. Sanderson RT. *Polar Covalence*. New York: Academic Press; 1983.
31. Rick SW, Stuart SJ. In: Lipkowitz KB, Boyd DB, editors. *Electronegativity Equalization Models*, *Reviews in Computational Chemistry*. New York: Wiley; 2002. p 106.
32. Allen LC. J Am Chem Soc 1989;111:9003.
33. Gyftopoulos EP, Hatsopoulos GN. Proc Natl Acad Sci U S A 1968;60:786.
34. Koopmans TA. Physica 1933;1:104.
35. Pearson RG. J Am Chem Soc 1963;85:3533.
36. Pearson RG. Science 1966;151:172.
37. Pearson RG. J Chem Educ 1968;45:581.
38. Pearson RG. J Chem Educ 1968;45:643.
39. Pearson RG, Chattaraj PK. Chemtracts-Inorg Chem 2008;21:1.
40. Pearson RG. J Chem Educ 1987;64:561.
41. Pearson RG. Coord Chem Rev 1990;100:403 and references therein.
42. Pearson RG. J Chem Sci 2005;117:369.
43. Yang W, Parr RG. Proc Natl Acad Sci U S A 1985;82:6723.
44. Maynard AT, Huang M, Rice WG, Covel DG. Proc Natl Acad Sci U S A 1998;95: 11578.
45. Chattaraj PK, Giri S, Duley S. Chem Rev 2010;111:PR43.
46. Gázquez JL, Cedillo A, Vela A. J Phys Chem A 2007;111: 1966.
47. Chattaraj PK, Chakraborty A, Giri S. J Phys ChemA 2009;113:10068.
48. van Vleck J. *The Theory of Electric and Magnetic Susceptibilities*. Oxford: Oxford University Press; 1932.
49. Dalgarno A. Adv Phys 1962;11:281.
50. Pearson RG. Proc Natl Acad Sci U S A 1986;83:8440.
51. Ghanty TK, Ghosh SK. J Phys Chem 1993;97:4951.
52. Ghanty TK, Ghosh SK. J Phys Chem 1996;100:12295.
53. Pal S, Chandra AK. J Phys Chem 1995;99:13865.
54. Politzer P. J Chem Phys 1987;86:1072.
55. Fuentealba P, Reyes O. J Mol Struct (Theochem) 1993;282:65.
56. Ghosh SK. J Phys Chem 1993;97:4951.
57. Chattaraj PK, Poddar A. J Phys Chem A 1998;102:9944.
58. Vela A, Gázquez JL. J Am Chem Soc 1990;112:1490.
59. Simon-Manso Y, Fuentealba P. J Phys Chem A 1998;102:2029.
60. Chattaraj PK, Arun Murthy TVS, Giri S, Roy DR. J Mol Struct (Theochem) 2007;813:63.

61. Parr RG, Yang W. Ann Rev Phys Chem 1995;46:701.
62. Ayers PW, Levy M. Theor Chem Acc 2000;103:353.
63. Fukui K, Yonezawa T, Shingu H. J Chem Phys 1952;20:722.
64. Fukui K, Yonezawa T, Nagata C, Shingu H. J Chem Phys 1954;22:1433.
65. Fukui K. Science 1982;218:747.
66. Fukui K. *Theory of Orientation and Stereoselection*. Berlin: Springer; 1975.
67. Yang W, Parr RG, Pucci R. J Chem Phys 1984;81:2862.
68. Chattaraj PK, Pérez P, Zevallos J, Toro-Labbé A. J Phys Chem A 2001;105:4272.
69. Chattaraj PK, Gutiérrez-Oliva S, Jaque P, Toro-Labbé A. Mol Phys 2003;101:2841.
70. Yang W, Mortier WJ. J Am Chem Soc 1986;108:5708.
71. Chattaraj PK, Cedillo A, Parr RG. J Chem Phys 1995;103:7645.
72. Berkowitz M, Parr RG. J Chem Phys 1988;88:2554.
73. Chattaraj PK, Cedillo A, Parr RG. J Chem Phys 1995;103:10621.
74. Klopman G. J Am Chem Soc 1968;90:223.
75. Berkowitz M, Ghosh SK, Parr RG. J Am Chem Soc 1985;107:6811.
76. Ghosh SK, Berkowitz M. J Chem Phys 1985;83:2976.
77. Berkowitz M. J Am Chem Soc 1987;109:4823.
78. Gázquez JL, Méndez F. J Phys Chem 1994;98:4591.
79. Li Y, Evans JNS. J Am Chem Soc 1995;117:7756.
80. Damoun S, Van de Woude G, Méndez F, Geerlings P. J Phys Chem A 1997;101:886.
81. Chattaraj PK. J Phys Chem A 2001;105:511.
82. Anderson JSM, Melin J, Ayers PW. J Chem Theoy Comput 2007;3:358.
83. Méndez F, Gázquez JL. J Am Chem Soc 1994;116:9298.
84. Li Y, Evans JNS. Proc Natl Acad Sci U S A 1996;93:4612.
85. Pérez P, Simon-Manso Y, Aizman A, Fuentealba P, Contreras R. J Am Chem Soc 2000;122:4756.
86. Melin J, Aparicio F, Subramanian V, Galvan M, Chattaraj PK. J Phys Chem A 2004;108:2487.
87. Chattaraj PK, Roy DR, Geerlings P, Torrent-Sucarrat M. Theor Chem Acc 2007;118:923.
88. De Proft F, Liu SB, Geerlings P. J Chem Phys 1998;108:7549.
89. Torrent-Sucarrat M, De Proft F, Geerlings P, Ayers PW. Chem Eur J 2008;14:8652.
90. Torrent-Sucarrat M, De Proft F, Ayers PW, Geerlings P. Phys Chem Chem Phys 2010;12:1072.
91. Chattaraj PK, Maiti B, Sarkar U. J Phys Chem A 2003;107:4973.
92. Roy DR, Parthasarathi R, Padmanabhan J, Sarkar U, Subramanian V, Chattaraj PK. J Phys Chem A 2006;110:1084.
93. Parthasarathi R, Padmanabhan J, Elango M, Subramanian V, Chattaraj PK. Chem Phys Lett 2004;394:225.
94. Roy DR, Subramanian V, Chattaraj PK. Ind J Chem A 2006;45A:2369.
95. Morell C, Grand A, Toro-Labbe A. J Phys Chem A 2005;109:205.
96. Roy DR, Sarkar U, Chattaraj PK, Mitra A, Padmanabhan J, Parthasarathi R, Subramanian V, Van Damme S, Bultinck P. Mol Divers 2006;10:119.

97. Politzer P, Weinstein H. J Chem Phys 1979;71:4218.
98. Parr RG, Bartolotti L. J Am Chem Soc 1982;104:3801.
99. Nalewajski R. J Phys Chem 1985;89:2831.
100. Yang W, Lee C, Ghosh SK. J Phys Chem 1985;89:5412.
101. Chattaraj PK. Curr Sci 1991;61:391.
102. Datta D. J Phys Chem 1986;90:4216.
103. Chattaraj PK, Nandi PK, Sannigrahi AB. Proc Indian Acad Sci (Chem Sci) 1991; 103:583.
104. Chattaraj PK, Ayers PW, Melin J. Phys Chem Chem Phys 2007;9:3853.
105. Chattaraj PK, Giri S, Duley S. J Phys Chem Lett 2010;1:1064.
106. Pérez P, Toro-Labbe' A, Aizman A, Contreras R. J Org Chem 2002;67:4747.
107. Chamorro E, Chattaraj PK, Fuentealba P. J Phys Chem A 2003;107:7068.
108. Pearson RG. Inorg Chem 1972;11:3146.
109. Nalewajski RF. J Am Chem Soc 1984;106:944.
110. Chattaraj PK, Lee H, Parr RG. J Am Chem Soc 1991;113: 1855.
111. Gázquez JL. In: Sen KD, editor. Volume 80, *Chemical Hardness: Structure and Bonding*. New York: Springer-Verlag; 1993. p. 27.
112. Chattaraj PK, Schleyer PvR. J Am Chem Soc 1994;116:1067.
113. Chattaraj PK, Maiti B. J Am Chem Soc 2003;125:2705.
114. Parr RG, Chattaraj PK. J Am Chem Soc 1991;113: 1854.
115. See, for example, p 747–748 and especially Fig. 31–32 of the following: Samuelson PA, Nordhaus WD. *Economics*. 13th ed. New York: McGraw-Hill; 1989.
116. Gázquez JL. J Phys Chem A 1997;101:4657.
117. Pearson RG. Acc Chem Res 1993;26:250.
118. Pearson RG. J Chem Educ 1999;76:267.
119. Zhou Z, Parr RG, Garst JF. Tetrahedron Lett 1988;29:4843.
120. Zhou Z, Parr RG. J Am Chem Soc 1989;111:7371.
121. Zhou Z, Parr RG. J Am Chem Soc 1990;112:5720.
122. Chattaraj PK, Sengupta S. J Phys Chem 1996;100:16126.
123. Chattaraj PK, Sengupta S. J Phys Chem A 1997;101:7893.
124. Chattaraj PK, Roy DR, Giri S. Comput Lett 2007;3:223.
125. Tanwar A, Pal S, Roy DR, Chattaraj PK. J Chem Phys 2006;125:056101.
126. Frisch MJ, Trucks GW, Schlegel HB, Scuseria GE, Robb MA, Cheeseman JR, Montgomery JA Jr. Vreven T, Kudin KN, Burant JC, Millam JM, Iyengar SS, Tomasi J, Barone V, Mennucci B, Cossi M, Scalmani G, Rega N, Petersson GA, Nakatsuji H, Hada M, Ehara M, Toyota K, Fukuda R, Hasegawa J, Ishida M, Nakajima T, Honda Y, Kitao O, Nakai H, Klene M, Li X, Knox JE, Hratchian HP, Cross JB, Adamo C, Jaramillo J, Gomperts R, Stratmann RE, Yazyev O, Austin AJ, Cammi R, Pomelli C, Ochterski JW, Ayala PY, Morokuma K, Voth GA, Salvador P, Dannenberg JJ, Zakrzewski VG, Dapprich S, Daniels AD, Strain MC, Farkas O, Malick DK, Rabuck AD, Raghavachari K, Foresman JB, Ortiz JV, Cui Q, Baboul AG, Clifford S, Cioslowski J, Stefanov BB, Liu G, Liashenko A, Piskorz P, Komaromi I, Martin RL, Fox DJ, Keith T, Al-Laham MA, Peng CY, Nanayakkara A, Challcombe M, Gill PMW, Johnson B, Chen W, Wong MW, Gonzalez C, Pople JA. *Gaussian 03, Revision B.03*. Pittsburgh (PA): Gaussian Inc.; 2003.

127. Huheey JE, Keiter EA, Keiter RL. *Introduction to Cluster Chemistry*. In: Mingos DMP, Wales DJ, editors. *Inorganic Chemistry: Principles of Structure and Reactivity*. 4th ed. New York: Harper-Collins; 1990.
128. Gonzalez-Moraga G. *Cluster Chemistry: Introduction to the Chemistry of Transition Metal and Main Group Element Molecular Clusters*; 1993.
129. Cotton FA, Wilkinson G, Murillo CA, Bochmann M. *Advances in Inorganic Chemistry*. 6th ed. New York: Wiley Interscience; 1999.
130. Kealey TJ, Pauson PL. Nature 1951;168:1039.
131. Wilkinson G, Rosenblum M, Whiting MC, Woodward RB. J Am Chem Soc 1952;74:2125.
132. Eiland PF, Pepinsky R. J Am Chem Soc 1952;74:4971.
133. Coriani S, Haaland A, Helgaker T, Jørgensen P. Chem Phys Chem 2006;7:245.
134. Kudinov AR, Rybinskaya I. Russ Chem Bull 1999;48:1615.
135. Kudinov AR, Loginov DA, Starikova ZA, Petrovskii PV. J Organomet Chem 2002;649:136.
136. Beck V, O'Hare D. J Organomet Chem 2004;689:3920.
137. Qian-Shu L, Heng-Tai Y, Au-chin T. Theor Chim Acta 1986;70:379.
138. Padma Malar EJ. Theor Chem Acc 2005;114:213 and references therein.
139. Li X, Kuznetsov AE, Zhang H-F, Boldyrev AI, Wang LS. Science 2001;291:859.
140. Kekulé A. Bull Soc Chim Fr (Paris) 1865;3:98.
141. Minkin VI, Glukhovtsev MN, Simkin BY. *Aromaticity and Antiaromaticity: Electronic and Structural Aspects*. New York: John Wiley & Sons; 1994.
142. Schleyer PvR. (Guest editor) Chem. Rev. (Special Issue on Aromaticity.) 2001;101(5).
143. Fowler PW, Havenith RWA, Steiner E. Chem Phys Lett 2001;342:85.
144. Fowler PW, Havenith RWA, Steiner E. Chem Phys Lett 2002;359:530.
145. Juselius J, Straka M, Sundholm D. J Phys Chem A 2001;105:9939.
146. Zhan CG, Zheng F, Dixon DA. J Am Chem Soc 2002;124:14795.
147. Schleyer PvR, Maerker C, Dransfeld A, Jiao H, Hommes NJRVE. J Am Chem Soc 1996;118:6317.
148. Li X, Zhang HF, Wang LS, Kuznetsov AE, Cannon AN, Boldyrev AI. Angew Chem Int Ed 2001;40:1867.
149. Kuznetsov AE, Boldyrev AI, Li X, Wang LS. J Am Chem Soc 2001;123:8825.
150. Mercero JM, Ugalde JM. J Am Chem Soc 2004;126:3380.
151. Boldyrev AI, Wang LS. Chem Rev 2005;105:3716 and references therein.
152. Tsipis C. Coord Chem Rev 2005;249:2740.
153. Chattaraj PK, Roy DR, Elango M, Subramanian V. J Mol Struct (Theochem) 2006;759:109.
154. Mallajosyula SS, Datta A, Pati SK. J Phys Chem B 2006;110:20098.
155. Chattaraj PK, Roy DR. J Phys Chem A 2007;111:4684.
156. Chattaraj PK, Sarkar U, Roy DR. J Chem Educ 2007;84:354.
157. Datta A, Mallajosyula SS, Pati SK. Acc Chem Res 2007;40:213.
158. Chi XX, Liu Y. Int J Quantum Chem 2007;107:1886.

159. Roy DR, Chattaraj PK. J Phys Chem A 2008;112:1612.
160. Chattaraj PK, Giri S. J Mol Struct (Theochem) 2008;865:53.
161. Chattaraj PK, Roy DR, Duley S. Chem Phys Lett 2008;460:382385.
162. Roy DR, Duley S, Chattaraj PK. Proc Indian Natl Sci Acad A 2008;74:11.
163. Chattaraj PK, Giri S. Int J Quantum Chem 2009;109:2373.
164. Duley S, Giri S, Chakraborty A, Chattaraj PK. J Chem Sci (S. K. Rangarajan Special Issue) 2010;121:849.
165. Chakraborty A, Giri S, Chattaraj PK. J Mol Struct (Theochem) 2009;913:70.
166. Giri S, Roy DR, Duley S, Chakraborty A, Parthasarathi R, Elango M, Vijayaraj R, Subramanian V, Islas R, Merino G, Chattaraj PK. J Comput Chem 2009;31:1815.
167. Giri S, Chakraborty A, Duley S, Merino G, Subramanian V, Chattaraj PK. ICCMSE, AIP Issue; 2009. In Press.
168. Duley S, Chakraborty A, Giri S, Chattaraj PK. J Sulf Chem 2010;31:231.
169. Duley S, Goyal P, Giri S, Chattaraj PK. Croat Chem Acta 2009;82:193.
170. Chattaraj PK, Duley S, Das R. Chem Educ 2010;15:474.
171. Chattaraj PK, Roy DR, Elango M, Subramanian V. J Phys Chem A 2005;109:9590.
172. Chattaraj PK, Giri S. J Phys Chem A 2007;111:11116.
173. Khatua S, Roy DR, Chattaraj PK, Bhattacharjee M. Chem Commun 2007;2:135.
174. Kuznetsov AE, Boldyrev AI. Chem Phys Lett 2004;388:452.
175. Middleton R, Klein J. Phys Rev A 1999;60:3786.
176. Kaplan IG, Dolgounitcheva O, Watts JD, Ortiz JV. J Chem Phys 2002;117:3687.
177. Liu Y, Chi XX, Wang XB. N03: PNIP (Poster No 4); International Conference on Computational Science; 2007 May 27–30; Beijing. Available at http://www.iccs-meeting.org/.
178. Oscar J, Jimenez-Halla C, Matito E, Blancafort L, Robles J, Sola M. J Comput Chem 2009;30:2764, (ibid. 2011;32:372.
179. Resa I, Carmona E, Gutierrez-Puebla E, Monge A. Science 2004;305:1136.
180. Yong L, Chi X. J Mol Struct (Theochem) 2007;818:93.
181. Parkin G. Acc Chem Res 1992;25:455.
182. Parkin G. Chem Rev 1993;93:887.
183. Parkin G. In: Atwood JL, Steed JW, editors. *Encyclopedia of Supramolecular Chemistry*. Boca Raton (FL): Taylor & Francis; 2004.
184. Jemmis ED, Kumar PNVP, Sastry GN. J Organomet Chem 1994;478:29.
185. Boatz JA, Gordon MS. Organometallics 1996;15:2118.
186. Rohmer MM, Bénard M. Chem Soc Rev 2001;30:340.
187. Ladinger JACR. Chimie 2002;5:235.
188. Vijay D, Sastry GN. J Mol Struct (Theochem) 2005;714:199.
189. Zdetsis AD. J Chem Phys 2007;127:014314.
190. Satpati P, Sebastian KL. Inorg Chem 2008;47:2098.
191. Giri S, Abhijith Kumar RPS, Chakraborty A, Roy DR, Duley S, Parthasarathi R, Elango M, Vijayraj R, Subramanian V, Merino G, Chattaraj PK. In: Chattaraj PK, editor. *Aromaticity and Metal Clusters*. Boca Raton (FL): Taylor & Francis; 2011.
192. Hoffmann R, Schleyer PvR, Schaefer HF III. Angew Chem Int Ed 2008;47:7164.

193. Gupta A, Chakraborty A, Giri S, Subramanian V, Chattaraj PK. Int J Chemoinf Chem Eng 2010;1:61.
194. Roy DR, Giri S, Chattaraj PK. Mol Divers 2009;13:551.
195. Mildred D, Crabtree G, Buchanan M. Basic Research Needs for the Hydrogen Economy. Argonne National Laboratory, U.S. Department of Energy, Office of Science Laboratory; 2003 May 15. Available at http://www.sc.doe.gov/bes/hydrogen.pdf.
196. Chakraborty A, Giri S, Chattaraj PK. New J Chem 2010;34: 1936.
197. Wang Q, Sun Q, Jena P, Kawazoe Y. ACS Nano 2009;3:621.
198. Shevlina SA, Guo ZX. Appl Phys Lett 2006;89(1–3):153104.
199. Srinivasu K, Chandrakumar KRS, Ghosh SK. Phys Chem Chem Phys 2008;10: 58332, and references 2–12 therein.
200. Peng Q, Chen G, Mizuseki H, Kawazoe Y. J Chem Phys 2009;131:214505.
201. Wu G, Wang J, Zhang X, Zhu L. J Phys Chem C 2009;113:7052.
202. Wagemans RWP, van Lenthe JH, de Jongh PE, van Dillen AJ, de Jong KP. J Am Chem Soc 2005;127:16675.
203. McNelles P, Naumkin FY. Phys Chem Chem Phys 2009;11:2858.
204. Giri S, Chakraborty A, Chattaraj PK. J Mol Model 2011;17:777.

8

ELECTRON DENSITY AND MOLECULAR SIMILARITY

N. SUKUMAR

8.1 THE MOLECULAR SIMILARITY PRINCIPLE IN DRUG DESIGN

The importance of molecular similarity measures [1] for drug design is summarized in the similarity principle, which states that similar molecules should exhibit a corresponding similarity in their biological effects [2, 3]. In other words, chemically similar molecules should exhibit similar patterns of activity toward protein targets [2]. The molecular similarity principle is a fundamental assumption implicit in most quantitative structure–activity relationship (QSAR) modeling. Such correlations have indeed been observed for many simple physicochemical properties, but owing to the complex nature of the activity landscape associated with biological assays, many deviations from the similarity principle have also been observed [2, 3], leading very similar molecules to exhibit very different activities in some assays. Such regions of the structure–activity landscape have been termed activity cliffs [3] and represent the most interesting regions of the structure–activity relationship for purposes of drug design. While similar molecules may not always exhibit similar activities in individual biological assays, similar molecules do display similar broad patterns of biological activities across a range of related targets [4, 5].

Changes in biological activities resulting from changes in molecular structure are described by chemists through structure–activity relationships. The classical QSAR approach, based on physicochemical characterizations of molecules, was pioneered by Hammett [6, 7] and developed by Hansch et al. [8], Taft [9], and

A Matter of Density: Exploring the Electron Density Concept in the Chemical, Biological, and Materials Sciences, First Edition. Edited by N. Sukumar.

others. Hammett introduced numerical descriptors to represent substituent groups R on a molecular scaffold in the context of a linear free energy relationship

$$\log(K/K_0) = \sigma\rho. \tag{8.1}$$

where K is an equilibrium constant and K_0 the equilibrium constant for the reference reaction with R = H, σ is the substituent constant characteristic of meta- or para-substitution and depends only on R, and ρ is the reaction constant characteristic of the given reaction. Systematic study with different R groups led to the explanation of both equilibrium and kinetic substituent effects, such as acidity and reaction rates. Nowadays, empirically observed correlations between molecular structures and biological activities are supplemented by statistically discovered quantitative correlations (QSAR) through the aid of sophisticated machine learning methods. All correlations between molecular structure and chemical or biological properties are ultimately based on the dependence of the property in question on the molecular electron density distribution. The detailed nature of this dependence is in general unknown, but the existence of a relationship can be justified by recourse to the Hohenberg–Kohn theorem. Structure–activity relationships are of great value in drug discovery. While the introduction of a new drug has often come about after a lengthy process of laboratory experimentation, lead compound discovery, animal testing, and preclinical and clinical trials, this process can take over a decade before a viable drug is ready for marketing. With the availability of large quantities of experimental data through automated high throughput screening techniques and the availability of cheap computational power, virtual high throughput screening (VHTS) and QSAR modeling offer an attractive strategy for accelerating the process of drug discovery. Furthermore, since nearly 90% of promising drug leads fail, often at an advanced stage of development, on account of unfavorable absorption, distribution, metabolism, elimination, and toxicity (ADMET) effects in the human body, virtual screening using ADMET filters are often also employed to weed out compounds likely to exhibit adverse side effects, thereby identifying the "losers" early. The most promising compounds that survive this "fail early, fail cheap" strategy are then selected for laboratory synthesis and preclinical testing.

The quantitative characterization of molecular structure as a set of numerical (analog or boolean) descriptors enables the computational implementation of pattern recognition or data mining schemes to search for compounds with specific properties or to search for molecules most similar to a given query molecule. This immediately brings up the question of how molecular similarity is to be assessed [10, 11]. There is no unique answer to this question and thus no ideal similarity measure. Chemical similarity includes constitutional similarity (similarity of atoms in the molecule), structural similarity (similarity of substructures comprising the molecule), similarity of three-dimensional shape, similarity of chemical properties, or similarity of effects on the human (or animal) body owing to binding to similar proteins. The concept of molecular similarity can be illustrated through the eastern parable of the six blind men and the elephant.

As the tale goes, six blind men went to feel an elephant and then described the experiences to each other. The first approached the broad side of the elephant and claimed that the elephant was similar to a wall. The second man felt the tusk and declared that the elephant was like a spear. To the third, who took the squirming trunk in his hands, the elephant was like a snake. The fourth man felt about the knee of the beast and said it was similar to a tree. The fifth chanced to touch the ear, and swore that the elephant most resembled a fan. Finally, the sixth man seized the swinging tail and said the animal was similar to a rope. None of these descriptions can be faulted. All are correct in their own way, but none is a complete or unique characterization. Any two (or more) molecules may likewise be similar in some ways, but not in others. The most appropriate similarity measure in any given situation is thus very much a domain-specific problem.

Structure–activity or structure–property relationships are commonly envisioned in an abstract high dimensional space of numerical descriptors or Boolean fingerprints. This space, commonly referred to as *chemistry space*, is often used to cluster molecules into similarity classes. It is the combinatorial and configurational space spanned by all possible molecules (i.e., those combinations of atoms allowed by the rules of valence in energetically stable spatial arrangements). It is estimated that the total number of possible small organic molecules populating chemistry space could exceed 10^{60}—a number that exceeds the total number of atoms in the known universe and is vastly greater than the number of molecules that have actually been isolated or synthesized. The choice of descriptors or fingerprints defining coordinates within chemistry space and the choice of the similarity metric determine the partitioning of the space into regions corresponding to local structural similarity. These are the regions (known as domains of applicability [12]) most likely to be successfully modeled by a structure–activity relationship. Any characterization of similarity thus depends both on the chemical space (molecular descriptor/fingerprint) representation and on the similarity assessment metric employed within that space.

While there are numerous ways of assessing molecular similarity, all are derived in some way or other from the electron density, since the electron density determines all molecular properties, by virtue of the Hohenberg–Kohn theorem (Chapter 5). Similar molecules must have similar electron density distributions. Some of the most theoretically satisfying molecular similarity measures are thus those that are derived directly from the molecular electron density distributions, but they often suffer from the disadvantage of requiring more intensive computational effort for their determination than some of the simpler similarity measures. In this chapter, we are only concerned with molecular similarity measures and molecular descriptors directly derived from the electron density. Similarity measures based on the integration of the total electron density or electron-density-derived properties over at points in space will be the subject matter of the next section. Thereafter (Section 8.3), we will treat similarity measures generated from the electron density at some special points in space, namely, the bond critical points between nuclei of Bader's theory of atoms in molecules (Chapter 4). Section 8.4 deals with electron-density-derived properties on the molecular van der Waals surface. It has been argued [13, 14] that the use of molecular descriptors and similarity measures based on local surface properties, such as those derived from the electron density, are likely to provide more generalizable QSAR models, applicable across diverse regions of chemistry space (a feature known as *scaffold hopping*), even though such descriptors do not encode the chemical constitution directly. In Section 8.5, we discuss some high throughput algorithms to encode molecular shape and electronic properties into compact signatures for similarity comparisons. The progression of concepts through the successive sections of this chapter proceeds from the most rigorous and computationally expensive to faster, cheaper, and high throughput techniques, while trying to preserve the most relevant chemical information. We conclude the chapter with the construction of network graphs from molecular similarity analysis.

8.2 ELECTRON-DENSITY-BASED ATOMIC AND MOLECULAR SIMILARITY ANALYSIS

From the realization that similar molecules must have similar electron density distributions, Carbó and coworkers pioneered the use of the electron density for the development of molecular quantum similarity measures (MQSM) for drug design and for pharmacological and toxicological modeling. MQSM encapsulate the principle that the more similar the electron distributions of any pair of molecules, the more similar will their properties be [2, 11, 12]. The simplest electron-density-based quantum similarity measure between two systems A and B is obtained from the superposition of the respective density distributions integrated over all space

$$Z_{AB} = \int \rho_A(\mathbf{r})\rho_B(\mathbf{r})d\mathbf{r}. \tag{8.2}$$

This measure is related to the Cartesian distance between the electron densities of A and B, since

$$\int |\rho_A(\mathbf{r}) - \rho_B(\mathbf{r})|^2 d\mathbf{r} = \int \rho_A^2(\mathbf{r})d\mathbf{r} + \int \rho_B^2(\mathbf{r})d\mathbf{r} - 2\int \rho_A(\mathbf{r})\rho_B(\mathbf{r})d\mathbf{r}. \tag{8.3}$$

Thus, the electron density functions are more similar when the overlap integral Z_{AB} between them is maximized. In general, a similarity measure can be constructed from a convolution of the electron density distributions using any positive definite similarity kernel $\Omega(\mathbf{r}_1, \mathbf{r}_2)$, also called a *separation operator*

$$Z_{AB}(\Omega) = \int d\mathbf{r}_1 \int d\mathbf{r}_2 \rho_A(\mathbf{r}_1)\, \Omega(\mathbf{r}_1, \mathbf{r}_2)\, \rho_B(\mathbf{r}_2). \tag{8.4}$$

The overlap form (Eq. 8.2) is obtained from the choice of the Dirac delta function $\Omega(\mathbf{r}_1, \mathbf{r}_2) = \delta(\mathbf{r}_1 - \mathbf{r}_2)$ as the similarity kernel in Equation (8.4). Other popular choices include the Coulomb operator r_{12}^{-1}

$$Z_{\mathrm{AB}}(\mathrm{Coul}) = \int d\mathbf{r}_1 \int d\mathbf{r}_2 \frac{\rho_{\mathrm{A}}(\mathbf{r}_1)\rho_{\mathrm{B}}(\mathbf{r}_2)}{|\mathbf{r}_1 - \mathbf{r}_2|}. \tag{8.5}$$

These similarity measures are normalized to provide a Carbó similarity index ranging between 0 and 1

$$C_{\mathrm{AB}} = \frac{Z_{\mathrm{AB}}}{\sqrt{Z_{\mathrm{AA}} Z_{\mathrm{BB}}}}. \tag{8.6}$$

Values of the Carbó similarity index closer to 1 indicate higher similarity. Using the shape function

$$\sigma_{\mathrm{A}}(\mathbf{r}) = \frac{\rho_{\mathrm{A}}(\mathbf{r})}{N_{\mathrm{A}}}, \tag{8.7}$$

and Equations 8.2 and 8.6, the overlap Carbó similarity index can be rewritten in a form independent of N_{A} and N_{B}

$$C_{\mathrm{AB}} = \frac{\int \sigma_{\mathrm{A}}(\mathbf{r})\sigma_{\mathrm{B}}(\mathbf{r}) d\mathbf{r}}{\sqrt{\int \sigma_{\mathrm{A}}^2(\mathbf{r}) d\mathbf{r} \int \sigma_{\mathrm{B}}^2(\mathbf{r}) d\mathbf{r}}}, \tag{8.8}$$

showing that this index depends only on the shapes of the electron density distributions and not on the number of electrons. Hodgkin and Richards [15] defined an alternative index

$$H_{\mathrm{AB}} = \frac{2\int \rho_{\mathrm{A}}(\mathbf{r})\rho_{\mathrm{B}}(\mathbf{r}) d\mathbf{r}}{\int \rho_{\mathrm{A}}^2(\mathbf{r}) d\mathbf{r} + \int \rho_{\mathrm{B}}^2(\mathbf{r}) d\mathbf{r}} = \frac{2Z_{\mathrm{AB}}}{Z_{\mathrm{AA}} + Z_{\mathrm{BB}}}, \tag{8.9}$$

which, in contrast to the Carbó index, depends on both the shapes and the extent of the electron density distributions.

Of course, the electron density $\rho(\mathbf{r})$ determines the external potential and all hence the energy and all molecular electronic properties (Hohenberg–Kohn theorem). In 2000, Ayers [16] proved a remarkable result that for a Coulombic system, the shape function $\sigma(\mathbf{r})$ determines the external potential and the total number of electrons and hence contains all the information contained in the electron density function. We can thus write

$$E = E[\sigma(\mathbf{r})], \tag{8.10}$$

analogous to Equation 5.2. We also saw in Chapter 5 how through the holographic electron density theorem of Riess and Münch [17] and Mezey [18, 19], the ground-state density of any finite volume or subdomain uniquely determines the ground-state properties of this or any other subdomain of the system (or of the whole system). Combining this insight with that of Ayers, we can conclude that knowledge of the shape function in a finite but otherwise arbitrary subdomain of a Coulombic system suffices to determine all properties of the system. This result has been termed the holographic shape theorem [19].

When applied to Hartree-Fock atomic densities across the periodic table (with spherically averaged densities in the case of incompletely filled atomic subshells), the Carbó atomic quantum similarity indices were highest for atom pairs closest to each other in the periodic table [20, 21]. Atomic quantum similarity indices mask periodicity information. However, the shape function can be used to define an information discrimination or entropy deficiency function

$$\Delta S_{\mathrm{A}}(\mathbf{r}) = \int d\mathbf{r}\sigma_{\mathrm{A}}(\mathbf{r}) \log \frac{\sigma_{\mathrm{A}}(\mathbf{r})}{\sigma_{\mathrm{A}}^{0}(\mathbf{r})}, \tag{8.11}$$

where $\sigma_{\mathrm{A}}(\mathbf{r})$ is the shape function of atom A and $\sigma_{\mathrm{A}}^{0}(\mathbf{r})$ that of the noble gas in the preceding row in the periodic table. This information discrimination function reveals periodicity information [21], as shown in Figure 8.1.

Quantum similarity measures in momentum space [22–29] may be defined in an analogous way using momentum space densities

$$Z_{\mathrm{AB}}(\mathbf{p}) = \int \pi_{\mathrm{A}}(\mathbf{p})\pi_{\mathrm{B}}(\mathbf{p})d\mathbf{p}. \tag{8.12}$$

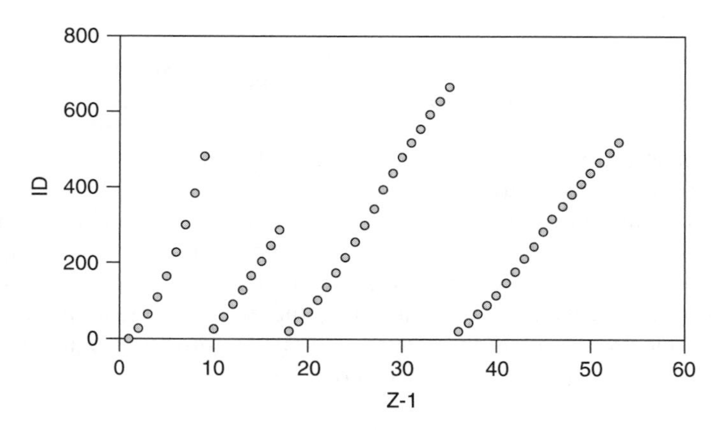

Figure 8.1 Information discrimination or entropy deficiency function versus atomic number for atomic densities with the noble gas of the previous row as reference [21]. (Reproduced with permission from Borgoo A, Godefroid M, Sen KD, De Proft F, Geerlings P. Chem Phys Lett 2004;399:363–367, Copyright 2004 Elsevier.)

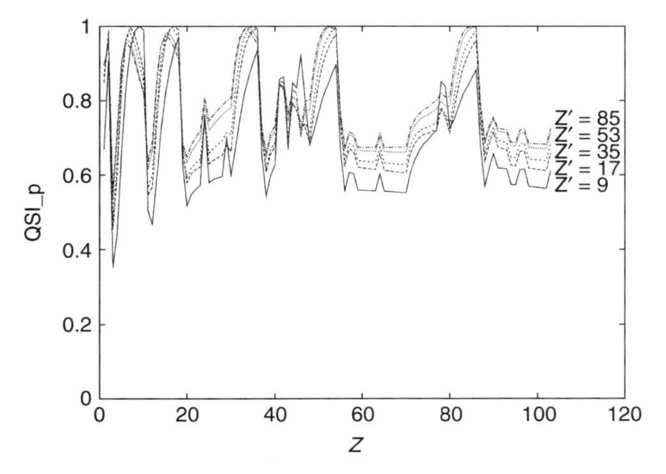

Figure 8.2 Momentum space similarity indices reveal periodicity trends for atoms [30]. (Reproduced with permission from Angulo JC, Antolín J. J Chem Phys 2007;126:044106, Copyright 2007 American Institute of Physics.)

As shown in Figure 8.2, the momentum space similarity indices (and the corresponding phase space quantities) reveal periodicity trends for atoms, in contrast to the position space Carbó indices [30].

It is also possible to use other quantities besides the density to define similarity measures. For instance, the molecular electrostatic potential (MEP)

$$v_e(\mathbf{r}) = \sum\nolimits_{\alpha} \frac{Z_\alpha}{|r - R_\alpha|} - \int \frac{\rho(\mathbf{r}')\mathrm{d}\mathbf{r}'}{|\mathbf{r} - \mathbf{r}'|}, \tag{8.13}$$

has long been used as a descriptor of chemical reactivity [31–36] and a molecular similarity measure [15]. Equation (8.13) is the solution of the Poisson equation for the electron density $\rho(\mathbf{r})$; the first term on the right-hand side is the external potential $v(\mathbf{r})$ due to the nuclei, while the second term is the interelectronic Coulomb repulsion. Since $v(\mathbf{r})$ is uniquely mapped to the electron density $\rho(\mathbf{r})$ by the Hohenberg–Kohn theorem, so is the MEP $v_e(\mathbf{r})$. Both $\rho(\mathbf{r})$ and $v_e(\mathbf{r})$ are experimentally accessible from high resolution X-ray diffraction data (Chapter 3).

MEP is also a principal ingredient in the popular comparative molecular field analysis (CoMFA) method [37] for virtual screening. In addition to the electrostatic fields, CoMFA also employs steric fields generated by a probe atom or group of atoms, on a grid of points around the set of molecules to be compared. The molecules are aligned in 3D space by maximizing the steric and electrostatic overlap between them. CoMFA is used to identify chemical features favoring or disfavoring a biological activity. These features are then employed in a 3D QSAR equation to build a partial least squares (PLS) model for the biological activity of interest. A set of molecules whose activities have been experimentally determined form the training set. MEP computed from molecular mechanics force fields, rather than those from the electron density, are generally sufficient for the purpose when dealing with biological activities that depend primarily on noncovalent interaction. The predictive quality of a CoMFA model depends critically on the quality of alignment between the molecules. Such field analysis techniques are useful in comparing molecules that have a common molecular scaffold. Molecules with very low similarity, that is, lacking a common skeleton, cannot be meaningfully superimposed. Silverman and Platt [38] have proposed a variant of the method, termed comparative molecular moment analysis (CoMMA) that uses moments of the mass and charge distributions, eliminating the need for molecular alignment.

Girones et al. [39] developed a kinetic energy-density-based MQSM

$$Z_{\mathrm{AB}}(\mathrm{KE}) = \int K_{\mathrm{A}}(\mathbf{r})K_{\mathrm{B}}(\mathbf{r})\mathrm{d}\mathbf{r}, \tag{8.14}$$

where $K_{\mathrm{A}}(\mathbf{r})$ and $K_{\mathrm{B}}(\mathbf{r})$ are the kinetic energy densities of the molecules A and B, respectively. A hardness-based similarity index has also been introduced using the chain rule relation between the hardness [40] (Chapter 5) and the Fukui function [40–45]

$$\begin{aligned}\eta &= \left[\frac{\partial^2 E}{\partial N^2}\right]_{v(\mathbf{r})} \\ &= \int \mathrm{d}\mathbf{r} \int \mathrm{d}\mathbf{r}' \frac{\delta^2 E[\rho(\mathbf{r})]}{\delta\rho(\mathbf{r})\delta\rho(\mathbf{r}')} \left[\frac{\partial\rho(\mathbf{r})}{\partial N}\right]_{v(\mathbf{r}')} \left[\frac{\partial\rho(\mathbf{r}')}{\partial N}\right]_{v(\mathbf{r}')} \\ &= \int \mathrm{d}\mathbf{r} \int \mathrm{d}\mathbf{r}' f(\mathbf{r})\eta(\mathbf{r},\mathbf{r}')f(\mathbf{r}'),\end{aligned} \tag{8.15}$$

where $\eta(\mathbf{r},\mathbf{r}')$ is the hardness kernel [46, 47]

$$\eta(\mathbf{r},\mathbf{r}') = \frac{\delta^2 F[\rho(\mathbf{r})]}{\delta\rho(\mathbf{r})\delta\rho(\mathbf{r}')}, \tag{8.16}$$

$$f(\mathbf{r}) = \left[\frac{\partial\rho(\mathbf{r})}{\partial N}\right]_{v(\mathbf{r}')} \tag{8.17}$$

is the Fukui function and $F[\rho(\mathbf{r})]$ is the Hohenberg–Kohn universal density functional (Chapter 5)

$$F[\rho(\mathbf{r})] = T[\rho(\mathbf{r})] + J[\rho(\mathbf{r})] + E_{\mathrm{xc}}[\rho(\mathbf{r})], \tag{8.18}$$

Equation (8.15) may be interpreted as a self-similarity of the Fukui function with the hardness kernel $\eta(\mathbf{r},\mathbf{r}')$ as the corresponding separation operator. Simple models for the hardness kernel, such as the Dirac delta form $\eta(\mathbf{r},\mathbf{r}') \approx (\delta r - r')$ and the Coulomb form $\eta(\mathbf{r},\mathbf{r}') \approx |\mathbf{r}-\mathbf{r}'|^{-1}$, have been shown to give good approximations for the global hardness and reasonable trends for the quantum similarities of atoms and molecules [48–50]. The Coulomb form is, in fact, the leading contribution to the hardness kernel, arising from the dominant, classical Coulomb repulsion term $J[\rho(\mathbf{r})]$ in the expression (Eq. 8.18) for $F[\rho(\mathbf{r})]$

$$J[\rho(\mathbf{r})] = 1/2\frac{\rho(\mathbf{r})\rho(\mathbf{r}')}{|\mathbf{r}-\mathbf{r}'|}. \tag{8.19}$$

Boon et al. [51] introduced a similarity index based on the local softness

$$S(\mathbf{r}) = \left[\frac{\partial\rho(\mathbf{r})}{\partial N}\right]_{v(\mathbf{r})}\left[\frac{\partial N}{\partial\mu}\right]_{v(\mathbf{r})}, \tag{8.20}$$

where

$$S = \left[\frac{\partial N}{\partial\mu}\right]_{v(\mathbf{r})} \tag{8.21}$$

is the global softness (Chapter 5). This gives a Carbó-type similarity index

$$C_{\mathrm{AB}}^{\mathrm{S}} = \frac{\int s_{\mathrm{A}}(\mathbf{r})s_{\mathrm{B}}(\mathbf{r})\,\mathbf{dr}}{\sqrt{\int s_{\mathrm{A}}^2(\mathbf{r})\,\mathbf{dr}\int s_{\mathrm{B}}^2(\mathbf{r})\,\mathbf{dr}}}, \tag{8.22}$$

$$= \frac{\int f_{\mathrm{A}}(\mathbf{r})f_{\mathrm{B}}(\mathbf{r})\,\mathbf{dr}}{\sqrt{\int f_{\mathrm{A}}^2(\mathbf{r})\,\mathbf{dr}\int f_{\mathrm{B}}^2(\mathbf{r})\,\mathbf{dr}}}, \tag{8.23}$$

which depends only on the Fukui functions. The corresponding Hodgkin and Richards-type similarity index

$$H_{\mathrm{AB}}{}^{\mathrm{S}} = \frac{2\int s_{\mathrm{A}}(\mathbf{r})s_{\mathrm{B}}(\mathbf{r})\,\mathbf{dr}}{\int s_{\mathrm{A}}^2(\mathbf{r})\,\mathbf{dr}+\int s_{\mathrm{B}}^2(\mathbf{r})\,\mathbf{dr}}, \tag{8.24}$$

depends on both the Fukui functions and the global softness. These reactivity-based similarity indices capture different and complementary information to that based on the electron densities alone. Further applications of reactivity-based indices are described in Chapter 7.

All these molecular quantum similarity measures depend on the relative positions and orientations of the molecules in three-dimensional space. The problem of molecular alignment thus requires serious consideration in molecular similarity assessment. The molecular alignment protocol is not only arbitrary, algorithm dependent, and thus not unique, but also computationally intensive for all but the smallest of molecules. The simplest commonly used protocols include superposition of the center of charge, or for small molecules, superimposing only the heavy atoms. The algorithm dependence of the molecular quantum similarity measures can be illustrated for the case of similarity analysis of the (R)- and (S)-enantiomers of CHFClBr, as given in Table 8.1. Superimposing different atoms

TABLE 8.1 Global Similarity Analysis of R and S Enantiomers of CHFClBr Using the Hirshfeld Partitioning Scheme[a]

Superimposed Atoms	Carbó Index C_{RS}
Cl, C, Br	0.990
F, C, Br	0.915
H, C, Br	0.906
F, C, Cl	0.098
H, C, Cl	0.089
H, C, F	0.054

Superimposing different atoms leads to different similarity measures, with the highest global similarity obtained when the atoms with the most electrons (Cl and Br) are superimposed.

[a] Adapted from References 20 and 53.

leads to very different similarity measures, with the highest global similarity obtained when the atoms with the most electrons (Cl and Br) are superimposed [52, 53]. Alignment based on a structural motif or pharmacophore (a 3D descriptor designed to represent a spatial combination of features responsible for a drug's geometry-dependent biological activity), such as the topogeometrical superposition algorithm [54], is useful for systems with a common functional group or substructure. None of these alignment schemes is more "correct" than another, just as no similarity index is more "correct" than another. It is important to choose an alignment protocol based on the goals of the analysis. Since electron densities are heavily concentrated around the nuclei, similarity indices based on the density can give exaggerated weight to small mismatches in regions of space with high electron density, for example, in the vicinity of nuclei, when comparing molecules with slightly different geometries. Bultinck et al. [55] introduced the QSSA algorithm that aligns molecules by maximizing their quantum similarity measure. This method gives internally consistent, globally maximal similarities, but this is achieved by superimposing the heaviest atoms and not necessarily the most similar functional groups.

Spatial autocorrelation functions [56] can be used to circumvent the need for molecular alignment, thus eliminating one of the major computational bottlenecks in molecular quantum similarity analysis

$$A(R_{xy}) = (1/n)\Sigma_{x,y}\mathrm{P}_x\mathrm{P}_y, \tag{8.25}$$

where P_x and P_y are properties of points x and y on the molecular van der Waals surface, binned by the distance R_{xy} between them. Alternatively, topological autocorrelation functions [57] can be defined through the relation

$$A(d) = \Sigma_i\mathrm{P}_i\mathrm{P}_{i+d}, \tag{8.26}$$

where P_i and P_{i+d} are atomic properties of two atoms i and $i + d$ separated by d bonds. Boon et al. [58] studied both types of autocorrelation functions using the electron density $\rho(\mathbf{r})$, the MEP $v_e(\mathbf{r})$, and the local softness $s(\mathbf{r})$. Similarity measures may be constructed from these autocorrelation functions either by computing the Euclidean distance D between a pair of molecules

$$D_{AB} = \sqrt{\sum_d [A_A(d) - A_B(d)]^2} \tag{8.27}$$

or by constructing Carbó or Hodgkin-Richards-type similarity indices from them

$$C'_{AB} = \frac{\Sigma_d A_A(d) A_B(d)}{\sqrt{\Sigma_d A_A^2(d) \Sigma_d A_B^2(d)}}, \tag{8.28}$$

$$H'_{AB} = \frac{2\Sigma_d A_A(d) A_B(d)}{\Sigma_d A_A^2(d) + \Sigma_d A_B^2(d)}. \tag{8.29}$$

In many applications in chemistry and biology, it is the similarity of certain parts of the molecules, rather than their global similarity, that is of interest. Since information on the global electron density is contained in any finite subdomain Ω, one can define local similarity indices, such as the local Carbó overlap similarity index

$$C_{AB}^{(\Omega)} = \frac{\int_\Omega \rho_A(\mathbf{r})\rho_B(\mathbf{r})\,d\mathbf{r}}{\sqrt{\int_\Omega \rho_A^2(\mathbf{r})\,d\mathbf{r} \int_\Omega \rho_B^2(\mathbf{r})\,d\mathbf{r}}}. \tag{8.30}$$

Investigation of $C_{AB}^{(\Omega)}$ for the (R)- and (S)-enantiomers of CHFClBr [52, 53] with Hirshfeld atomic subdomains (Chapter 4) reveals that the local electron density is not superimposable ($C_{AB}^{(\Omega)} \neq 1$), even in atomic regions other than the asymmetric carbon. Information about chirality can thus be obtained from the electron density in any arbitrary finite subdomain, as required by the holographic electron density theorem.

Girones and Ponec [59] used the domain-averaged Fermi hole density $g_\Omega(\mathbf{r})$ [60, 61]:

$$g_\Omega(\mathbf{r}_1) = N_\Omega \rho(\mathbf{r}_1) - 2\int_\Omega \rho(\mathbf{r}_1, \mathbf{r}_2) d\mathbf{r}_2, \tag{8.31}$$

$$\text{where } N_{(\Omega)} = \int_\Omega \rho_A(\mathbf{r}) d\mathbf{r} \tag{8.32}$$

is the mean number of electrons in the domain Ω and $\rho_A(\mathbf{r}_1,\mathbf{r}_2)$ is the pair density, to define the fragment molecular quantum self-similarity measure for fragment A

$$Z_{AA}^{(\Omega)} = \int_\Omega g_\Omega^A(\mathbf{r}) g_\Omega^A(\mathbf{r}) d\mathbf{r}, \tag{8.33}$$

Matta [62] proposed using the integral of the Laplacian of the electron density (chapter 4) to define a similarity index for a pharmacophore that reflects the reactivity of the molecule without being biased by small nuclear cusp mismatches:

$$R_{\Omega\cup\Omega'} = \frac{\displaystyle\int\limits_{\substack{\Omega\cup\Omega' \\ \nabla^2\rho(\boldsymbol{r})\geq 0}} \nabla^2\rho_\Omega \nabla^2\rho'_\Omega \mathrm{d}v}{\sqrt{\displaystyle\int\limits_{\substack{\Omega \\ \nabla^2\rho(\boldsymbol{r})\geq 0}} \nabla^2\rho_{\Omega'} \mathrm{d}v \int\limits_{\substack{\Omega' \\ \nabla^2\rho(\boldsymbol{r})\geq 0}} \nabla^2\rho_\Omega \mathrm{d}v}}. \tag{8.34}$$

8.3 MOLECULAR SIMILARITY MEASURES FROM CRITICAL POINTS OF THE ELECTRON DENSITY

Determination of similarity indices by superposition of densities (or density-derived properties) at all points in space is computationally expensive. Furthermore, as discussed above, such superposition is arbitrary, not uniquely defined and severely biased by the density contributions from the atomic core regions. Paul Popelier [63, 64] proposed a molecular similarity measure based on the topology of the electron density field. Popelier's scheme, called quantum topological molecular similarity (QTMS) [65, 66], utilizes properties (such as the electron density ρ_b, Laplacian $\nabla^2\rho_b$, bond ellipticity ε_b, and kinetic energy K_b) of the bond critical points of the electron density, thereby avoiding both the computationally intensive spatial integrations and molecular alignment problems of Carbó-type similarity indices; it is also not dominated by contributions from the core densities. These bond critical point properties can be used to compute a Cartesian distance measure between any pair of critical points (and thus between any pair of molecules sharing a common skeleton) in an abstract bond critical point space [63, 64] or utilized as descriptors in a PLS regression equation [65, 66] to construct a QSAR model for the property of interest. Popelier's first analysis [63] simply mapped all the critical points in a set of molecules onto an abstract 3D Cartesian space, called the *BCP space* (Fig. 8.3), the axes of which were the bond critical point properties ρ_b, $\nabla^2\rho_b$, and ε_b. It was demonstrated that a simple Cartesian distance measure in this BCP space yielded a good regression for the Hammett [6, 7] σ parameter of para-substituted benzoic acids. In subsequent analyses, Popelier et al. used the PLS method [67–69] to construct latent variables from the bond critical point descriptors that maximize the variance in the set of data. These latent variables were then regressed against the molecular property to be predicted, such as the Hammett σ parameter [65], the bond dissociation energy [66], the acid dissociation constant pK_a [70], or the proton affinity pK_b [71]. This method, termed *QTMS* [65], is also successful at identifying a molecular fragment containing the

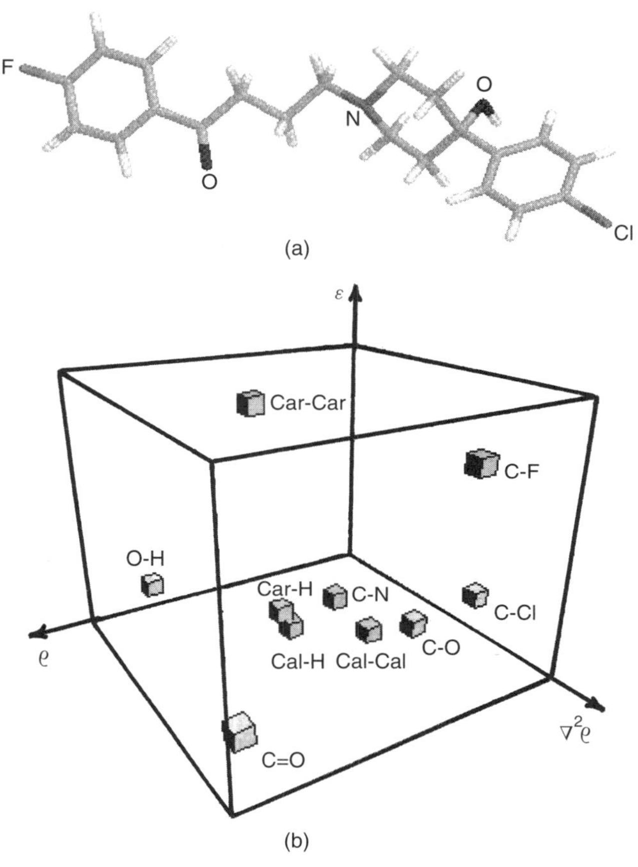

Figure 8.3 The critical points in a set of molecules mapped onto the BCP space [63]. (Reproduced with permission from Popelier PA. J Phys Chem A 1999;103:2883–2890, Copyright 1999 American Chemical Society.)

active center responsible for the QSAR. In other words, the molecular fragment responsible for a given biological activity is not determined beforehand but is obtained from the ranking of features in the PLS analysis. For instance, Figure 8.4 shows a variable importance plot of features from a PLS analysis for the Hammett σ parameter of trisubstituted benzoic acids [65]—the most important features arise from the BCP descriptors of bonds in the carboxyl group. Similarity metrics employing BCP properties are most useful when the molecules to be compared have a common structural core containing a common number of critical points.

It should be reiterated that critical points obtained from experimental electron density distributions depend on the level of crystallographic resolution. Topological analysis of molecular electron density maps at atomic resolution

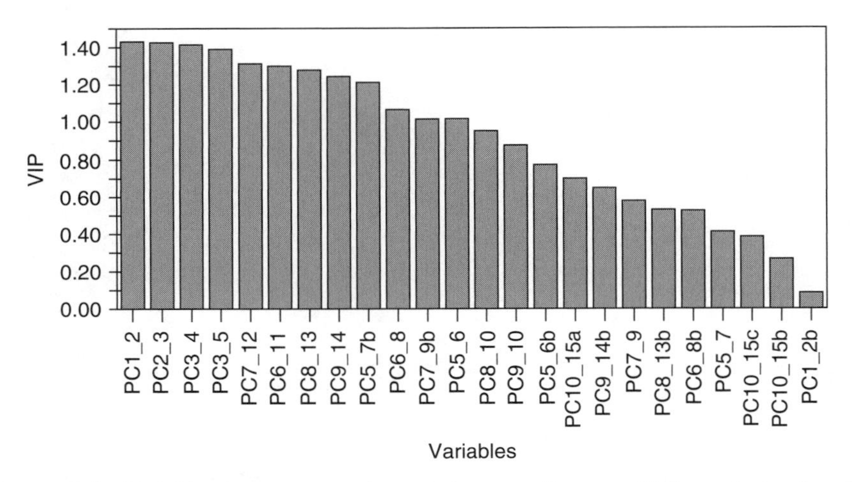

Figure 8.4 Variable importance plot of features from a PLS analysis for the Hammett σ parameter of trisubstituted benzoic acids [65]. (Reproduced with permission from O'Brien SE, Popelier PA. J Chem Inf Model 2001;41:764–775, Copyright 2001 American Chemical Society.)

shows peaks and [3, −1] saddle points at the positions of atomic nuclei and chemical bonds, respectively (Chapter 4). At lower crystallographic resolutions, as commonly obtained from X-ray diffraction data for biological macromolecules, these critical points can merge and the critical points seen can be representative of groups of atoms rather than the atoms themselves. For example, Figure 8.5 shows the critical point graphs obtained from topological analysis at (a) 2.5 Å and (b) 3.0 Å resolution maps of the electron density for four benzodiazepine-type molecules. Leherte et al. [72] developed a method for similarity comparison of molecular electron density distributions at various levels of crystallographic resolution from critical point analysis. In this method, the critical points are converted to fully connected graphs and each graph represented by a 2D matrix, whose diagonal elements are the electron density values at the critical points (both peaks and saddle points, with different kinds of critical points appropriately weighted) and the off-diagonal elements are the Cartesian distances between pairs of critical points. The root mean squared deviation between these matrices for each pair of molecules is then minimized, using either genetic algorithms or Monte Carlo simulated annealing, to determine the best alignment. They found that at 3.0 Å resolution, molecules with similar affinities to a biological receptor adopt a similar topology of their pharmacophore elements, leading to the conclusion that recognition of a ligand by a receptor is a medium resolution phenomenon. Molecules capable of binding to a receptor present common topological features at various levels of resolution.

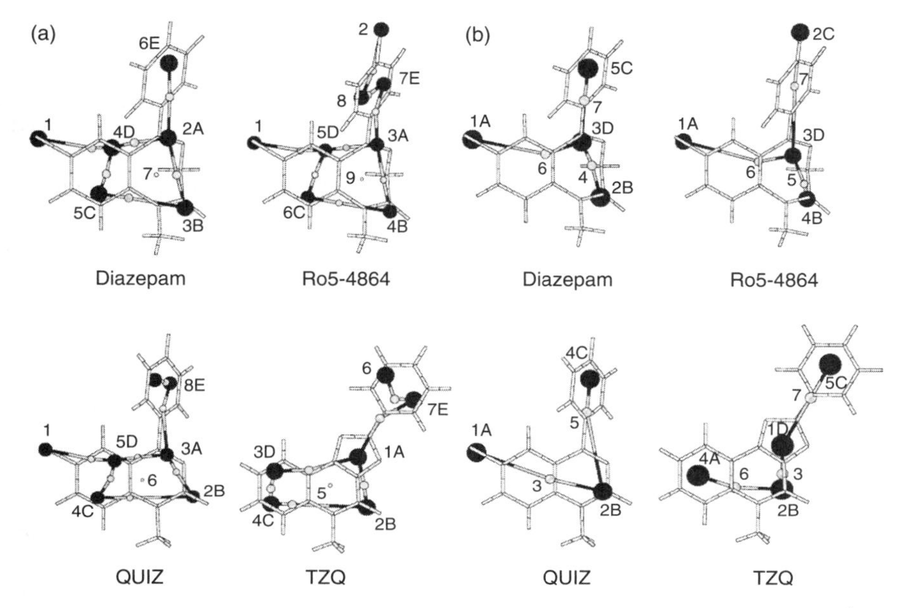

Figure 8.5 The critical point graphs obtained from topological analysis at (a) 2.5 Å and (b) 3.0 Å resolution maps of the electron density for four benzodiazepine-type molecules [72]. (Reproduced with permission from Leherte L, Meurice N, Vercauteren DP. J Chem Inf Comput Sci 2000; 40: 816–832, Copyright 2000 American Chemical Society.)

8.4 ELECTRON-DENSITY-DERIVED MOLECULAR SURFACE DESCRIPTORS

The primary drawback of molecular descriptors and similarity indices derived from electron density distributions is the intensive computational effort required to generate them through *ab initio* quantum chemical calculations. The transferable atom equivalent (TAE) method [73–75] circumvents this computational bottleneck through the use of Bader's Atoms in Molecules formalism [76] (Chapter 4). By precomputing an extensive library of transferable atomic fragment densities (the TAE library) from *ab initio* wave functions, the RECON algorithm [74, 75] exploits the theory of atoms in molecules for rapid, high throughput computation of molecular electronic properties from the atomic charge density fragments stored in the TAE library. Recall that atomic fragments that satisfy Bader's virial partitioning prescription have well-defined properties that are approximately additive and transferable from one molecule to another. Molecular descriptors can be constructed in most cases by simple arithmetic operations on the respective atomic descriptors stored in the TAE library, making the method well suited for virtual high throughput screening applications.

Rather than using the density itself for molecular comparisons, TAE descriptors encode the distributions of electron-density-based molecular properties, such as kinetic energy densities [76], local average ionization potentials [77–80], electrostatic potentials [31–36], Fukui functions [40–45], electron density gradients, and second derivatives or Laplacian distributions [76]. These density-derived properties are much more sensitive indicators of the local chemical environment than is the density itself. For applications involving biomolecular recognition and other noncovalent interactions, it is most useful to employ descriptors mapped onto the van der Waals surface of the molecule. In computational applications, the van der Waals surface is well approximated by an electron density isosurface with $\rho(\mathbf{r}) = 0.2$ electrons per cubic Bohr. Surface integrals, extrema, and histogram bins of density-derived surface properties have been employed as descriptors. In addition, integrated atomic properties (integrated over the atomic basins) are also employed, often as autocorrelation functions (see below). TAE descriptors are generally used in conjunction with modern machine learning (classification or regression) techniques [69, 81–90], a training set of molecules with known activity, feature selection algorithms [91, 92], and rigorous validation protocols [93–96] to generate predictive models within a well-defined domain of applicability [12, 97]. Applications to the prediction of small molecule toxicity [75], polymer glass-transition temperatures [75], chromatographic column retention times and selectivity [98–102], antibody binding sites on antibiotics [36], transcription factor binding sites on DNA sequences [103], and classification of odorants [104] have been reported.

One of the most commonly used and well-known electron-density-derived descriptors is the MEP (Eq. 8.8). The van der Waals surface integrals, extrema, and histogram bins of $\upsilon_{\mathrm{e}}(\mathbf{r})$ have been commonly employed as descriptors [32–36, 83, 84, 98–102]. Quantitative correlations of the values of electrostatic potential minima with the carcinogenic activity of molecules [31] showed the promise of this descriptor for biological and environmental applications and paved the way for its extensive use [33] in QSAR and drug design. Histogram bins of $\upsilon_{\mathrm{e}}(\mathbf{r})$ also map the polar and hydrophobic regions of a molecule. Electropositive and electronegative regions of the molecular surface are represented by the high and low histogram bins of MEP, respectively, while the middle bins correspond to hydrophobic regions. The electrostatic potential is associated with many other molecular and intermolecular phenomena, including acid–base interactions, solvation behavior, and pK_{a} correlations [33–35], as well as protein–ligand and protein–DNA binding.

Perhaps the simplest electron-density-derived descriptor, besides the density itself, is the gradient of the electron density normal to the molecular van der Waals surface ($\nabla\rho\cdot\mathbf{n}$, where n is the surface normal). $\nabla\rho\cdot\mathbf{n}$ has been employed to distinguish soft, polarizable regions of the electron density from regions where the electron density is more tightly bound. The values of $\nabla\rho\cdot\mathbf{n}$ are much smaller over electron-rich π systems and aromatic rings than over polarized or electron-deficient alkyl carbons. Since $\rho(\mathbf{r})$ decreases away from the attractors, $\nabla\rho{\cdot}\mathbf{n}$ is always negative; large negative values of $\nabla\rho{\cdot}\mathbf{n}$ indicate that the electron density of the underlying molecular region is more tightly held and less likely to extend very far from the molecule.

In Chapter 4, we have seen that the electronic kinetic energy density distribution can be represented in either the gradient form

$$G(\mathbf{r}) = -\nabla\psi^* \cdot \nabla\psi, \tag{8.35}$$

or the Schrödinger form

$$K(\mathbf{r}) = -(\psi^*\nabla^2\psi + \psi\nabla^2\psi^*). \tag{8.36}$$

Either or both of these quantities may be employed as molecular descriptors. Imbalance between K and G is responsible for nonzero values of the Laplacian ($\nabla^2\rho$), the trace of the second-derivative matrix of the electron density at any point in space

$$L(\mathbf{r}) = -\frac{1}{4}\nabla^2\rho(\mathbf{r}) = K(\mathbf{r}) - G(\mathbf{r}). \tag{8.37}$$

The Laplacian has been extensively studied by Bader et al. [76], who identified this quantity as a useful descriptor in quantifying donor/acceptor interactions and selectivity toward electrophilic aromatic substitution [76, 105]. Since the Laplacian reflects overall normalization of the molecular electron density ($\int \mathrm{L}(\mathbf{r})d\mathbf{r} = 0$), Laplacian peaks in the outer core and valence regions are often matched by "shadows" of these internal Laplacian extrema, of opposite sign, on the molecular van der Waals surface. Thus, less negative regions of surface values of K often indicate the presence of Brönsted bases. The rate of change of the electronic kinetic energy density normal to and away from the molecular surface ($\nabla K\mathrm{n}$) has also been used to describe differences in the polarizability and hydrophobicity of molecular regions [76, 106, 107].

We have already encountered the application of the Fukui reactivity indices (Eq. 8.17 above and Chapter 7) to molecular similarity analysis. Since the number of electrons N can vary in integer increments, the derivative in Equation (8.17) is discontinuous and thus not uniquely defined. One can define three different Fukui indices depending on whether one takes the limit as $\delta N \to 0^+$, $\delta N \to 0^-$ or the average of the two limiting expressions. Using the Koopmans theorem, these indices may be approximated as

$$f^-(\mathbf{r}) = \left[\frac{\partial\rho(\mathbf{r})}{\partial N}\right]_{\mathrm{v}(\mathbf{r});\ \delta N\to 0-} \approx \rho_{\mathrm{HOMO}}(\mathbf{r}), \tag{8.38}$$

$$f^+(\mathbf{r}) = \left[\frac{\partial\rho(\mathbf{r})}{\partial N}\right]_{v(\mathbf{r});\ \delta N\to 0+} \approx \rho_{\mathrm{LUMO}}(\mathbf{r}), \tag{8.39}$$

and

$$f^0(\mathbf{r}) \approx 1/2[\rho_{\mathrm{HOMO}}(\mathbf{r}) + \rho_{\mathrm{LUMO}}(\mathbf{r})], \tag{8.40}$$

which describe reactivity toward electrophilic, nucleophilic, and radical attack, respectively.

One of the most interesting descriptors derived from the electron density is the local average ionization potential, $\overline{I}$, of Politzer et al. [77–80]:

$$\overline{I}(\mathbf{r}) = \sum_i \frac{\rho_i(\mathbf{r})|\varepsilon_i|}{\rho(\mathbf{r})}, \tag{8.41}$$

where the summation runs over the occupied orbitals, ε_i are the orbital energies, and the molecular surface is thus encoded with energy-weighted orbital densities ρ_i. $\overline{I}(\mathbf{r})$ identifies hard regions of the electron density (represented by maxima or high bins of local average ionization potential on the molecular surface) and soft, electron-donor, or hydrogen-bond-acceptor regions (corresponding to minima or low bins of $\overline{I}(\mathbf{r})$ on the surface). $\overline{I}(\mathbf{r})$ has been found to be important in describing differential solubility, donor/acceptor, and hydrophobic/hydrophilic interaction tendencies, as well as intermolecular binding modes, such as induced-dipole and protein–DNA interactions. In a variant of the TAE method, $\overline{I}(\mathbf{r})$ and MEP have been employed to reconstruct the chemical properties of DNA sequences through electron density characterization of DNA fragments consisting of three stacked base pairs [103]. *Ab initio* electronic structure calculations were first performed on all possible sets of three stacked base pairs, with the central base pair residing in the electronic environment of the flanking base pairs. The local electronic environments induced by neighboring base pairs were observed to have a strong influence on electronic properties, such as $v_{\mathrm{e}}(\mathbf{r})$ and $\overline{I}(\mathbf{r})$. The surface electron density properties of the central base pair of each triplet constituted a library of base pairs (with their local electronic environment), which was then employed to reconstruct the properties of any arbitrary DNA sequence. Figure 8.6 shows a rectangular grid of pixels on the major groove surface of DNA (the "Dixel" representation), colored by discretized values of $\overline{I}(\mathbf{r})$ and MEP.

In a similar manner to the local ionization potential, the local electron affinity may be defined as

$$\overline{E}(\mathbf{r}) = \frac{-\sum_{i=\mathrm{LUMO}}^{N} \rho_i(\mathbf{r})|\varepsilon_i|}{\sum_{i=\mathrm{LUMO}}^{N} \rho_i(\mathbf{r})} \tag{8.42}$$

This and other descriptors from conceptual DFT (Chapter 7), such as the electronegativity

$$\chi(\mathbf{r}) \approx 1/2[\overline{I}(\mathbf{r}) + \overline{E}(\mathbf{r})], \tag{8.43}$$

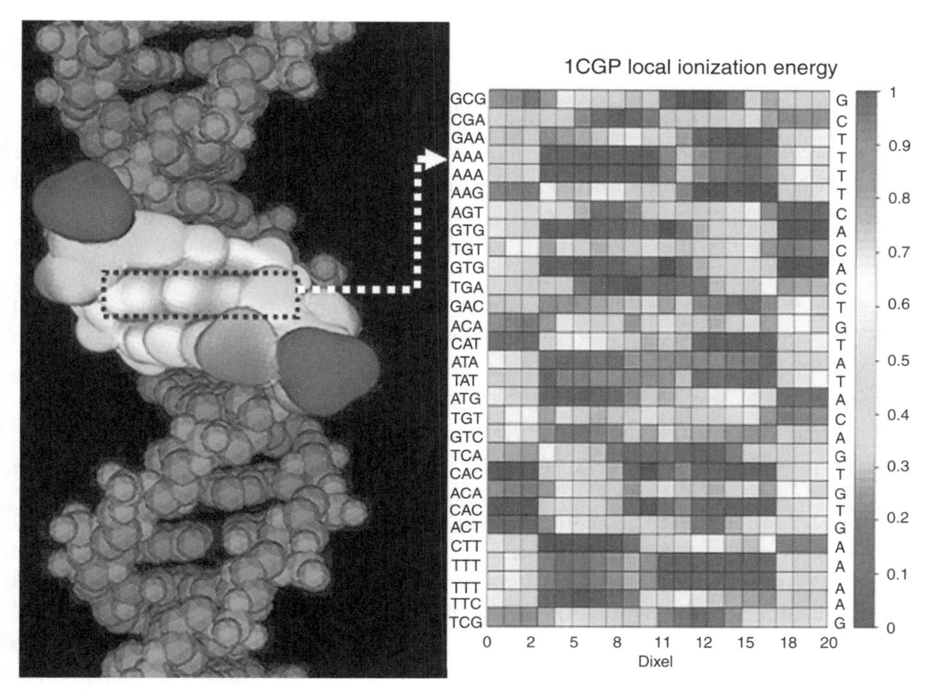

Figure 8.6 The "Dixel" representation: a rectangular grid of pixels on the major groove surface of DNA, colored by discretized values of $\overline{I}(\mathbf{r})$ and MEP [103]. (Reproduced with permission from Sukumar N, Krein M, Breneman CM. Curr Opin Drug Discov Dev 2008;11(3):311–319, Copyright 2008 Thomson.) *See insert for color representation of the figure.*

local hardness

$$\eta(\mathbf{r}) \approx 1/2[\overline{I}(\mathbf{r}) - \overline{E}(\mathbf{r})], \tag{8.44}$$

and local polarizability

$$\alpha(\mathbf{r}) = \frac{-\sum_{i=1}^{N} \rho_i'(\mathbf{r}) q_i \overline{a}_i(\mathbf{r})}{\sum_{i=1}^{N} \rho_i'(\mathbf{r}) q_i}, \tag{8.45}$$

have been employed within the framework of semiempirical MO theory [13, 14]. Ehresmann found that the local electron affinity, local hardness, and local polarizability showed little correlation with other descriptors in common use and thus these descriptors effectively extend the variance of the descriptor set [13, 14]. The use of such local electron-density-derived descriptors increases the likelihood of scaffold hopping (i.e., switching from one structural type to another) in QSAR and virtual screening applications and can thus lead to more robust and general QSPR models.

8.5 ALIGNMENT-FREE MOLECULAR SHAPE AND ELECTRONIC PROPERTY DESCRIPTORS

Histogram distributions of molecular surface properties generated by TAE reconstruction from fragments are insensitive to molecular conformation, while the corresponding properties from *ab initio* methods are much more computationally intensive to generate. Molecular interaction fields are not only computationally intensive but also alignment dependent. Three-dimensional shape information can, however, be incorporated within the TAE RECON formalism in a computationally inexpensive implementation (RECON Autocorrelation Descriptors) [75] through property autocorrelation functions of the type (Eq. 8.25), where P_x and P_y are atomic properties of atoms x and y, binned by the distance R_{xy} between them for each TAE property P. An analogous scheme using autocorrelation functions of sequence-derived properties has also been employed for amino acid sequences characterizing proteins [108], employing Equation (8.26), where P_i and P_{i+d} are the values of a particular property for a pair of amino acids i and $i + d$ separated by d residues along the protein sequence. Autocorrelation descriptors measure the correlation of a property with itself measured along the sequence (topological, conformation-insensitive autocorrelations) or through 3D space (spatial, conformation-sensitive autocorrelations).

Constructing a shape signature of electron-density-derived properties on the molecular surface is another means of retaining conformation and shape sensitivity without the need for molecular alignment. The Zauhar shape signatures method [109, 110] uses a ray-tracing procedure within the interior of the molecular envelope (defined by either the van der Waals or solvent-accessible surface). Ray-length and angle-of-reflection information are recorded at each point of intersection of the ray with the surface (Fig. 8.7). This encoding of molecular shape through the distribution of ray lengths rapidly generates a compact fingerprint (shape signature) for each molecule without computationally intensive 3D alignments between molecules. The property-encoded surface translator (PEST) [111] method extends this formalism by recording TAE surface property information at each point of intersection of the ray with the molecular surface. Descriptors are encoded as two-dimensional histograms and wavelet coefficients [112] and used directly in machine learning algorithms for both similarity assessment and QSAR/QSPR. Descriptors incorporating both shape and electronic property information may thus be generated not only for whole molecules but also for molecular fragments and binding pockets. Property-encoded shape distributions (PESD) [113–115] have been used to compare similarities in shape and property distributions on the surfaces of protein binding sites in a sequence and fold-independent manner. Shapes and electronic properties of protein binding sites are often conserved without significant conservation of the amino acid residues in the sequence. PESD signatures (Fig. 8.8) employed MEP and MLP (molecular lipophlicity potential) to find binding sites of high similarity even in the absence of sequence similarity and to predict the binding affinities of protein–ligand complexes.

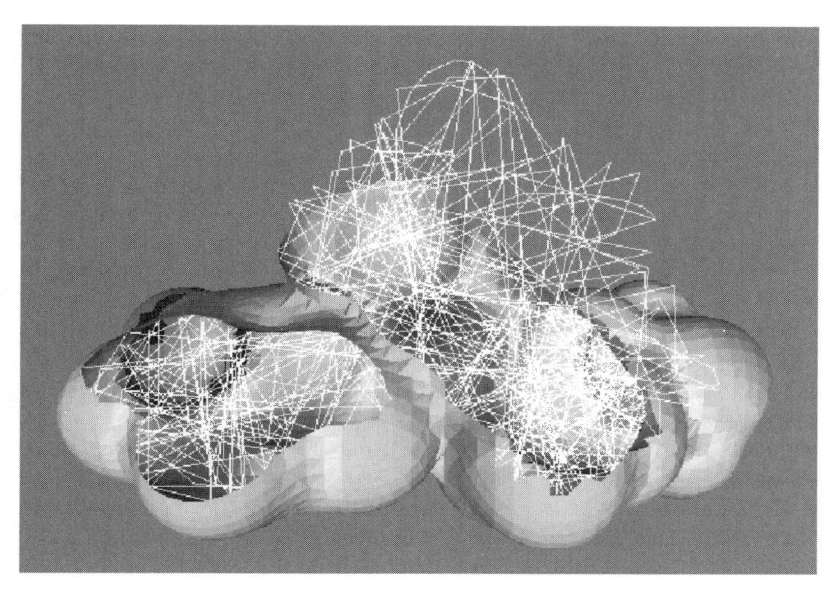

Figure 8.7 PEST ray trace showing ray length and angle-of-reflection information recorded at each point of intersection of the rays with the molecular surface [111]. (Reproduced with permission from Breneman CM, Sundling CM, Sukumar N, Shen L, Katt WP, Embrechts MJ. J Comput Aided Mol Des 2003;17:231–240, Copyright 2003 Springer.) *See insert for color representation of the figure.*

Property moments may also be used to define descriptors combining shape and electronic property information [75]. The ultrafast shape recognition (USR) similarity search tool developed by Ballester and Richards [116–118] employs molecular shape moments with respect to a small set of well-defined points within a molecule: namely the centroid (*ctd*), the closest atom to the *ctd* (*cst*), the farthest atom from the *ctd* (*fct*), and the farthest atom from the *fct* (*ftf*). The USR algorithm is also alignment-free and extremely fast, well suited for high throughput applications and has been shown to perform well at shape classification [118]. Combining electrostatic information with shape recognition methods [75, 119, 120] incorporates electrostatic complementarity into the description of molecular interactions. In RECON [75], TAE property moments with respect to *ctd, cst, fct*, and *ftf*

$$\mathrm{USP}_{mk} = \sum_i \mathrm{P}_i \mathrm{R}_{ik}^m \tag{8.46}$$

are computed for each TAE property P, where the summation i runs over all atoms in the molecule, $m = 1, 2, 3$ (corresponding to first, second, and third moments), and $k = ctd, cst, fct, ftf$, thereby generating rapid shape-electronic-property hybrid descriptors for high throughput screening applications. Adding both chirality and electrostatic complementarity to USR has been shown to result

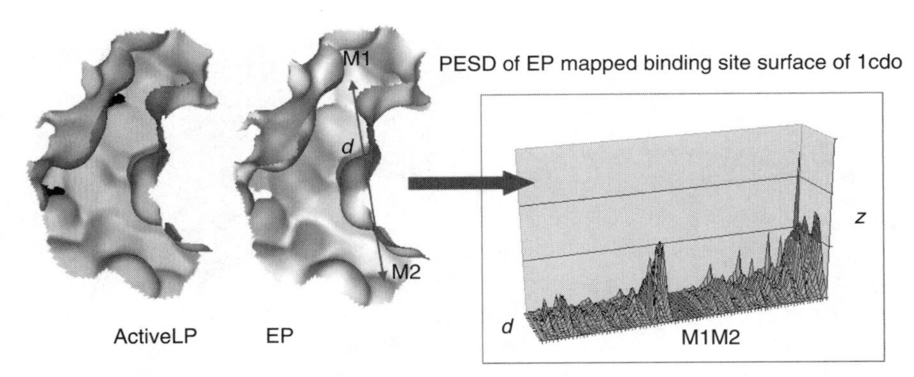

Figure 8.8 Construction of PESD signatures employing MEP and MLP [114]. (Reproduced with permission from Das S, Krein MP, Breneman CM. J Chem Inf Model 2010; 50 (2): 298–308, Copyright 2010 American Chemical Society.) *See insert for color representation of the figure*.

in significant enrichment in virtual screens [119, 120]. The ability to distinguish enantiomers is thus of utmost importance in the study of biomolecule interactions and in drug design. Nine of the top ten drugs on the market today have chiral active ingredients. There are several drugs (such as the β-blocker Propanolol), where one enantiomer is several orders of magnitude more potent than the other, and even some drugs (such as L-Dopa, used to treat Parkinson's disease) where the enanctiomer (R-Dopa) is toxic! Armstrong's chiral shape recognition (CSR) method builds on USR without significant computational overhead by including moments with respect to a fourth centroid, defined through a cross product operation: parity inversion changes the signs of all coordinates except that of the fourth centroid, so that the moments with respect to this fourth centroid are different for any chiral molecule and its enantiomer.

8.6 NETWORK GRAPHS FROM MOLECULAR SIMILARITY

Let us conclude this chapter by bringing the discussion back to where we started, namely the molecular similarity principle in drug design, which provides much of the motivation for the study of molecular similarity. As we have seen, the enormous size of chemistry space makes its thorough exploration impossible. Therefore, a key consideration in drug design is devising strategies to optimally direct research efforts toward regions of chemistry space that are most likely to contain molecules with useful biological activity. The regions of chemistry space that have been explored through experimental investigations—or even through detailed computations—are extremely limited and constitute an obviously biased sample. Chemists isolate, synthesize, and study molecules for a variety of intellectual, sociological, and practical reasons. Thus it is not even clear whether different regions of chemistry space or chemistry spaces constructed using different descriptors or different similarity metrics need have any common characteristics or whether the network topology of chemistry spaces should resemble that of biological networks or of social networks.

Any of the molecular similarity measures discussed in this chapter can be employed to construct a network representation of chemistry space, where each molecule is a node of the network and a discretized similarity measure is used to define the edges. The degree distribution $P(k)$ of a network is the probability that a given node in a network has exactly k links or connections to other nodes. Investigations of a number of chemistry space networks using a variety of similarity measures, including electron-density-based ones with the TAE method [121], reveal a heavy tail degree distribution [121–123] characteristic of a small-world network, where the probability that a node has k links approximates a power-law degree distribution $P(k) \sim k^{-\gamma}$. Such distributions appear linear on a plot of log $P(k)$ versus log k (Fig. 8.9), and nodes whose degrees deviate significantly from the average degree are extremely rare. The properties of such a scale-free network are often determined by a relatively small number of highly connected nodes (called *hubs*). Hubs in chemistry space are represented by molecules with high

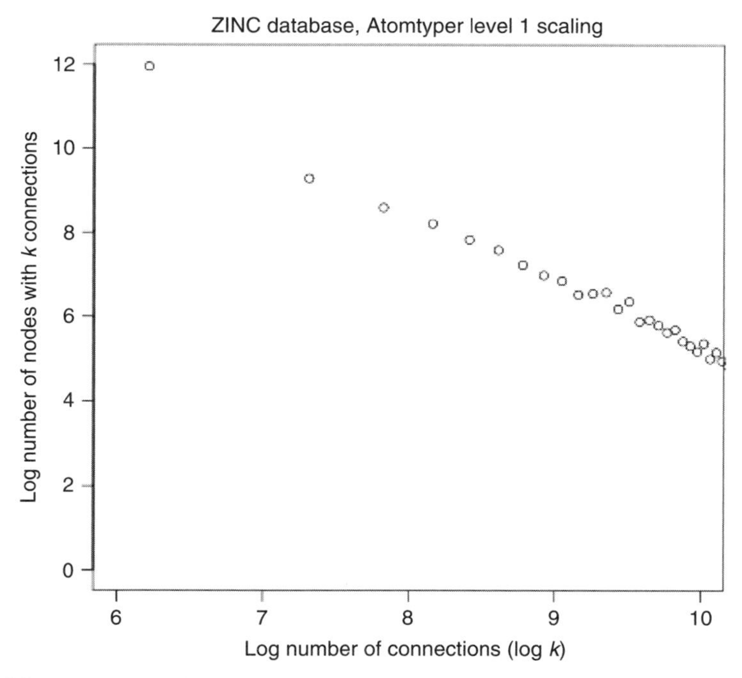

Figure 8.9 The probability that a node of a chemical space network has k links approximates a power-law degree distribution $P(k) \sim k^{-\gamma}$, as shown by the linear plot of log $P(k)$ versus log k [121]. (Reproduced with permission from Krein MP, Sukumar N. J Phys Chem A 2011;11:6, Copyright 2011 American Chemical Society.)

leverage in structure–activity relationship models. Such molecules are important for maintaining the diversity of a chemical library and for ensuring good predictive performance of structure–activity relationship models across a wide domain of applicability. This ability to identify diverse structures spanning very different bond frameworks or structural scaffolds with similar activities (scaffold hopping) is of great importance for drug design.

Relating chemical similarity to similarity in biological activity produced by the molecules introduces another level of complication [124]. The similarity principle, that similar molecules should exhibit similar activities in biological assays, is a fundamental assumption implicit in QSAR studies. Deviations from the similarity principle [2] lead to the failure of QSAR models [125]. Maggiora [3] postulated that such deviations arise on account of the complex nature of the activity landscape associated with biological assays, and he coined the term "activity cliffs" to characterize such regions of the structure–activity landscape. In Maggiora's topographical metaphor, smooth regions of the structure–activity landscape (either flat, like Kansas, or like the rolling hills of England) are those that best satisfy the similarity principle. Activity cliffs may be characterized using measures such as the structure–activity landscape index (SALI) [126, 127]

$$\mathrm{SALI}_{i,j} = \frac{|A_i - A_j|}{\{1 - \mathrm{sim}(i, j)\}}, \tag{8.47}$$

where A_i and A_j are the activities of the ith and the jth molecules and sim(i, j) is the similarity coefficient between the two molecules. This quantifies the change in biological activity produced by a given change in chemical structure. Utilizing a cutoff value of the index, one can represent sets of molecules through network graphs that highlight abrupt changes in biological activity associated with the steepest cliffs. Steep activity cliffs (Bryce canyonlike regions), associated with high SALI values, represent the most challenging regions of a structure–activity relationship to model with a QSAR, but they are also the most interesting regions for purposes of drug design because small structural modifications in a molecule can lead to a drug with vastly improved potency (a process known as *lead optimization*).

Molecular similarity analysis is thus of central importance in understanding the chemistry of materials and in drug design. The electron density determines all molecular properties by virtue of the Hohenberg–Kohn and holographic electron density theorems and it is thus *the* key determinant of molecular similarity. The density may be used directly in molecular similarity analysis by means of Carbó-type similarity indices or indirectly through electron-density-derived descriptors employed in classification or regression analyses to predict the properties of molecules yet to be synthesized. Such descriptors may also be used to map out the topology of chemistry space and to concentrate research efforts in the search for molecules with specific properties or specific biological activities. Further examples of the application of electron-density-based descriptors from conceptual DFT in QSAR were encountered in Chapter 7.

REFERENCES

1. Carbó R, editor. *Molecular Similarity and Reactivity: From Quantim Chemical to Phenomenological Approaches*. Amsterdam: Kluver; 1995.
2. Martin YC, Kofron JL, Traphagen LM. J Med Chem 2002;45:4350–4358.
3. Maggiora GM. J Chem Inf Model 2006;46(4):1535.
4. Klabunde T. Br J Pharmacol 2007;152(1):5–7.
5. Rognan D. Br J Pharmacol 2007;152:38–52.
6. Hammett LP. Chem Rev 1935;17:125–136.
7. Hammett LP. J Am Chem Soc 1937;59:96–103.
8. Hansch C, Muir RM, Fujita T, Maloney PP, Geiger F, Streich M. J Am Chem Soc 1963;85:2817–2824.
9. Taft RW. J Am Chem Soc 1952;74:3120–3128.
10. Johnson M, Maggiora G. *Concepts and Applications of Molecular Similarity*. New York: John Wiley and Sons; 1990.
11. Carbó R, Arnau J, Leyda L. Int J Quantum Chem 1980;17:1185–1189.
12. Nikolova N, Jaworska J. QSAR Comb Sci 2003;22:1006–1026.
13. Ehresmann B, Martin B, Horn AHC, Clark T. J Mol Model 2003;9:342–347.
14. Ehresmann B, de Groot MJ, Alex A, Clark T. J Chem Inf Comput Sci 2004;44(2): 658–668.
15. Hodgkin EE, Richards WG. Int J Quantum Chem Quantum Biol Symp 1987; 14:105–110.
16. Ayers PW. Proc Natl Acad Sci U S A 2000;97(5):1959–1964.
17. Riess J, Münch W. Theor Chem Acta 1981;58:295–300.
18. Mezey PG. Mol Phys 1990;96:169–178.
19. Mezey PG. J Chem Inf Comput Sci 1999;39:224–230.
20. Robert D, Carbó-Dorca R. Int J Quantum Chem 2000;77:685–692.
21. Borgoo A, Godefroid M, Sen KD, De Proft F, Geerlings P. Chem Phys Lett 2004; 399:363–367.
22. Cooper DL, Allen NL. J Am Chem Soc 1992;114:4773–4776.
23. Allen NL, Cooper DL. J Chem Inf Comput Sci 1992;32:587–590.
24. Allen NL, Cooper DL. In: Sen KD, editor. Volume 173, *Molecular Similarity*, *Topics in Current Chemistry*. Berlin, New York: Springer,; 1995. p 86–111.
25. Cooper DL, Allen NL. In: Carbó R, editor. *Molecular Similarity and Reactivity: From Quantum Chemical to Phenomenological Approaches*. The Netherlands: Kluwer Academic; 1995. p 31–55.
26. Measures PT, Mort KA, Allen NL, Cooper DL. J Comput Aided Mol Des 1995; 9:331–340.
27. Measures PT, Allen NL, Cooper DL. Adv Mol Simil 1996;1:61.
28. Allen NL, Cooper DL. J Math Chem 1998;23:51–60.
29. Amat L, Carbó-Dorca R, Cooper DL, Allen NL. Chem Phys Lett 2003;367: 207–213.
30. Angulo JC, Antolín J. J Chem Phys 2007;126:044106.

31. Politzer P, Daiker KC. In: Deb BM, editor. *The Force Concept in Chemistry*. New York: Van Nostrand Reinhold; 1981, Chapter 6.
32. Politzer P, Sukumar N, Jayasuriya K, Ranganathan S. J Am Chem Soc 1988;110: 3425–3430.
33. Politzer P, Truhlar DG, editors. *Chemical Applications of Atomic and Molecular Electrostatic Potentials*. New York: Plenum Press; 1981.
34. Murray JS, Sukumar N, Ranganathan S, Politzer P. Int J Quantum Chem 1990;38: 611–629.
35. Gonzalez OG, Murray JS, Peralta-Inga Z, Politzer P. Int J Quantum Chem 2001;83: 115–121.
36. Kulshrestha P, Sukumar N, Murray JS, Giese RF, Wood TD. J Phys Chem. A 2009; 113(4):756–766.
37. Cramer I, Richard D, Patterson DE, Bunce JD. J Am Chem Soc 1988;110: 5959–5967.
38. Silverman BD, Platt DE. J Med Chem 1996;39(11):2129–2140.
39. Gironés X, Gallegos A, Carbó-Dorca R. J Chem Inf Comput Sci 2000;40: 1400–1407.
40. Parr RG, Yang W. *Density-Functional Theory of Atoms and Molecules*. New York: Oxford University Press; 1989.
41. Fukui K, Yonezawa T, Shingu H. J Chem Phys 1952;20:722–725.
42. Fukui K, Yonezawa T, Nagata C, Shingu H. J Chem Phys 1954;22:1433–1442.
43. Fukui K. *Theory of Orientation and Stereoselection*. Berlin: Springer-Verlag; 1975.
44. Parr RG, Yang W. J Am Chem Soc 1984;106:4049–4050.
45. Fukui K. Science 1987;218:747–754.
46. Ghosh SK. Chem Phys Lett 1990;172:77–82.
47. Chattaraj PK, Cedillo A, Parr RG. J Chem Phys 1995;103:7645.
48. Borgoo A, Torrent-Sucarrat M, De Proft F, Geerlings P. J Chem Phys 2007;126: 234104.
49. Liu S, De Proft F, Parr RG. J Phys Chem A 1997;101:6991–6997.
50. Torrent-Sucarrat M, Duran M, Sola M. J Phys Chem A 2002;106:4632–4638.
51. Boon G, De Proft F, Langenaeker W, Geerlings P. Chem Phys Lett 1988;295: 122–128.
52. Geerlings P, Boon G, Van Alsenoy C, De Proft F. Int J Quantum Chem 2005;101: 722–732.
53. Boon G, Van Alsenoy C, De Proft F, Bultinck P, Geerlings P. J Phys Chem A 2003;107:11120–11127.
54. Gironés X, Robert D, Carbó-Dorca R. J Comput Chem 2001;22:255–263.
55. Bultinck P, Kuppens T, Gironés X, Carbó-Dorca R. J Chem Inf Comput Sci 2003;43:1143–1150.
56. Wagener M, Sadowski J, Gasteiger J. J Am Chem Soc 1995;117:7769–7775.
57. Moreau G, Broto P. Nouv J Chim 1980;4:359–360, 757–764.
58. Boon G, Langenaeker W, De Proft F, De Winter H, Tollenaere JP, Geerlings P. J Phys Chem A 2001;105:8805–8814.
59. Gironés X, Ponec R. J Chem Inf Model 2006;46(3):1388–1393.

60. Ponec R. J Math Chem 1997;21:323–333.
61. Ponec R. J Math Chem 1998;23:85–103.
62. Matta C. J Phys Chem A 2001;105(49):11088–11101.
63. Popelier PA. J Phys Chem A 1999;103:2883–2890.
64. O'Brien SE, Popelier PA. Can J Chem 1999;77(1):28–36.
65. O'Brien SE, Popelier PA. J Chem Inf Model 2001;41:764–775.
66. Singh N, Loader RJ, O'Malley PJ, Popelier PA. J Phys Chem A 2006;110: 6498–6503.
67. Wold H. *Encyclopedia of Statistical Sciences*. New York: John Wiley and Sons; 1985. p 581–591, Chapter 6.
68. Wold S, Sjöström M, Eriksson L. Chemom Intell Lab Syst 2001;58:109–130.
69. Livingstone DJ. *Data Analysis for Chemists*. Oxford: Oxford University Press; 1995.
70. Chaudry UA, Popelier PA. J Org Chem 2004;69(2):233–241.
71. Hawe GI, Alkorta I, Popelier PA. J Chem Inf Model 2010;50:87–96.
72. Leherte L, Meurice N, Vercauteren DP. J Chem Inf Comput Sci 2000;40:816–832.
73. Breneman CM, Rhem M. J Comput Chem 1997;18(2):182–197.
74. Whitehead CE, Breneman CM, Sukumar N, Ryan MD. J Comput Chem 2003;24: 512–529.
75. Sukumar N, Breneman CM. QTAIM in drug discovery and protein modeling. In: Matta CF, Boyd RJ, editors. *The Quantum Theory of Atoms in Molecules: From Solid State to DNA and Drug Design*. Hoboken (NJ): Wiley-VCH; 2007. p 471–498.
76. Bader RFW. *Atoms in Molecules: A Quantum Theory*. Oxford: Oxford Press; 1990.
77. Murray JS, Politzer P, Famini GR. J Mol Struct (Theochem) 1998;454:299–306.
78. Sjoberg P, Murray JS, Brinck T, Politzer PA. Can J Chem 1990;68:1440–1443.
79. Politzer PA, Murray JS, Grice ME, Brinck T. J Chem Phys 1991;95:6699–6704.
80. Murray JS, Abu-Awwad F, Politzer PA. J Mol Struct (Theochem) 2000;501–502: 241–250.
81. Kewley RH, Embrechts MJ, Breneman CM. In: Volume 8, Dagli C, Akay M, Philip Chen CL, editors. *Intelligent Engineering Systems through Artificial Neural Networks*. New York: ASME Press; 1998. p 391–396.
82. Kewley RH, Embrechts MJ, Breneman CM. IEEE Trans Neural Netw 2000;11(3): 668–679.
83. Song M, Breneman CM, Bi J, Sukumar N, Bennett KP, Cramer S, Tugcu N. J Chem Inf Comput Sci 2002;42:1347–1357.
84. Breneman CM, Bennett KP, Embrechts MJ, Cramer SM, Song M, Bi J, Sukumar N. In: Cawse JN, editor. *Experimental Design for Combinatorial and High Throughput Materials Development*. New York: John Wiley and Sons; 2002. p 203–238.
85. Bennett KP, Embrechts MJ. In: Suykens JAK, Horváth G, Basu S, Micchelli C, Vandewalle J, editors. *Advances in Learning Theory: Methods, Models and Applications, NATO ASI Ser.: Computer and Systems Sciences*. Amsterdam: IOS Press; 2003. p 227–246.
86. Embrechts MJ, Arciniegas FA, Ozdemir M, Kewley RH. Int J Smart Eng Syst Des 2003;5:225–239.

87. Embrechts MJ, Szymanski B, Sternickel M. In: Ovaska SJ, editor. *Computationally Intelligent Hybrid Systems*. Wiley Interscience; 2004, Chapter 10. p 317–363.
88. Bennett KP, Parrado-Hernandez E. J Mach Learn Res 2006;7:1265–1281.
89. Embrechts MJ, Bress RA, Kewley RH. Stud Fuzz 2006;207:447–462.
90. Embrechts MJ, Ekins S. Drug Metab Dispos 2007;35(3):325–327.
91. Embrechts MJ, Bress RC, Kewley RH. In: Guyon I, Gunn S, editors. *Feature Selection in Machine Learning*. Berlin: Springer Verlag; 2004.
92. Momma M, Bennett K. In: Guyon I, Gunn S, Nikravesh M, Zadeh L, editors. *Feature Extraction, Foundations and Applications*. Berlin: Springer Verlag; 2006.
93. Golbraikh A, Tropsha A. J Mol Graph Model 2002;20(4):269–276.
94. Golbraikh A, Tropsha A. J Chem Inf Comput Sci 2003;43(1):144–154.
95. Tropsha A. Mol Inform 2010;29(6–7):476–488.
96. Breneman CM. Predictive Cheminformatics: Best Practices for Determining Model Domain Applicability, Sanibel Conference. Florida; 2007. Available at http://reccr.chem.rpi.edu/Presentations/Sanibel2007_BestPractices.pdf. Accessed 2011.
97. Tetko IV, Sushko I, Pandey AK, Zhu H, Tropsha A, Papa E, Oberg T, Todeschini R, Fourches D, Varnek A. J Chem Inf Model 2008;48(9):1733–1746.
98. Mazza CB, Sukumar N, Breneman CM, Cramer S. Anal Chem 2001;73:5457–5461.
99. Tugcu N, Song MH, Breneman CM, Sukumar N, Bennett KP, Cramer SM. Anal Chem 2003;75:3563–3572.
100. Ladiwala A, Rege K, Breneman CM, Cramer SM. Proc Natl Acad Sci U S A 2005;102(33):11710–11715.
101. Liu J, Yang T, Ladiwala A, Cramer SM, Breneman CM. Sep Sci Technol 2006; 41(14):3079–3107.
102. Chen J, Luo Q, Breneman CM, Cramer SM. J Chromatogr A 2007;1139(2): 236–246.
103. Sukumar N, Krein M, Breneman CM. Curr Opin Drug Discov Dev 2008;11(3): 311–319.
104. Lavine BK, Davidson CE, Breneman C, Katt W. Chemoinformatics 2004;275: 399–425.
105. Bader RFW, MacDougall PJ, Lau CDH. J Am Chem Soc 1984;106:1594–1605.
106. Bader RFW, MacDougall PJ. J Am Chem Soc 1985;107:6788–6795.
107. Bader RFW, Chang C. J Phys Chem 1989;93:2946–2956.
108. Bergeron C, Hepburn T, Sundling M, Sukumar N, Katt WP, Bennett KP, Breneman CM. Prediction of peptide bonding affinity: Kernel methods for nonlinear modeling. 2011. Available at http://arxiv.org/abs/1108.5397. Accessed 2011.
109. Zauhar R, Moyna G, Tian L, Li Z, Welsh WJ. J Med Chem 2003;46:5674–5690.
110. Nagarajan K, Zauhar R, Welsh WJ. J Chem Inf Model 2005;45:49–57.
111. Breneman CM, Sundling CM, Sukumar N, Shen L, Katt WP, Embrechts MJ. J Comput Aided Mol Des 2003;17:231–240.
112. Sundling M, Sukumar N, Zhang H, Embrechts MJ, Breneman CM. Wavelets in chemistry and cheminformatics. In: Boyd D, Lipkowitz K, editors. Volume 22, *Reviews in Computational Chemistry*. Hoboken (NJ): Wiley-VCH; 2006. p 295–329.

113. Das S, Kokardekar A, Breneman CM. J Chem Inf Model 2009;49(12):2863–2872.
114. Das S, Krein MP, Breneman CM. J Chem Inf Model 2010;50(2):298–308.
115. Das S, Krein MP, Breneman CM. Bioinformatics 2010;26(15):1913–1914.
116. Ballester PJ, Richards WG. Proc R Soc London, Ser A 2007;463(2081):1307–1321.
117. Ballester PJ, Richards WG. J Comput Chem 2007;28:1711–1723.
118. Ballester PJ, Finn PW, Richards WG. J Mol Graph Model 2009;27:836–845.
119. Armstrong MS, Morris GM, Finn PW, Sharma R, Richards WG. J Mol Graph Model 2009;28:368–370.
120. Armstrong MS, Morris GM, Finn PW, Sharma R, Moretti L, Cooper RI, Richards WG. J Comput Aided Mol Des 2010;24:789–801.
121. Krein MP, Sukumar N. J Phys Chem A 2011;11:6.
122. Benz RW, Swamidass J, Baldi P. J Chem Inf Model 2008;48:1138–1151.
123. Tanaka N, Ohno K, Niimi T, Moritomo A, Mori K, Orita M. J Chem Inf Model 2009;49, 703(12):2677–2686.
124. Bajorath J, Peltason L, Wawer M, Guha R, Lajiness MS, Van Drie JH. Drug Discov Today 2009;14(13–14):698–705.
125. Kubinyi H. Why Models Fail. Available at http://www.kubinyi.de/san-francisco-09-06.pdf. Accessed 2012.
126. Guha R, Van Drie JH. J Chem Inf Model 2008;48:646–658.
127. Guha R, Van Drie JH. J Chem Inf Model 2008;48(8):1716–1728.
128. Peltason L, Bajorath J. J Med Chem 2007;50:5571–5578.
129. Rhem MO. RECON: An Algorithm for Molecular Charge Density Reconstruction Using Atomic Charge Density Fragments [PhD Dissertation]. Rensselaer Polytechnic Institute, Troy; 1996.
130. Thom R. *Structural Stability and Morphogenesis*. Reading (MA): Benjamin; 1975.
131. Vapnik VN. *The Nature of Statistical Learning*. New York: Springer-Verlag; 1996; *Statistical Learning Theory*. New York: John Wiley and Sons; 1998.
132. Wawer M, Peltason L, Weskamp L, Teckentrup A, Bajorath J. J Med Chem 2008;51:6075–6084.

9

ELECTROSTATIC POTENTIALS AND LOCAL IONIZATION ENERGIES IN NANOMATERIAL APPLICATIONS

PETER POLITZER, FELIPE A. BULAT, JAMES BURGESS, JEFFREY W. BALDWIN, AND JANE S. MURRAY

9.1 THE ELECTRONIC DENSITY

The fundamental importance of the electronic density $\rho(\mathbf{r})$ is well established, for example, by the famous Hohenberg–Kohn theorem [1]. In terms of understanding and predicting chemical reactive behavior, however, the total electronic density is not very informative. A two-dimensional plot of $\rho(\mathbf{r})$ for a molecule typically shows merging circular isodensity contours centered on the nuclei. This is because the molecular electronic density differs only slightly (but crucially) from the superimposed atomic densities. See, for instance, Hazelrigg and Politzer [2] and Iwasaki and Saito [3].

More interesting is the density difference $\Delta\rho(\mathbf{r})$, obtained by subtracting from the $\rho(\mathbf{r})$ of the molecule the electronic densities of its constituent atoms, placed at the same positions as they occupy in the molecule. A plot of $\Delta\rho(\mathbf{r})$ shows very clearly the electronic charge buildup in certain regions and depletion in others that accompanies the formation of the molecule, and emphasizes the subtlety of these changes. The study of $\Delta\rho(\mathbf{r})$ was at one time an area of considerable activity [2, 4, 5], and $\Delta\rho(\mathbf{r})$ continues to be used in crystallography [3, 6]. There is an element of ambiguity, however, in deciding whether to subtract the densities of the atoms in their ground states or in their valence states. This ambiguity

A Matter of Density: Exploring the Electron Density Concept in the Chemical, Biological, and Materials Sciences, First Edition. Edited by N. Sukumar.

can be avoided by looking at the Laplacian of $\rho(\mathbf{r})$, $\nabla^2\rho(\mathbf{r})$, which also reveals electronic charge buildup and depletion [7, 8].

However, any attempt to analyze and predict chemical reactivity from the electronic density alone is ignoring the role of the positively charged nuclei. A reactant A that approaches a molecule B encounters not only the negative electrostatic effect of the electrons of B but also the positive one of its nuclei. Both need to be explicitly taken into account in making predictions.

In this chapter, brief overviews of two complementary molecular properties that provide, together, a realistic basis for understanding molecular reactivity are presented. These are the electrostatic potential V($\mathbf{r}$) and the average local ionization energy $\overline{I}(\mathbf{r})$. As shall be seen, both are formally related to $\rho(\mathbf{r})$, but they go beyond $\rho(\mathbf{r})$ in terms of insight into the electrostatics and the energetics of reactive behavior.

9.2 THE ELECTROSTATIC POTENTIAL

According to Coulomb's law, a stationary point charge Q_1 creates a potential $V(R)$ at any distance R that is given simply by

$$V(R) = \frac{1}{4\pi\varepsilon_0}\frac{Q_1}{R}, \tag{9.1}$$

where ε_0 is the permittivity of free space, a constant. $V(R)$ can be regarded as the "potential" of Q_1 to interact with another point charge Q_2 placed at the distance R. Their interaction energy would be

$$\Delta E = \frac{1}{4\pi\varepsilon_0}\frac{Q_1 Q_2}{R}. \tag{9.2}$$

If Q_1 and Q_2 have the same sign, whether positive or negative, ΔE is positive and the interaction is repulsive, and thus destabilizing. If they have opposite signs, ΔE is negative and the interaction is attractive, and thus stabilizing.

Equation 9.1 can easily be extended to atoms and molecules. Invoking the Born–Oppenheimer approximation, the nuclei can be treated as stationary point charges, and Equation 9.1 can be summed over each one. The electrons are not stationary, but the electron density function $\rho(\mathbf{r})$ gives the average number of them in each volume element d$\mathbf{r}$, and Equation 9.1 can be applied again, but now integrating (rather than summing) over the infinite number of volume elements d$\mathbf{r}$. Thus, the potential created at any point $\mathbf{r}$ by the nuclei and electrons of the molecule is given by

$$V(\mathbf{r}) = \frac{1}{4\pi\varepsilon_0}\left[\sum_{\mathrm{A}}\frac{Z_{\mathrm{A}}e}{|\mathbf{R}_{\mathrm{A}} - \mathbf{r}|} - e\int\frac{\rho(\mathbf{r}')\mathrm{d}\mathbf{r}'}{|\mathbf{r}' - \mathbf{r}|}\right]. \tag{9.3}$$

In Equation 9.3, Z_A is the atomic number of nucleus A and e is the magnitude of the charge on an electron or proton. Thus, $Z_A e$ is the charge on nucleus A, located at $\mathbf{R}_A$ and $|\mathbf{R}_A - \mathbf{r}|$ is its distance from the point $\mathbf{r}$. The quantity $e\rho(r')\mathrm{d}\mathbf{r}'$ is the amount of electronic charge in the volume element $\mathrm{d}\mathbf{r}'$, at a distance $|\mathbf{r}' - \mathbf{r}|$ from $\mathbf{r}$.

For convenience, $V(\mathbf{r})$ is typically written in terms of atomic units (au). It then takes the form

$$V(\mathbf{r}) = \sum_A \frac{Z_A}{|\mathbf{R}_A - \mathbf{r}|} - \int \frac{\rho(\mathbf{r}')\mathrm{d}\mathbf{r}'}{|\mathbf{r}' - \mathbf{r}|}. \tag{9.4}$$

Going back to Equations 9.1 and 9.2, it is seen that the potential at R has the same sign and magnitude as would the interaction energy with a unit positive point charge. This is true as well for Equation 9.4; $V(\mathbf{r})$ can be viewed as the interaction energy with a +1 point charge placed at $\mathbf{r}$. For this reason, $V(\mathbf{r})$ is customarily quoted in units of energy rather than potential. One atomic unit of energy = 1 hartree = 27.21 eV = 627.5 kcal/mol; 1 kcal/mol = 4.184 kJ/mol.

The sign of $V(\mathbf{r})$ in any region depends on whether the positive contribution of the nuclei or the negative one of the electrons is dominant there. Note that Equation 9.4 can be applied to atoms as well as molecules; the summation will then be over just the one nucleus.

An important feature of the electrostatic potential is that it is a real physical property, an observable. It can be determined experimentally, by diffraction techniques [9, 10], as well as computationally. This is in marked contrast to arbitrarily defined properties such as atomic charges, for which there is no rigorous physical basis. They depend very much on the procedure used to assign them. For instance, six different proposed atomic charge definitions predict the carbon in CH_3NO_2 to have charges ranging from −0.478 to +0.564 [11]!

Molecular electrostatic potentials can be presented in various formats, for instance, as two-dimensional plots in selected molecular planes. In the context of reactivity, however, an effective approach is to compute $V(\mathbf{r})$ on a molecular surface, as defined by an outer contour of the molecule's electronic density $\rho(\mathbf{r})$ [12]. The $\rho(\mathbf{r}) = 0.001$ au (electrons/bohr^3) contour is commonly used for this purpose; it generally encompasses at least 96% of the molecule's electronic charge [12]. Taking a contour of $\rho(\mathbf{r})$ to define the surface has the significant advantage that it reflects features specific to that molecule, for example, lone pairs, π electrons, strained bonds, etc. The 0.001 au contour is normally beyond the van der Waals radii of the constituent atoms (except for hydrogen) [13], and the $V(\mathbf{r})$ on this surface shows realistically what another entity "sees" as it approaches the molecule.

The electrostatic potential on a 0.001 au molecular surface is denoted $V_S(\mathbf{r})$. Its most positive and most negative values (maxima and minima) are labeled $V_{S,\max}$ and $V_{S,\min}$; there may be several of each on a given molecular surface. (Note that while the only maxima in $V(\mathbf{r})$ occur at the nuclei [14], there can exist

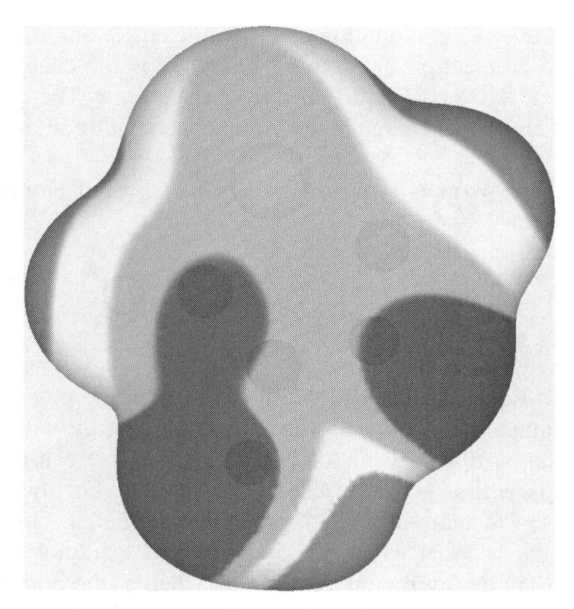

Figure 9.1 Color-coded diagram of the electrostatic potential on the molecular surface of 4-hydroxy-1,3-thiazole computed at the B3PW91/6-31G(d,p) level. The locations of the atomic nuclei are visible through the surface; sulfur is at the top, and nitrogen is at the lower right. Color ranges, in kilocalories per mole, are red > 15.0 > yellow > 0.0 > green > − 8.0 > blue. (*See insert for color representation of the figure*.)

maxima in $V_S(\mathbf{r})$ because it is defined on the molecular surface, a two-dimensional subdomain of three-dimensional real space.)

Our experience has been that most organic molecules have weakly positive potentials over much of their surfaces, becoming stronger in the vicinities of acidic hydrogens. Negative $V_S(\mathbf{r})$ values are associated primarily with lone pairs, π electrons of unsaturated molecules, and strained C–C bonds [15]. To illustrate, Figure 9.1 displays $V_S(\mathbf{r})$ on the 0.001 au surface of 4-hydroxy-1,3-thiazole, **1**, computed at the density functional B3PW91/6-31G(d,p) level [16] with the WFA-Surface Analysis Suite [17].

1

Figure 9.1 shows strongly negative potentials associated with the nitrogen and oxygen lone pairs and with the π electrons. Their $V_{S,\,min}$, are, respectively, −24.8, −24.4, and −11.4 kcal/mol. The nitrogen lone pair would normally have a more negative $V_{S,\,min}$, but it is partially neutralized by the proximity of the hydroxyl hydrogen. The most positive regions are near the hydrogens, the strongest being due to the hydroxyl hydrogen, $V_{S,\,max} = 47.9$ kcal/mol. It is to be noted that the sulfur has both positive and negative potentials on its surface. This exemplifies the fallacy, discussed in detail elsewhere [18, 19], of treating atoms in molecules as having only positive or only negative charges.

While the emphasis in this chapter is on chemical reactivity, the electrostatic potential is of much greater and more fundamental significance. For example, the energies of atoms and molecules can be expressed *rigorously* in terms of their electrostatic potentials, specifically at the positions of the nuclei [20, 21]. Realistic covalent radii can be determined from the minimum of $V(\mathbf{r})$ along the internuclear axis of two bonded atoms [22, 23]. These and other aspects of atomic and molecular electrostatic potentials are discussed in detail in several reviews [19, 20, 24].

9.3 THE AVERAGE LOCAL IONIZATION ENERGY

The energy I_i required to remove an electron i from an atom or molecule is a *global* property of the atom or molecule as a whole and is often associated with a particular atomic or molecular orbital φ_i. However chemical reactivity is *local* and site specific. For analyzing and understanding reactive behavior, it would therefore be useful to be able to identify and rank particular sites where the electrons are least strongly held, where they are most available, as well as where they are tightly bound. The average local ionization energy, $\overline{I}(\mathbf{r})$, serves this purpose. It focuses on the point in the space of the atom or molecule, rather than on a particular orbital.

How can one formulate a local ionization energy? Within the framework of Hartree–Fock theory, the ionization energy of an electron i in orbital φ_i of a molecule X is given by

$$I_i = E(\mathrm{X}_i{}^{+}) + E(e^{-}) - E(\mathrm{X}), \tag{9.5}$$

where $E(\mathrm{X})$ and $E(X_i{}^{+})$ are the energies of X and the positive ion formed from X by loss of an electron from orbital φ_i. From Equation 9.5 and the Hartree–Fock expressions for $E(\mathrm{X})$ and $E(X_i^{+})$, it readily follows that

$$I_i = |\varepsilon_i|, \tag{9.6}$$

where ε_i is the energy of an electron in φ_i. Equation 9.6 is based on a major assumption, which is that the occupied orbitals of X are unaffected by the loss of electron i. This assumption finds some support in Koopmans' theorem [25, 26] and also benefits from a fortuitous partial cancellation of errors; this has been discussed in detail by Politzer et al. [27]. In practice, Equation 9.6 has been found to be reasonably accurate at the Hartree–Fock level for estimating the first ionization energy of an atom or molecule, that is, when ϕ_i is the highest occupied orbital, but less effective for the lower ones [27].

If $\rho_i(\mathbf{r})$ is the electron density corresponding to Hartree–Fock orbital $\varphi_i(\mathbf{r})$, and $\rho(\mathbf{r})$ is the total electronic density, then the average orbital energy $\overline{\varepsilon}(\mathbf{r})$ at a point $\mathbf{r}$ is

$$\overline{\varepsilon}(\mathbf{r}) = \frac{\sum_i \rho_i(\mathbf{r})\varepsilon_i}{\rho(\mathbf{r})}, \tag{9.7}$$

the summation being over all occupied orbitals. Then if Equation 9.6 is assumed to be valid, Equation 9.7 can be rewritten as

$$\overline{I}(\mathbf{r}) = \frac{\sum_i \rho_i(\mathbf{r})|\varepsilon_i|}{\rho(\mathbf{r})}. \tag{9.8}$$

Equation 9.8 defines the average local ionization energy at point $\mathbf{r}$. It was originally introduced, in the Hartree–Fock framework, by Sjoberg et al. [28]; its relationship to Equation 9.7 was pointed out later by Nagy et al. [29, 30].

Can the formula for $\overline{I}(\mathbf{r})$, Equation 9.8, be applied within the context of Kohn–Sham density functional theory, which is now the dominant approach in molecular electronic structure calculations? Various aspects of this question are discussed by Politzer et al. [27], but the general answer has been yes. The magnitudes of Kohn–Sham orbital energies usually underestimate experimental ionization energies, while the Hartree–Fock energies overestimate them.

$$|\varepsilon_i|,\ \text{Kohn–Sham} < I_i,\ \text{experimental} < |\varepsilon_i|,\ \text{Hartree–Fock} \tag{9.9}$$

However, the key point is that the relative values and the trends, in both the Hartree-Fock and the Kohn-Sham $\overline{I}(\mathbf{r})$, have been found to be physically meaningful; see, for example, Sjoberg et al. [28], Murray et al. [31], and Politzer et al. [32, 33]. Exchange-only Kohn–Sham $\overline{I}(\mathbf{r})$ has been compared to Hartree–Fock $\overline{I}(\mathbf{r})$, both theoretically [34] and computationally [34, 35]. It was found that the difference is given approximately by the response portion of the Kohn–Sham potential [34]. Furthermore, for all atoms H through Kr, the difference was shown to be very small when computed with nearly exact spherically symmetric wavefunctions [35].

Just as with the electrostatic potential, when analyzing reactivity, we normally compute $\overline{I}(\mathbf{r})$ on the 0.001 au molecular surface, yielding $\overline{I}_s(\mathbf{r})$ [17]. This is shown in Figure 9.2 for 4-hydroxy-1,3-thiazole, **1**, computed with the B3PW91/6-31G(d,p) procedure. In Figure 9.2, our primary interest is in the regions having the lowest $\overline{I}_s(\mathbf{r})$, that is, the minima $\overline{I}_{S,\min}$. These indicate the locations of the

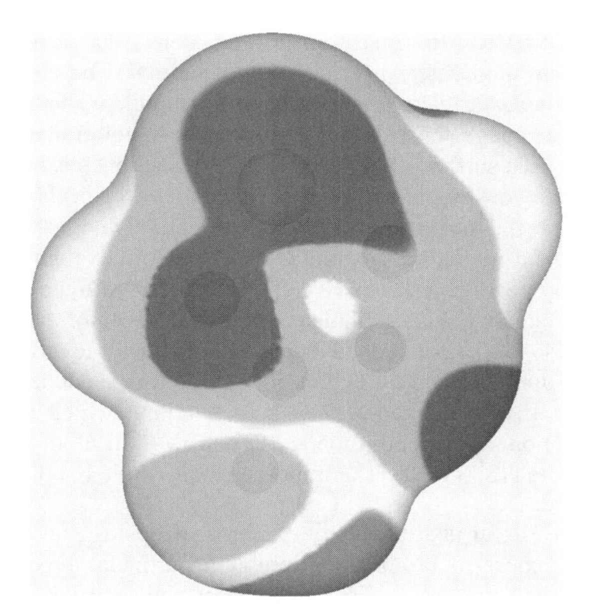

Figure 9.2 Color-coded diagram of the average local ionization energy on the molecular surface of 4-hydroxy-1,3-thiazole computed at the B3PW91/6-31G(d,p) level. The locations of the atomic nuclei are visible through the surface; sulfur is at the top, and nitrogen is at the lower right. Color ranges, in electron volts, are red $>$ 14.0 $>$yellow $>$ 12.0 $>$green $>$ 10.0 $>$blue. (*See insert for color representation of the figure*.)

least tightly held, most reactive electrons. In **1**, these are the π electrons ($\overline{I}_{S,min}$ = 8.7 eV) and the nitrogen lone pair ($\overline{I}_{S,min}$ = 9.0 eV). It may be surprising that the oxygen lone pair, despite having a significantly negative $V_{S,\ min}$, has a relatively high $\overline{I}_{S,min}$ of 10.77 eV. This is just one of many examples showing that $\overline{I}_s(\mathbf{r})$ and $V_S(\mathbf{r})$ do not necessarily show the same trends.

As with the electrostatic potential, the fundamental significance of the average local ionization energy goes well beyond chemical reactivity. It is linked to local kinetic energy density, atomic shell structure, electronegativity, local polarizability/hardness, etc. For overviews of these aspects of $\overline{I}(\mathbf{r})$, see Politzer et al. [27, 36].

9.4 REACTIVITY

9.4.1 Overview

Both the electrostatic potential and the average local ionization energy have been used extensively in analyzing and predicting reactive behavior. This work has been reviewed elsewhere, both for $V(\mathbf{r})$ [19, 20, 24] and for $\overline{I}(\mathbf{r})$ [27, 36]. Hence, we now limit ourselves to a brief overview and then proceed to discuss three specific areas in greater detail.

With respect to $V(\mathbf{r})$, the first point to consider is that it is computed using the average electronic density $\rho(\mathbf{r})$ of the molecule X. The close approach of a reactant Y will necessarily perturb the $\rho(\mathbf{r})$ of X, and so the $V(\mathbf{r})$ obtained for isolated X will no longer be valid. This effect can be minimized by determining $V(\mathbf{r})$ on the 0.001 au surface of X, that is, $V_S(\mathbf{r})$, since this lies beyond the van der Waals radii of the constituent atoms of X (except hydrogen) [13]. However, this means that $V(\mathbf{r})$ is the most useful, as $V_S(\mathbf{r})$, at relatively large X—Y separations, that is, noncovalent interactions or the early stages of covalent ones.

Thus, strongly positive hydrogen $V_{S,\,max}$ identify good hydrogen bond donors, while strongly negative $V_{S,\,min}$ indicate potential acceptors. Hydrogen $V_{S,\,max}$ and basic site $V_{S,\,min}$ have indeed been demonstrated to correlate well with hydrogen-bond-donating and hydrogen-bond-accepting tendencies [37]. Another electrostatically driven interaction is σ-hole bonding, which involves a region of positive $V_S(\mathbf{r})$ on a covalently bonded Group IV–VII atom interacting with a negative $V_S(\mathbf{r})$ site [18, 38, 39]. The importance of σ-hole bonding is increasingly being recognized.

In some areas, as in pharmacology, patterns of positive and negative surface potentials that appear to promote or inhibit a particular type of activity may provide insight into the early stages of large systems "recognizing" each other [10, 40–42]. Drug–receptor and enzyme–substrate interactions can sometimes be elucidated in this manner. However, the electrostatic potential on a molecular surface contains a great deal of information beyond its general pattern of positive and negative regions and its maxima and minima [17]. Some of this can be extracted by characterizing $V_S(\mathbf{r})$ in terms of quantities such as its average deviation, positive and negative average values, positive and negative variances, etc. It has been found that a variety of condensed phase physical properties that depend on noncovalent interactions can be represented analytically in terms of subsets of these quantities [43, 44]. These properties include heats of phase transitions, boiling points and critical constants, solubilities and solvation energies, partition coefficients, viscosities, diffusion coefficients, surface tensions, etc.

$V_S(\mathbf{r})$ can be a very useful guide to noncovalent interactions. However it is not reliable, except as an adjunct, for processes involving the formation of a covalent bond. It was already pointed out, in Figures 9.1 and 9.2, that the strengths of negative potentials do not necessarily correlate with the reactivities of electrons. The latter requires $\overline{I}_s(\mathbf{r})$; its lowest values, the $\overline{I}_{S,\,min}$, reveal the sites of the most readily available electrons. For example, in benzene derivatives such as phenol, the most negative $V_{S,\,min}$ is generally associated with the substituent. Yet electrophilic substitution occurs on the aromatic ring. This is correctly predicted by the minima of $\overline{I}_s(\mathbf{r})$, the $\overline{I}_{S,min}$, as is the directing effect (*ortho*/*para* or *meta*) and whether the substituent activates or deactivates the ring [27, 28, 31–33].

$V_S(\mathbf{r})$ and $\overline{I}_s(\mathbf{r})$ can thus be viewed as complementing each other. Indeed, the regions of negative potential sometimes provide long-range guidance of an approaching electrophile toward locations of low $\overline{I}_s(\mathbf{r})$ where reaction can occur. This is illustrated by protonation [45], and by electrophilic attack on furan and pyrrole [33].

The $\overline{I}_{S,min}$ locate the most reactive electrons, and hence the positions most vulnerable to electrophiles or free radicals. $\overline{I}_{S,min}$ have also been found to characterize free radical sites, localized multiple bonds, and strained C–C bonds. Such applications have been summarized by Politzer et al. [27, 36]. Do the $\overline{I}_{S,max}$ show where nucleophilic attack is favored? There are some indications that this may be so [17], but it has not been confirmed.

9.4.2 Noncovalent Interactions: Chemical Vapor Sensing

As mentioned earlier, the molecular electrostatic potential is a reliable guide to noncovalent interactions [17, 20, 33, 41–45]. In this section, we discuss how the potential computed on molecular surfaces can be utilized in exploring possible interactions between gas-phase analytes and sorbent groups in chemical sensing [46]. The sorbent groups responsible for selectivity are chemically bound to the surface of a nanoelectromechanical resonator. In particular, low-cost complementary metal oxide semiconductor (CMOS) nanomechanical resonators have been functionalized to attract analytes of interest by Baldwin et al. [47–49]. Their extremely small masses and high surface/volume ratios have demonstrated femtogram (10^{-15} g) sensitivity and such functionalized nanomechanical resonators emerge as a competitive and complementary method for vapor sensing applications. These types of low-power chemical microsensors have an important role in helping to counter, reduce, and eliminate the threats and effects of chemical and biological agents and high-yield explosives. Note, by way of comparison, that the canine nose—the most successful detector used at present for vapor sensing of explosives—has detection limits of 10^{-12}–10^{-13} g [50].

Figure 9.3 shows a typical CMOS-integrated bridge resonator with an approximate length of 5 μm, with all-electrical actuation and readout. The shaded area was chemically functionalized with sorbent monolayers that interact favorably (and selectively) with analyte molecules in the gas phase. The adsorption of these molecules onto the surface changes the effective mass of the resonator, and hence its resonant frequency. Such frequency shifts can then be used to determine the amount of mass that was added and thus the concentration of the analyte in the gas phase.

The sensitivity and selectivity toward individual analytes in these microsensors is dictated by specific molecular interactions that determine sorption thermodynamics and kinetics [46, 51]. Chemical functionalization with highly selective sorbent groups is the basis for sensor technology involving nanomechanical resonators [47–49]. A key issue to address is the problem of designing functional sorbent groups to favor the adsorption of target molecules in preference to interferents present in the environment (e.g., water and vehicle exhaust.). The choices of functional groups of interest can be narrowed through a detailed understanding of the different types of possible interactions (e.g., acid-base, π–π, hydrogen bonding, and σ-hole bonding). Weak interactions (in the 2–20 kcal/mol range) between the functionalized surface and the analyte result in reversible binding, so that exposure to the analyte vapor produces a measurable frequency response,

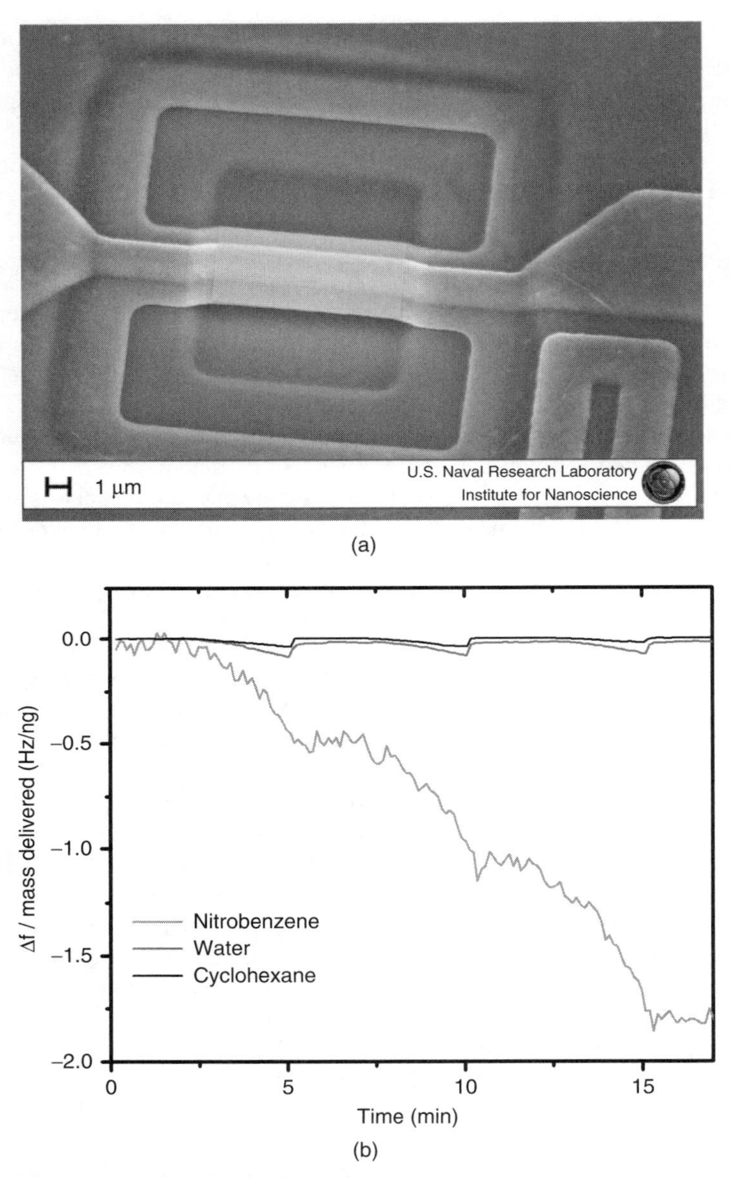

Figure 9.3 (a) Chemically functionalized CMOS nanomechanical resonator (shaded bridge) with electrical actuation and transduction. (b) Exposure of CMOS nanomechanical resonator to nitrobenzene (analyte), water (interferent), and cyclohexane (interferent) shows the selectivity of sensor. On the abscissa, f = frequency. (*See insert for color representation of the figure*.)

as seen in Figure 9.3. Reversibility ensures that a fast purge cycle or a quick heat pulse would be sufficient to release the analyte, thus resetting the sensor for the next vapor exposure. The functionalized resonator response (calculated as Δ frequency/mass delivered) differs for water, cyclohexane, and nitrobenzene (Fig. 9.3). Note in Figure 9.3 that the resonator response is much greater for nitrobenzene (a simulant of TNT, 2,4,6-trinitrotoluene) than for the other analytes. The interaction with nitrobenzene is in fact so strong that it requires a heating cycle to initiate desorption from the surface.

In this section, we show how the molecular electrostatic potential can be a valuable tool for elucidating the molecular interactions between sorbent groups on the resonator surface and the analytes and/or interferents of interest. In particular, we focus on a specific sorbent group used to functionalize nanomechanical resonator surfaces: an aliphatic chain that is chemically bonded to the silicon resonator surface and is terminated by a hexafluoroisopropanol (HFIPA) group (Fig. 9.4). In our calculations, we simply use HFIPA, **2** to model this sorbent. We consider its interactions with 2,4-dinitrotoluene (DNT, **3**), TNT, **4**, and triethylamine (TEA, **5**). DNT, a decomposition product of TNT, is often found in the headspace over explosives [52].

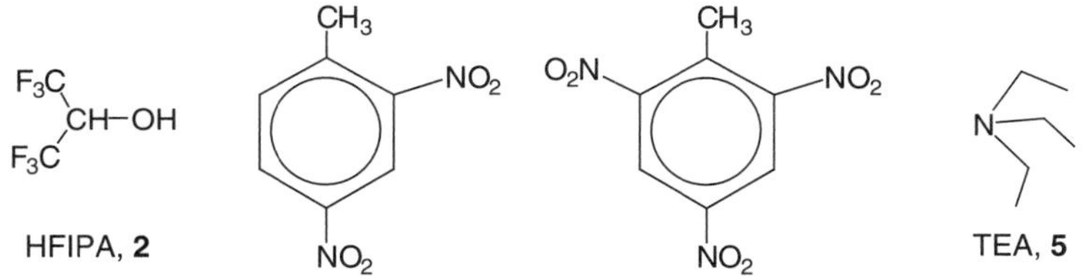

HFIPA, **2** DNT, **3** TNT, **4** TEA, **5**

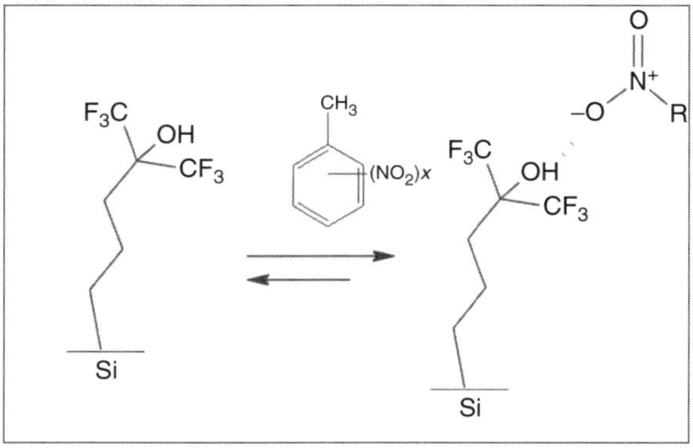

Figure 9.4 Schematic view of the interaction of molecules containing nitro groups with a silicon surface functionalized with HFIPA.

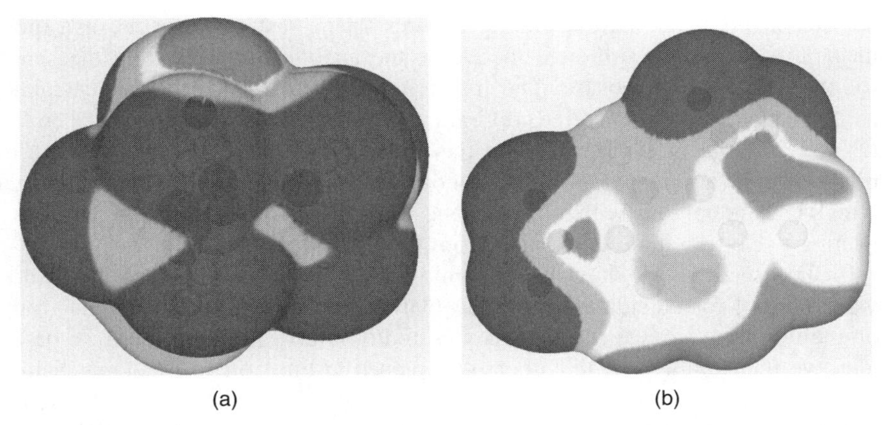

Figure 9.5 Calculated molecular electrostatic potentials on the 0.001 au molecular surfaces of HFIPA (a) and DNT (b), from B3LYP/6-311G(d,p) wavefunctions. In HFIPA, the hydroxyl group is at the top; in DNT, the methyl group is at the right. Color ranges, in kilocalories per mole, are (a, HFIPA) red > 30 > yellow > 17.0 > green > 0.0 > blue, and (b, DNT) red > 20 > yellow > 13.0 > green > 0.0 > blue. (*See insert for color representation of the figure*.)

We have computed the electrostatic potentials $V_S(\mathbf{r})$ on the molecular surfaces of HFIPA, DNT, TNT, and TEA; these surfaces are defined as the 0.001 au contours of the molecules' electronic densities. Optimized geometries and their corresponding wavefunctions were obtained with the Gaussian 09 Suite [16] using density functional methods: the B3LYP [53–55] energy functional and a triple-zeta basis set augmented with polarization functions, 6-311G(d,p). We performed frequency calculations at the same level to confirm that the optimized structures do indeed correspond to energy minima. The wavefunctions were then utilized to compute electrostatic potentials $V_S(\mathbf{r})$ on the molecular surfaces and to determine the surface minima/maxima using the WFA Suite [17].

The surface potentials of HFIPA and DNT are shown in Figure 9.5, color-coded according to the values of $V_S(\mathbf{r})$. HFIPA displays a very positive region (>30 kcal/mol) associated with the hydroxyl hydrogen atom and negative regions due to the electronegative fluorines in the fluoromethyl groups as well as the hydroxyl oxygen (in the background). The remainder of the aliphatic backbone (not shown) is weakly positive. DNT has a positive potential over most of the aromatic ring, indicating that the ring is deactivated because of the electron-withdrawing nature of the NO_2 groups. Negative $V_S(\mathbf{r})$ regions are due to the nitro oxygens, and weakly positive ones (although stronger than over the aromatic ring) are around the aromatic and aliphatic hydrogen atoms.

The features outlined above confirm a very obvious interaction possibility between HFIPA and DNT (Fig. 9.4), which is through the hydrogen bond facilitated by the positive $V_S(\mathbf{r})$ of the HFIPA hydroxyl hydrogen and the negative $V_S(\mathbf{r})$ regions around the nitro oxygens in DNT. Interestingly, another attractive interaction might be between the positive potentials of the methyl or aromatic hydrogens and the negative regions of the fluorines in HFIPA. These would not be typical hydrogen bonds because the hydrogens involved are not bonded to electronegative atoms such as oxygen or nitrogen, although clearly the deactivated aromatic ring plays an equivalent role. We provide evidence that such C-H—F satellite interactions assist the formation of stable complexes between HFIPA and DNT, complementing the primary O-H—-O hydrogen bonds.

In order to verify that the sorbent group HFIPA and the analyte DNT interact as described earlier, we have optimized the geometries of several possible supramolecular HFIPA–DNT complexes. We chose different starting geometries in optimizing their structures in order to probe various possibilities without biasing the results toward a specific one. Figure 9.6 shows the five most stable HFIPA–DNT complexes that were found. The most stable is S1DNT-A, with a calculated binding energy of 7.7 kcal/mol. This complex clearly exhibits an O-H—O hydrogen bond, but we also observe relatively close contacts between methyl hydrogens and fluorine atoms, representing weaker C-H—F hydrogen bonds. The computed H—O and H—F separations in Figure 9.6 are less than or approximately equal to the sums of the van der Waals radii of the respective atoms, which are 2.7 Å for both the H–O and H–F pairs [56]. Analogous interactions are observed in S1DNT-B (binding energy = 4.9 kcal/mol); however, the satellite C-H—F contacts are through a less positive aromatic hydrogen, which helps to explain the less favorable interaction. S1DNT-C (binding energy = 3.8 kcal/mol) and S1DNT-D (binding energy = 2.9 kcal/mol) involve only C-H—F interactions. S1DNT-E shows that when the negative potential around the fluorines points toward the center of the aromatic ring, which is moderately positive, some stabilization does take place and results in a rather weakly bound complex (1.8 kcal/mol). The structures in Figure 9.6 demonstrate the significant stabilizing effect that can be provided by C-H—F interactions.

The interactions of TNT with HFIPA display features very similar, as expected, to those described above for DNT and HFIPA. The electrostatic potential on the molecular surface of TNT, shown by Politzer et al. [57], has features analogous to that of DNT, taking into account that there is an additional nitro group present. The most stable HFIPA–TNT complex has a slightly lower binding energy of 6.9 kcal/mol. TEA, on the other hand, interacts significantly less strongly with HFIPA, binding energy = 4.2 kcal/mol. While TEA does form an O-H—N hydrogen bond, it does not have the possibility of further stabilization through C-H—F contacts. Finally, we mention that the interaction of HFIPA with cyclohexane is found to display an even lower stabilization of 2.1 kcal/mol, evidently because of the absence of any strong hydrogen bonding. More details on these HFIPA–analyte complexes can be found elsewhere [46].

9.4.3 Covalent Interactions: Graphene

Graphene is a single sheet of carbon atoms having sp^2 hybridization that form a lattice of fused six-membered rings, a conjugated and conducting

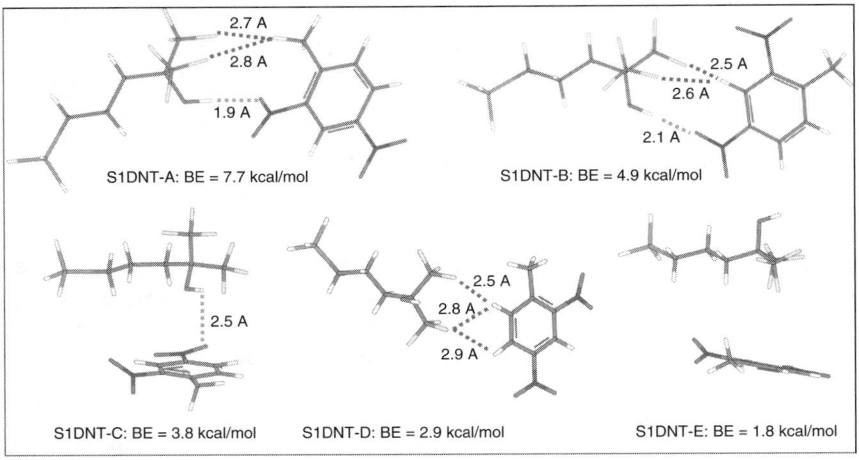

Figure 9.6 Various stable complexes formed between HFIPA and DNT, with binding energies ranging from 7.7 kcal/mol to 1.8 kcal/mol at the B3LYP/6-311G(d,p) level. Some key structural elements (distances between closest contacts) are shown. (*See insert for color representation of the figure*.)

two-dimensional material that has attracted considerable attention since first knowingly isolated by Geim et al. [58–60]. Owing to its peculiar band structure, with a linear dispersion relation around the Fermi level, graphene is a semimetal or a zero-bandgap semiconductor [61]. Chemical modifications, for example, oxidation, hydrogenation, and fluorination, are promising routes to fully exploiting applications of graphene in a wide variety of fields [62–66]. We show here that the average local ionization energy is a valuable tool for predicting and understanding the reactivity of graphene.

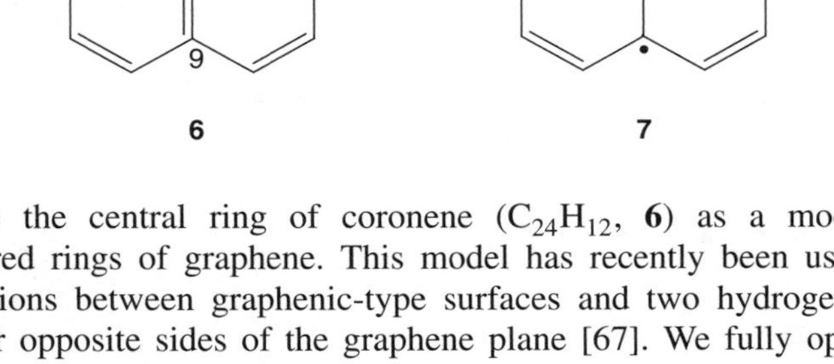

6 **7**

We take the central ring of coronene ($C_{24}H_{12}$, **6**) as a model for the six-membered rings of graphene. This model has recently been used to study the interactions between graphenic-type surfaces and two hydrogen atoms on the same or opposite sides of the graphene plane [67]. We fully optimized the molecular structures of coronene (**6**) and the 1-H-coronene radical ($C_{24}H_{13}$, **7**), which has a doublet spin state, at the B3PW91/6-311G(d,p) level using the Gaussian 09 suite [16]. The vibrational frequencies confirmed that the geometries correspond to energy minima, and molecular wavefunctions were obtained for these structures. The latter were then utilized to compute the average local ionization energy $\overline{I}_S(\mathbf{r})$ on the 0.001 au surfaces and to determine the local minima/maxima using the WFA-Surface Analysis Suite [17].

$\overline{I}_S(\mathbf{r})$ for coronene (**6**) is shown in Figure 9.7. There are $\overline{I}_{S,min}$ associated with each carbon in the central ring, 9.85 eV (above and below the molecular plane), and also with the six carbons linked to these (e.g., C5 and C8), 9.83 eV. However the lowest $\overline{I}_{S,min}$, 8.88 eV, are near the midpoints of the outermost bonds on the periphery of **6**, for example, C6–C7. This indicates that these outer bonds have some double bond character [68, 69], which is confirmed by their shorter lengths (1.37 A) compared to the other C–C bonds in **6**, which range from 1.41 A to 1.42 A.

Our present focus is on the central ring exclusively, since this is intended to be a reduced model for graphenic materials. It has been found to be quite reliable for this purpose [67]; trends observed for the central atoms of coronene have been seen to be equivalent to those obtained with much larger models, such as circumpyrene, $C_{42}H_{16}$. The carbon atoms in the central ring of **6** are all equivalent to each other by symmetry, which is of course broken when a hydrogen atom is added to one of these carbons (labeled C1) to form the 1-H-coronene radical (**7**). This creates a radical site at C9.

The $\overline{I}_S(\mathbf{r})$ plots for **7** on both sides of the molecular plane shown are in Figure 9.8. On the side opposite to the hydrogen atom (a), the ortho and para positions on the central ring are favored for electrophilic attack, as indicated by the $\overline{I}_{S,min}$ above C4 and the two equivalent C2 positions. The values of the $\overline{I}_{S,min}$ at the *ortho* carbons are the lowest associated with the central ring, 9.04 eV, a significant ~0.8 eV less than the corresponding $\overline{I}_{S,min}$ of coronene. The $\overline{I}_{S,min}$ at the para position is 9.75 eV, and we found no $\overline{I}_{S,min}$ near the meta carbons. These observations correlate very well with recently reported binding energies for a second hydrogen atom reacting at these sites [67], which indicates that the ortho and para are greatly favored; the binding energies are 2.89 eV and 1.74 eV, respectively, significantly stronger than at the meta positions (0.86 eV).

The reactivity of the 1-H-coronene radical on the side of the hydrogen atom (Fig. 9.8b) is obscured by its presence. (It bonds perpendicularly to the molecular plane.) There is an $\overline{I}_{S,min}$ at the para position, with a magnitude of 9.45 eV, and a second one in the general region of the hydrogen and the two ortho carbons, 9.44 eV. The similarity of these $\overline{I}_{S,min}$ is consistent with those of the reported binding energies of a second hydrogen atom at the para and ortho sites, 2.04 and 2.16 eV, respectively [67]. The meta positions display higher $\overline{I}_{S,min}$, 9.78 eV, and they are not favored for reaction, with binding energies of 0.76 eV. Hydrogenation as well as fluorination of graphene modeled by coronene and by periodic systems is further discussed elsewhere [70].

Figure 9.7 Diagram of the average local ionization energy on the molecular surface of coronene, computed at the B3PW91/6-311 G(d,p) level. Color ranges, in electron volts, are red >12 > yellow > 10 > green > 9 > blue. The light blue circles indicate the positions of the local minima found in the central ring. (*See insert for color representation of the figure.*)

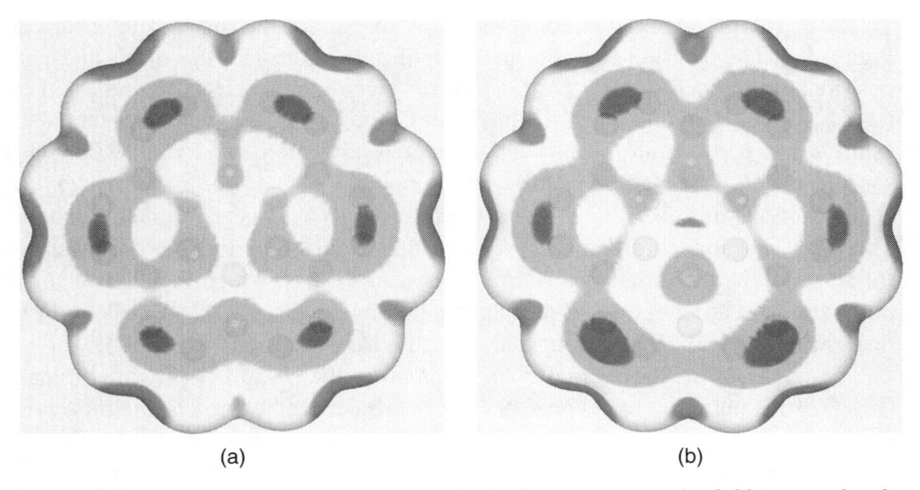

Figure 9.8 Diagram of the average local ionization energy on the 0.001 au molecular surface of the 1-H-coronene radical computed at the B3PW91/6-31G(d,p) level. The added hydrogen atom is in (b), coming out of the plane of the figure, in the lower portion of it. Color ranges, in electron volts, are red > 12 > yellow >10 > green >9 > blue. The light blue circles indicate the positions of the local minima found in the central ring and that associated with C9 (a). (*See insert for color representation of the figure.*)

9.4.4 Carbon Nanotubes

Carbon nanotubes share many structural, chemical, and electronic features with graphene. They can be thought of as rolled-up graphene sheets [71, 72], with the diameters of the resulting tubes and the orientations of their axes with respect to graphene's lattice dictating many of their properties through the effects of electronic confinement [73]. A very good approximation to their electronic band structure is obtained by "zone folding" the two-dimensional band structure of graphene, which gives reasonable predictions of the types of tubes that have metallic and semiconducting character [71]. The one-dimensional sub-bands obtained by the zone folding technique correspond to "cuts" through graphene's Brillouin zone, the orientation and spacing of which is dictated by the alignment of the tube's axis and diameter, respectively. This is because only specific values of the wave vector in the direction perpendicular to the tube's axis are allowed. Metallic tubes are obtained when the sub-bands intersect the K-points in the two-dimensional Brillouin zone of graphene. Semiconducting tubes result when none of these sub-bands cross the K-point, and a finite band gap is thus created. The gap is inversely proportional to the tube's diameter, since with greater diameters, more sub-bands must cut through the Brillouin zone.

Carbon nanotubes, first observed by Iijima [74], have spawned remarkable interest because of their potential significance in many areas. Their general properties (electrical, mechanical, chemical, thermal, optical, etc.) have been extensively studied experimentally and computationally [71–75]. They have found applications as chemical and biochemical sensors [76–78], reinforcement for nanocomposites [79], acoustic projectors (loudspeakers) [80, 81], mechanical sensors [82, 83], nonlinear optical materials [84–87], hydrogen storage [88, 89], etc. For many of these purposes, pristine carbon nanotubes are highly desirable, while for others, the presence of defects is critical and even the basis of their activities. Chemical modification, for example, fluorination [90], can alter and enhance many nanotube properties, such as electrical conductivity and chemical reactivity.

We will show that the electrostatic potential and the average local ionization energy are valuable tools for understanding and predicting nanotube behavior. They can offer insight into the notable features of charge delocalization in carbon nanotubes of certain types, and into the effects on their reactivities when defects disrupt the carbon networks. We shall begin by analyzing the general characteristics of the electrostatic potentials of model carbon nanotubes and emphasize some remarkable properties of (n,0) tubes that display notably more facile charge delocalization than other types. We shall then examine defective nanotubes using the average local ionization energy, which, as for graphene (discussed above), is a reliable indicator of the relative reactivities of different sites.

The electrostatic potentials of carbon nanotubes on their 0.001 au outer surfaces are usually quite uniform [72, 91], with little variation in comparison to typical organic molecules. For hydrogen-terminated carbon nanotubes, the most significant variations of their surface potentials occur at the ends, whereas for capped tubes, they are in the regions of greatest curvature, that is, at the caps

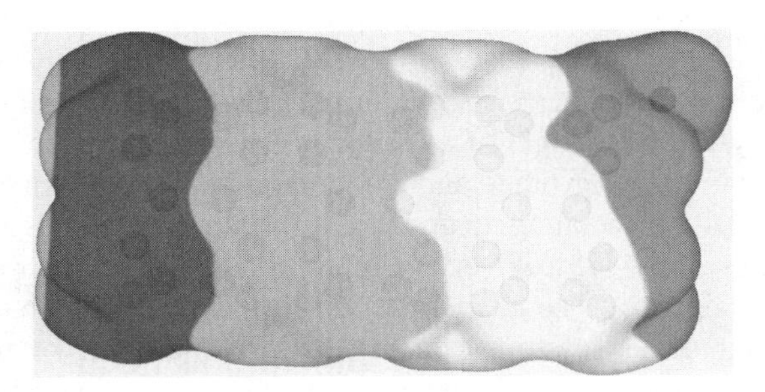

Figure 9.9 Diagram of the electrostatic potential on the 0.001 au surface of an NH_2-end-substituted (6,0) carbon nanotube, at the HF/STO-4G level. The NH_2 group is at the right. The tube is otherwise terminated with hydrogens. Color ranges, in kilocalories per mole, are red $>20 >$ yellow $>0 >$ green $> -20 >$ blue. (*See insert for color representation of the figure*.)

themselves [72, 91, 92]. The lateral surfaces generally show weakly positive (capped tubes) or weakly negative (hydrogen-terminated tubes) potentials, the latter as a consequence of electron withdrawal from the hydrogen atoms [91].

Figure 9.9 shows the surface potential of a hydrogen-terminated (6,0) tube, that has an NH_2 group substituted at one end. Instead of the bland and weak potentials normally found along a tube surface (Fig. 9.10a), Figure 9.9 reveals a striking and unexpected feature: there is a marked gradation of the potential from strongly positive at one end to strongly negative at the other. This remarkable ability to distribute charge along the entire tube length seems to be a feature of (n,0) nanotubes [72, 84, 85]; these are characterized by a large number of C–C bonds that are parallel to the tube axis. Gradations of surface potentials have also been observed in other end-substituted (n,0) tubes [84]. Other types of tubes, such as (5,5) and (n,1), show more localized responses to end substitution [84]. These features suggest that (n,0) carbon nanotubes could be used as "conjugated bridges" in donor-bridge-acceptor systems to yield materials with large nonlinear optical responses [84]. It was found, however, that a novel donor-nanotube paradigm [85] yields even larger nonlinear optical responses than typical donor-nanotube-acceptor systems. It has been further proposed that donor-nanotube-donor motifs may yield yet greater responses [86]. A notable consequence of these observations is the recognition that carbon nanotubes of the (n,0) type can act as reliable charge acceptors, perhaps because of their ability to delocalize the charge throughout their lengths.

Defects are important in nanotube chemistry because they enhance reactivity locally around the defect site, significantly activating the rather unreactive nanotube surface. This is important for the functionalization of carbon nanotubes [93]. Defects have also been instrumental in chemical sensing with nanotubes [77, 78], since adsorption at defect sites produces large electronic responses that increase the sensitivities of the devices [77]. An important type of defect is the so-called Stone–Wales [77, 94], in which four fused six-membered rings are

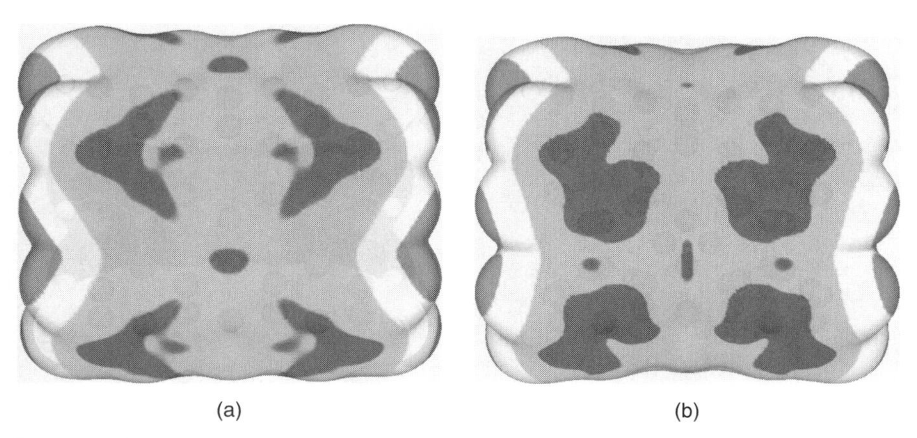

Figure 9.10 Diagram of the electrostatic potential on the 0.001 au surface of a pristine (a) and a Stone–Wales defective (b) (5,5) carbon nanotube, at the HF/STO-4G level. Both tubes are terminated with hydrogens. The defect is in the central portion of the tube facing the reader (b). Color ranges, in kilocalories per mole, are red > 9 > yellow > 0 > green > −5 > blue. (*See insert for color representation of the figure*.)

replaced by two five-membered and two seven-membered rings, following the rotation by 90° of the central bond:

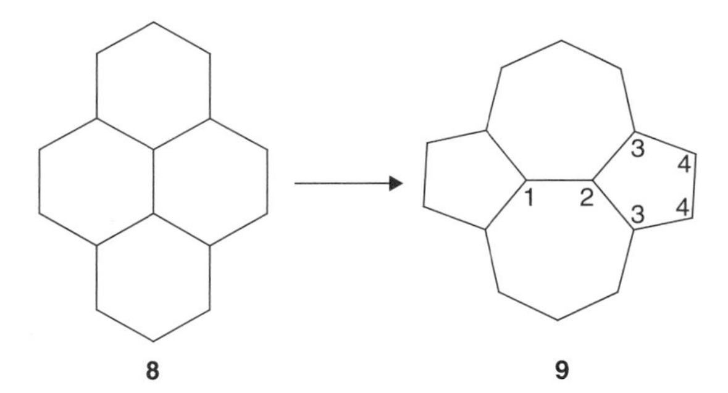

The electrostatic potentials for pristine and Stone–Wales defective (5,5) carbon nanotubes are shown in Figure 9.10. The surface potential for the pristine tube displays the general weakness (lateral sides) that typifies all-carbon or hydrogen-terminated systems. The tube with the Stone–Wales defect has some qualitative changes in the distribution of the most negative potentials around the rings involved in the defect; the bonds on its periphery become more negative. However the effects of the defect on $V_S(\mathbf{r})$ are extremely localized, and the potential on the remainder of the nanotube surface (including the other side, not shown in the figure) seems to be unaffected by its presence.

The average local ionization energy on the outer surface of the tube (Fig. 9.11) will provide a better understanding of the changes brought about by the Stone–Wales defect. The $\overline{I}_s(\mathbf{r})$ of a pristine (5,5) nanotube indicates that

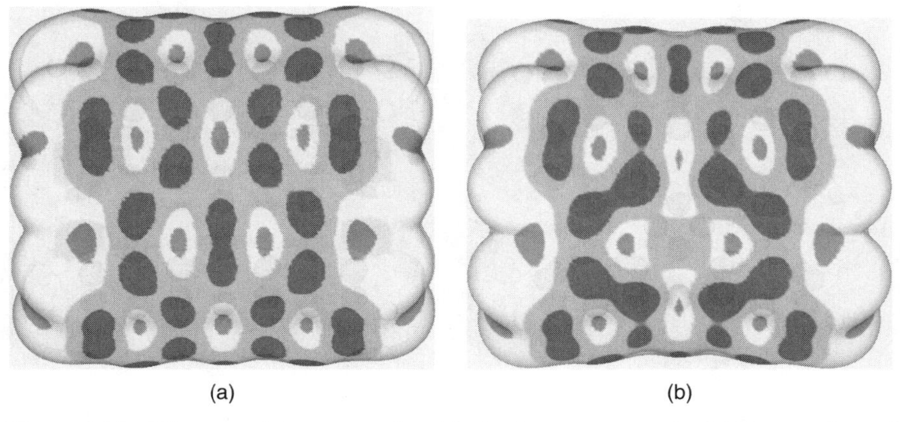

Figure 9.11 Diagram of the average local ionization energy on the 0.001 au surface of a pristine (a) and a Stone–Wales defective (b) (5,5) carbon nanotube at the HF/STO-4G level. Both tubes are terminated with hydrogens. The defect is in the central portion of the tube facing the reader (b). Color ranges, in electron volts, are red > 18.0 > yellow > 16.0 > green > 14.5 > blue. (*See insert for color representation of the figure*.)

all carbons are roughly equivalent, with $\overline{I}_{S,\,min}$ in the 13.6–13.9 eV range. For the nanotube containing a Stone–Wales defect, on the other hand, the carbon atoms involved in the defect show quite different reactivities, as can be seen from the values of the $\overline{I}_{S,min}$ associated with them. C3, which is simultaneously part of five-, six-, and seven-membered rings, has the lowest $\overline{I}_{S,\,min}$, ~13.2 eV. C4, which is a part of both five- and six-membered rings, displays the second lowest $\overline{I}_{S,\,min}$ at ~13.4 eV. C1 and C2, which form the bond common to the two seven-membered rings, have the highest $\overline{I}_{S,\,min}$ found for any carbon atom in either the pristine or the defective tubes (~14.6 eV). Thus, the center of the Stone–Wales defect (C1 and C2) seems to be quite unreactive compared to its periphery. One explanation for this is in terms of the lower curvature at the center of the defect; this results in less local strain, less negative potentials, and diminished reactivity [95].

As pointed out by Dinadayalane et al. [95], the values of the $\overline{I}_{S,\,min}$ on the outer surface of a (5,5) carbon nanotube containing a Stone–Wales defect correlate very well with hydrogen and fluorine chemisorption energies. The lower the $\overline{I}_{S,\,min}$ associated with each carbon, the more negative is the chemisorption interaction energy [95]. These observations emphasize the effectiveness of $\overline{I}_S(\mathbf{r})$ in predicting site reactivity. In this context, it should be noted that a single calculation produces all the $\overline{I}_{S,\,min}$ on the surface [17], whereas direct computation of interaction energies must be done separately at each site [67, 95].

9.5 SUMMARY

With the development of density functional methodology and ever-increasing processing capabilities, meaningful computational analyses of nanomaterials are now a realistic possibility. For instance, nanotubes with more than 70 carbons have been treated at levels that include electron correlation [82, 85, 95, 96]. In this chapter, we focused on two site-specific surface properties, the electrostatic potential $V_S(\mathbf{r})$ and the average local ionization energy $\overline{I}_s(\mathbf{r})$. These provide complementary insights into reactive behavior. $V_S(\mathbf{r})$ reflects the charge distribution, *both* nuclear and electronic, and thus is especially useful with respect to electrostatically driven noncovalent interactions; $\overline{I}_s(\mathbf{r})$ deals with the energetics of electron availability, for example, in covalent bond formation. An important feature, particularly for extended systems such as nanomaterials, is that a single calculation of $V_S(\mathbf{r})$ and $\overline{I}_s(\mathbf{r})$ will identify and rank all likely sites for noncovalent, electrophilic, and free radical reactions, without the need to compute the interaction energies at the various possible sites. We have given three examples of the applications of $V_S(\mathbf{r})$ and $\overline{I}_s(\mathbf{r})$ to nanomaterials. There will be many more in the future.

ACKNOWLEDGMENT

This work was partially supported by the Office of Naval Research.

REFERENCES

1. Hohenberg P, Kohn W. Phys Rev B 1964;136(3):864–871.
2. Hazelrigg MJ Jr., Politzer P. J Phys Chem 1969;73(4):1008–1011.
3. Iwasaki F, Saito Y. Acta Crystallogr B 1970;26(3):251–260.
4. Eisenstein M, Hirshfeld FL. Chem Phys 1979;42(3):465–474.
5. Feil D. Chem Scr 1986;26:395–408.
6. Cady HH. Acta Crystallogr 1967;23(4):601–609.
7. Bader RWF. *Atoms in Molecules: A Quantum Theory*. Oxford: Clarendon Press; 1990.
8. Cremer D, Gauss J. J Am Chem Soc 1986;108(24):7467–7477.
9. Stewart RF. Chem Phys Lett 1979;65(2):335–342.
10. Politzer P, Truhlar DG, editors. *Chemical Applications of Atomic and Molecular Electrostatic Potentials*. New York: Plenum Press; 1981.
11. Wiberg KB, Rablen PR. J Comput Chem 1993;14(12):1504–1518.
12. Bader RFW, Carroll MT, Cheeseman JR, Chang C. J Am Chem Soc 1987;109(26): 7968–7979.
13. Murray JS, Politzer P. Croatica Chim Acta 2009;82(1):267–275.
14. Pathak RK, Gadre SR. J Chem Phys 1990;93(3):1770–1773.
15. Murray JS, Lane P, Politzer P. Mol Phys 1998;93(2):187–194.
16. Frisch MJ, Trucks GW, Schlegel HB, Scuseria GE, Robb MA, Cheeseman JR, Scalmani G, Barone V, Mennucci B, Petersson GA, Nakatsuji H, Caricato M, Li X, Hratchian HP, Izmaylov AF, Bloino J, Zheng G, Sonnenberg JL, Hada M, Ehara M, Toyota K, Fukuda R, Hasegawa J, Ishida M, Nakajima T, Honda Y, Kitao O, Nakai H, Vreven T, Montgomery JA Jr, Peralta JE, Ogliaro F, Bearpark M, Heyd JJ, Brothers E, Kudin KN, Staroverov VN, Kobayashi R, Normand J, Raghavachari K, Rendell A, Burant JC, Iyengar SS, Tomasi J, Cossi M, Rega N, Millam JM, Klene M,

Knox JE, Cross JB, Bakken V, Adamo C, Jaramillo J, Gomperts R, Stratmann RE, Yazyev O, Austin AJ, Cammi R, Pomelli C, Ochterski JW, Martin RL, Morokuma K, Zakrzewski VG, Voth GA, Salvador P, Dannenberg JJ, Dapprich S, Daniels AD, Farkas Ö, Foresman JB, Ortiz JV, Cioslowski J, Fox DJ. *Gaussian 09, Revision A.1*. Wallingford (CT): Gaussian, Inc.; 2009.

17. Bulat FA, Toro-Labbé A, Brinck T, Murray JS, Politzer P. J Mol Model 2010;16(11): 1679–1691.
18. Politzer P, Murray JS, Concha MC. J Mol Model 2008;14(7):659–665.
19. Murray JS, Politzer P. Comput Mol Sci 2010;1(2):153–163.
20. Politzer P, Murray JS. Theor Chem Acc 2002;108(3):134–142.
21. Politzer P. Theor Chem Acc 2004;111(2–6):395–399.
22. Wiener JMM, Grice ME, Murray JS, Politzer P. J Chem Phys 1996;104(13): 5109–5111.
23. Politzer P, Murray JS, Lane P. J Comput Chem 2003;24(4):505–511.
24. Politzer P, Murray JS. In: Chattaraj PK, editor. *Chemical Reactivity Theory: A Density Functional View*. Boca Raton (FL): CRC Press; 2009. p 243–254.
25. Koopmans TA. Physica 1934;1(1–6):104–113.
26. Nesbet RK. Adv Chem Phys 1965;9:321–363.
27. Politzer P, Murray JS, Bulat FA. J Mol Model 2010;16(11):1731–1742.
28. Sjoberg P, Brinck T, Murray JS, Politzer P. Can J Chem 1990;68(8):1440–1443.
29. Nagy Á, Parr RG, Liu S. Phys Rev A 1996;53(5):3117–3121.
30. Gál T, Nagy Á. Mol Phys 1997;91(5):873–880.
31. Murray JS, Brinck T, Politzer P. J Mol Struct (Theochem) 1992;255(1):271–281.
32. Politzer P, Abu-Awwad F, Murray JS. Int J Quantum Chem 1998;69(4):607–613.
33. Politzer P, Murray JS, Concha MC. Int J Quantum Chem 2002;88(1):19–27.
34. Bulat FA, Levy M, Politzer P. J Phys Chem A 2009;113(7):1384–1389.
35. Politzer P, Shields ZP-I, Bulat FA, Murray JS. J Chem Theor Comput 2011;7(2): 377–384.
36. Politzer P, Murray JS. In: Toro-Labbé A, editor. *Theoretical Aspects of Chemical Reactivity*. Amsterdam: Elsevier; 2007. p 119–137.
37. Hagelin H, Murray JS, Brinck T, Berthelot M, Politzer P. Can J Chem 1995;73(4): 483–488.
38. Murray JS, Lane P, Politzer P. J Mol Model 2009;15(6):723–729.
39. Politzer P, Murray JS, Clark T. Phys Chem Chem Phys 2010;12(28):7748–7757.
40. Weiner PK, Langridge R, Blaney JM, Schaefer R, Kollman PA. Proc Natl Acad Sci USA 1982;79(12):3754–3758.
41. Politzer P, Laurence PR, Jayasuriya K. Environ Health Perspect 1985;61:191–202.
42. Naray-Szabó G, Ferenczy GG. Chem Rev 1995;95(4):829–847.
43. Murray JS, Politzer P. J Mol Struct (Theochem) 1998;425(1–2):107–114.
44. Politzer P, Murray JS. Fluid Phase Equilib 2001;185(1–2):129–137.
45. Brinck T, Murray JS, Politzer P. Int J Quantum Chem 1993;48(2):73–88.
46. Bulat FA, Burgess J, Baldwin JW, Murray JS, Politzer P. Assessment of the interactions governing analyte adsorption in functionalized silicon resonator surfaces. 2011. In preparation.

47. Baldwin JW, Zalalutdinov MK, Pate BB, Martin MJ, Houston BH. Proceedings NANO '08, 8th IEEE Conference Nanotechnology; Arlington, Texas, USA. 2008. pp 139–142.
48. Baldwin JW, Houston BH, Pate BB, Zalalutdinov MK. Patterned functionalization of nanomechanical resonators for chemical sensing. US patent Application 20100086735; Int'l patent #'s US20100086735-A1; WO2010040128-A; 2010.
49. Zalalutdinov MK, Cross JD, Baldwin JW, Ilic BR, Zhou W, Houston BH, Parpia JM. J MicroElectroMech Syst 2010;19(4):807–815.
50. Krausa M, Reznev AA, editors. *Vapour and Trace Detection of Explosives for Anti-terrorism Purposes*. Dordrecht: Kluwer; 2003.
51. Colton RJ, Russell JN. Science 2003;299(5611):1324–1325.
52. Grate JW. Chem Rev 2008;108(2):726–745.
53. Becke A. J Chem Phys 1993;98(7):5648–5652.
54. Lee C, Yang W, Parr RG. Phys Rev B 1988;37(2):785–789.
55. Vosko SH, Wilk L, Nusair M. Can J Phys 1980;58(8):1200–1211.
56. Bondi A. J Phys Chem 1964;68(3):441–451.
57. Politzer P, Murray JS, Koppes WM, Concha MC, Lane P. Central Eur J. Energ Mater 2009;6(2):165–180.
58. Novoselov KS, Geim AK, Morozov SV, Jiang D, Zhang Y, Dubonos SW, Grigorieva IV, Firsov AA. Science 2004;306(5696):666–669.
59. Geim AK, Novoselov KS. Nat Mater 2007;6(3):183–191.
60. Geim AK. Science 2009;324(5934):1530–1534.
61. Castro Neto AH, Guinea F, Peres NMR, Novoselov KS, Geim AK. Rev Mod Phys 2009;81(1):109–162.
62. Elias DC, Nair RR, Mohiuddin TMG, Monozov SV, Blake P, Halsall MP, Ferrari AC, Boukhvalov DW, Katsnelson MI, Geim AK, Novoselov KS. Science 2009;323(5914):610–613.
63. Luo Z, Yu T, Kim K, Ni Z, You Y, Lim S, Shen Z. ACS Nano 2009;3(7):1781–1788.
64. Robinson JT, Burgess JS, Junkermeier CE, Badescu SC, Reinecke TL, Perkins FK, Zalalutdniov MK, Baldwin JW, Cuthbertson JC, Sheehan PE, Snow ES. Nano Lett 2010;10(8):3001–3005.
65. Burgess JS, Matis BR, Robinson JT, Bulat FA, Perkins FK, Houston BH, Baldwin JW. Carbon 2011;49(13):4420–4426..
66. Burgess JS, Baldwin JW, Robinson JT, Bulat FA, Houston BH. In: Nelson DJ, Brammer CN, editor. *Fluorine-related Nanoscience with Energy Applications*, Volume 1064, *ACS Symposium Series*. American Chemical Society; 2011. p 11–30
67. Teillet-Billy D, Rougeau N, Ivanovskaya VV, Sidis V. Int J Quantum Chem 2010; 110(12):2231–2236.
68. Murray JS, Abu-Awwad F, Politzer P. J Mol Struct (Theochem) 2000;501–502(1): 241–250.
69. Peralta-Inga Z, Murray JS, Grice ME, Boyd S, O'Connor CJ, Politzer P. J Mol Struct (Theochem) 2001;549(1-2):147–158.
70. Bulat FA, Burgess JS, Matis BR, Baldwin JW Macaveiu L, Murray JS, Politzer P,. Hydrogenation and fluorination of graphene models: Analysis via the average local ionization energy. J Phys Chem A 2011.Submitted for publication.

71. Saito R, Dresselhaus G, Dresselhaus MS. *Physical Properties of Carbon Nanotubes*. London: Imperial College; 1998.
72. Politzer P, Murray JS, Lane P, Concha MC. Handbook of semiconductor nanostructures and nanodevices. In: Balandin AT, Wang KL, editors. Volume 2, *Nanofabrication and Nanoscale Characterization*. Stevenson Ranch: American Scientific Publishers; 2006. p 215–240.
73. Charlier J-C, Blase X, Roche S. Rev Mod Phys 2007;79(2):677–732.
74. Iijima S. Nature 1991;354(6348):56–58.
75. Kim OK, Je J, Baldwin JW, Pehrsson PE, Buckley LJ. J Am Chem Soc 2003; 125(15):4426–4427.
76. Wang J. Electroanalysis 2005;17(1):7–14.
77. (a) Snow ES, Perkins FK, Houser EJ, Badescu SC, Reinecke TL. Science 2005;307:1942–1945; (b) Robinson JA, Snow ES, Badescu SC, Reinecke TL, Perkins FK. Nano Lett 2006;6(8):1747–1751.
78. Snow ES, Novak JP, Lay MD, Houser EH, Perkins FK, Campbell PM. J Vac Sci Technol B 2004;22(4):1990–1994.
79. Thostensona ET, Renb Z, Choua T-W. Compos Sci Technol 2001;61(13):1899–1912.
80. Xiao L, Chen Z, Feng C, Liu L, Bai Z-Q, Wang Y, Qian L, Zhang Y, Li Q, Jiang K, Fan S. Nano Lett 2008;8(12):4539–4545.
81. Aliev AE, Lima MD, Fang S, Baughman RH. Nano Lett 2010;10(7):2374–2380.
82. Bulat FA, Couchman L, Yang W. Nano Lett 2009;9(5):1759–1763.
83. Lia X, Lia C, Lia X, Zhu H, Weia J, Wanga K, Wu D. Chem Phys Lett 2009; 481(4–6):224–228.
84. Politzer P, Murray JS, Lane P, Concha MC, Jin P, Peralta-Inga Z. J Mol Model 2005;11(4–5):258–264.
85. Xiao D, Bulat FA, Yang W, Beratan DN. Nano Lett 2008;8(9):2814–2818.
86. Ma F, Zhou Z-J, Li Z-R, Wu D, Li Y, Li Z-S. Chem Phys Lett 2010;488(4–6): 182–186.
87. Zhang L, Thomas J, Allen SD, Wang Y-X. Opt Eng 2010;49(6):063801.
88. Liu C, Fan YY, Liu M, Cong HT, Cheng HM, Dresselhaus MS. Science 1999; 286(5442):1127–1129.
89. Schlapbach L, Züttel A. Nature 2001;414(6861):353–358.
90. Pehrsson PE, Zhao W, Baldwin JW, Song CH, Liu J, Kooi S, Zheng B. J Phys Chem B 2003;107(24):5690–5695.
91. Peralta-Inga Z, Lane P, Murray JS, Boyd S, Grice ME, O'Connor CJ, Politzer P. Nano Lett 2003;3(1):21–28.
92. Politzer P, Lane P, Concha MC, Murray JS. Microelectr. Eng. 2005;81(2–4):485–493.
93. Mickelson ET, Huffman CB, Rinzler AG, Smalley RE, Hauge RH, Margrave JL. Chem Phys Lett 1998;296(1–2):188–194.
94. Suenaga K, Wakabayashi H, Koshino M, Sato Y, Urita K, Iijimi S. Nature Nanotechnol. 2007;2(6):358–360.
95. Dinadayalane TC, Murray JS, Concha MC, Politzer P, Leszczynski J. J Chem Theor Comput 2010;6(4):1351–1357.
96. Peralta-Inga Z, Boyd S, Murray JS, O'Connor CJ, Politzer P. Struct Chem 2003;14(5):431–443.

10

PROBING ELECTRON DYNAMICS WITH THE LAPLACIAN OF THE MOMENTUM DENSITY

Preston J. MacDougall and M. Creon Levit

"And now, for something completely different."

10.1 INTRODUCTION

The density concept is at once fundamentally important to all of physical science, and, perhaps subconsciously, already familiar to students who are just beginning to explore it. This is why one of the first experiments done in a physical science course, such as general chemistry, often involves measuring the mass density of an unknown. Building on familiar ground, when quantum mechanical models of the atom are introduced, and necessarily abstract entities such as atomic orbitals are presented, a key concept that is skipped at the instructor's peril is the electron density cloud. While the probabilistic nature of the atom is initially uncomfortable for many students, the familiarity of the density concept, combined with fuzzy clouds, helps the medicine go down.

Subsequently, and unfortunately, the simplicity and familiarity of the density concept often serves as a faulty crutch when yet more abstract and convoluted concepts are introduced. For instance, almost all organic chemistry textbooks now contain numerous and colorful electrostatic potential energy maps of oddly shaped organic molecules. Students are almost invariably told that "red areas

A Matter of Density: Exploring the Electron Density Concept in the Chemical, Biological, and Materials Sciences, First Edition. Edited by N. Sukumar.

are regions of high electron density and blue areas are regions of low electron density." Still later in the undergraduate chemistry curriculum, during a physical chemistry course, students may learn how to calculate local electron densities and how to integrate electrostatic potential energies, thus understanding the difference between these physical quantities, but by then the damage is done. Like a scalpel, a physical concept is most useful when it is sharply defined and used with precision.

Several other chapters in this book, which is dedicated to sharply defining and precisely applying the numerous facets of the density concept, have explored the rich vein of topological analysis of the total electronic charge density in position-space—denoted by $\rho(r)$. Instead of considering physicochemical information contained in the distribution of the total electron density as a function of the position of an electron (regardless of spin, binding energy, or momentum), this chapter considers physicochemical information contained in the distribution of the total electron density as a function of the (linear) momentum of an electron (regardless of spin, binding energy, or position)—denoted by $\Pi(p)$.

It is important to note that while a $3n$-dimensional n-electron wavefunction in position-space (neglecting spin) is directly related to its corresponding n-electron wavefunction in momentum-space, by a $6n$-dimensional Fourier–Dirac transformation [1], *no transformation exists between the corresponding three-dimensional densities in position- and momentum-space*. Indeed, by virtue of the Heisenberg uncertainty principle, knowledge of the probability density at a precisely defined point, r, necessitates correspondingly high uncertainty in the value of the momentum, p, of the located electrons. Similarly, knowledge of the probability density at the resolution in p that is presented here, necessitates that the uncertainty in the location of the electrons is larger than the entire van der Waals volume of the molecules studied.

In other words, in the figures presented below, the speed and direction of the electrons that give rise to the topological features observed is precisely known, but we do not know where in the molecule those electrons have a high probability of being located. Often, by comparing topological properties from chemically similar molecules, but with a single chemical substitution, the (dis)appearance of a feature may be inferred to have arisen from the part of the molecules in which the substitution was made. Even then, however, there is uncertainty as the effect of chemical substitution can be short and/or long range. The analysis of six-dimensional Husimi functions is also of interest and allows the simultaneous analysis of "fuzzy" probability distributions of both position and momentum densities [2], but the trade-off is loss of sharply defined topological features as well as the necessity to visualize six-dimensional functions instead of the more manageable three.

With regard to the nature of the physicochemical information that can be expected to be gleaned from analysis of the electron momentum density, it should correspondingly be exclusive of the type of information that has been demonstrated as recoverable from the more familiar face of the electron density, $\rho(r)$.

As has been shown in other chapters in this book, and a large body of interdisciplinary literature, the concepts of atoms, molecular structure, and sites of potential chemical reactivity, which intuitively depend on where electrons are, can be derived via topological analysis of $\rho(r)$ and its Laplacian, $\nabla^2\rho(r)$ [3]. In cases of near perfect transferability, such as a homologous series of hydrocarbons [4], or the side chains of the naturally occurring amino acids [5], there is very high fidelity between the corresponding transferable properties that are predicted by the quantum theory of atoms in molecules, and those that are at the core of the canon of experimental chemistry.

In highly resolved regions in momentum-space, we do not know where the electrons are, rather we know which way they are going and how quickly or slowly. The atomicity of molecules and matter is completely lost in momentum-space! It is, therefore, the electron transport properties of molecules and matter that are most directly probed in momentum-space. The new question is, are there definable features in the computed or measured electron momentum densities that can be correlated with the substance's electron transport properties, such as electrical conductivity and its anisotropy?

It would be reasonable to be skeptical of an affirmative answer. After all, electrical conductivity is a response function, conventionally explained within the context of solid-state (band) theory as resulting from excitation of a multitude of electrons within and/or into the conduction band. Prediction of such collective behavior, based on properties of a single-particle probability density, for a system in a stationary state, especially the ground state, seems too good to be true and is a difficult proposition to swallow. However, the reader has presumably already accepted an analogous proposition above, which has been effectively applied elsewhere in this book, and in a large body of interdisciplinary literature [3]. The topological properties of the Laplacian of the total electron density in position-space, $\nabla^2\rho(r)$, have been shown, both empirically [6] and formally [7], to reflect the number and arrangement of pairwise aggregations of electrons that were first postulated by Lewis in 1916 [8] and later expanded upon by Gillespie in his VSEPR model of molecular geometry [9]. These studies have demonstrated that while the coarse structure of the total single-particle probability density (in position-space) faithfully recovers the atomicity of matter, a result of the dominant electron–nuclear attractions, it gives no direct indication of the important electron correlations that are crucial to even a qualitative understanding of the electronic structure of matter. However, by accentuating the barely perceptible local fluctuations in the total density, resulting from more subtle factors, such as the exceedingly complex correlations in electron–electron interactions that arise from ensuring antisymmetry of the many-electron wavefunction, the topology of $\nabla^2\rho(r)$, which is still a single-particle property density, has a structure that surprisingly bears the imprint of collective behavior. The reflected correlations can range from the partial organization of the octet of valence electrons in the oxygen atom in water molecule into two bonding and two nonbonding pairs of electrons, to the pairwise aggregation of a multitude of valence electrons of the thousands of carbon atoms in a long chain of polyethylene.

In addition to partially reflecting the many-electron correlations in molecules and matter, the topology of $\nabla^2\rho(r)$ also displays "lumps and holes" that predict potential sites of chemical reactivity [3, 6, 10]. In essence, since the interactions of a molecule to approaching reactants is a response function, this proven utility of the Laplacian analysis of computed or measured single-particle probability densities, for molecules or matter in their stationary and ground states, is an example of a proposition that had earlier sounded too good to be true.

In this chapter, we briefly summarize previously published evidence, and present new computational support for the hypothesis that Laplacian analysis of the total electron density in momentum-space (referred to as the momentum density) is a simple and practical tool for probing, classifying, and predicting a wide range of electron transport properties of molecules and matter. Figure 10.1 is like déjà vu all over again. The appearance of the total electron density in position-space is very pedestrian, displaying a topology that reflects the dominant physical force in molecules. A tremendous amount of chemical insight is revealed by analysis of its Laplacian. In the free-electron theory of metals, all metals have basically the same total momentum distribution—a spherical step function with a rounded edge at the Fermi momentum. Again, the Laplacian analysis reveals much more structure. But how do we interpret it?

10.2 COMPUTATIONAL METHODS

All molecular computations reported or discussed in this chapter were obtained via ab initio calculations using Gaussian at the Hartree–Fock level of theory (6-311g** basis set) [11]. All geometries were fully optimized at this level, except when certain geometrical parameters were modified, as noted in the corresponding discussion. Selected calculations were repeated with second order Møller–Plesset perturbation theory, and there were no discernable differences in the topological properties of $\nabla^2\Pi$ reported here. Following calculation of the orbital-based wavefunctions, Fourier–Dirac transformations and a sum over orbital densities were executed to generate cubic grids of total momentum density at 0.02 au intervals, out to momenta of ± 1.0 au in each direction. For each of these grids, values of $\nabla^2\Pi$ were computed numerically using FAST, the Flow Analysis Software Toolkit developed for classical fluid dynamics by the NASA Ames Research Center [12].

10.3 A POSTULATE AND ITS EXISTING SUPPORT

In previous work, we have shown, with a single simple postulate, that the topological properties of the Laplacian of the electron momentum density, $\nabla^2\Pi(p)$, can be interpreted as a probe of the electron dynamics of atoms, molecules, or macroscopic systems [13]. On the basis of intuitive dynamical models, it is postulated that the only electron–electron interactions that are resistive (hinder electrical

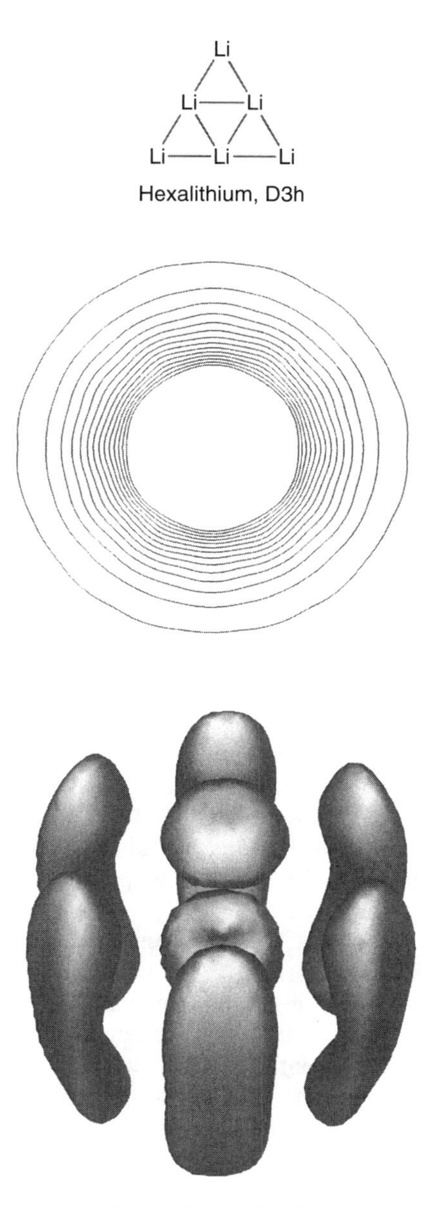

Figure 10.1 The contour plot is for $\Pi(p)$ calculated in one of the σ_v planes of symmetry of the hexalithium cluster. The data can be described as a plateau with a fluted and steep rim. The lower image is an isovalue surface for $\nabla^2\Pi$ in the same cluster. The vertical axis corresponds to the component of an electron's momentum perpendicular to the plane of the nuclei. All momenta coordinates inside the envelope are momenta for which the electron dynamics are locally laminar ($\nabla^2\Pi < 0$). The doughnut in the center is a momentum concentration at the origin in momentum-space. The momentum coordinates in both images extend out to ± 1.0 au.

conductivity) are those between electrons whose momenta are simultaneously within the same small sphere in momentum-space but are not equal [13].

The intuitive foundation of this postulate can be illustrated with an abstract example. Imagine an infinite one-dimensional chain of jellium (electrons embedded in a continuum of positive charge) with a Boltzmann distribution of electron momenta in either direction. No matter where they are, electrons that have exactly the same momentum have not collided in the immediate past and will not collide in the immediate future. For electrons that have greatly differing momenta, if and when they collide, they will do so with a very short interaction time, effectively an elastic collision. For electrons that differ only slightly in momentum, if they collide, they will do so with a long interaction time, which we term *a resistive interaction*. Maxwell's interpretation of the Laplacian of a scalar field [14], when applied to the Laplacian of the total electron momentum density (Eq. 10.1), helps make the connection between the intuitive model and the electron dynamics of a three-dimensional many-electron system.

$$\Pi(p)-\Pi_{\text{ave}}=-\tfrac{1}{10}p^2\nabla^2\Pi(p)+O(p^4). \tag{10.1}$$

In Equation 10.1, $\Pi(p)$ is the value of the total electron momentum density at a point p in momentum-space and integrated over all possible electron positions. The average value of the momentum density within a small sphere (with radius p), which is centered at p, is denoted Π_{ave}. If the sampling sphere is small, then the remaining terms in the expansion, beginning with a term including p^4, are negligible. In the case of interpreting $\nabla^2\rho(r)$, the local minima and maxima in the Laplacian correspond to lumps and holes, respectively, and the connection to sites of chemical reactivity was intuitive. From Equation 10.1, it is evident that regions in momentum-space where the Laplacian is negative (local concentrations) correspond to places in momentum-space where electrons, which can be anywhere in the molecule, are more likely to have the same momenta than to have slightly different momenta. The opposite is true for regions in momentum-space where the Laplacian is positive: electrons are less likely to have the same momenta than to have slightly different momenta. Also, the larger the magnitude of the Laplacian, the greater is the disparity between the probabilities of same momenta versus slightly different momenta.

Invoking the postulate above, when most of the electrons in a small sphere in momentum-space have the same momenta (regions where $\nabla^2\Pi < 0$), these electrons will experience few resistive interactions and the electron dynamics can be described as locally laminar. The situation is reversed when a small fraction of electrons have the same momenta (regions where $\nabla^2\Pi > 0$), and the electron dynamics can be described as locally nonlaminar or locally turbulent. Note that in this context, "locally" does not refer to a portion of a molecule but to a small region of momentum-space. We can therefore see that with a single, simple postulate, which is easily implemented on computed or measured data, all momentum-space is partitioned into regions where the electron dynamics are, to varying degrees, either locally laminar or locally turbulent. In addition, these regions balance out, since the overall integration of the Laplacian for a smooth, well-behaved function must be zero.

Under the influence of an external electric field, however, it is very clear that the overall electron dynamics of most forms of matter do not balance out. In fact, few physical properties of matter vary over as many orders of magnitude as does electrical conductivity. The additional factor that must be considered, in addition to whether electrons are locally laminar or turbulent, is whether or not they are weakly or strongly bound. Sagar et al. have reported shell structure in $\nabla^2\Pi$ for spherically averaged atoms and ions, much like that found for $\nabla^2\rho$ [15]. Alternating shells of momentum concentration and depletion are found for each quantum shell, including the core. Thus, core electrons have regions in momentum-space where they are locally laminar, but they will not contribute significantly to the electrical conductivity of a substance.

As one might expect since core electrons have greater binding energies, Sagar et al. reported that electrons in core orbitals contribute disproportionately to the shells of momentum concentration at higher momentum values. Conversely, valence electrons contribute disproportionately to the slow regime, which is approximately up to 0.5 au of momentum. We can relate the Laplacian probe of electron dynamics discussed above to the electron transport properties of molecules and matter by recognizing that primarily valence electrons contribute to electrical conductivity in the presence of an external electric field. Consequently, a substance that has very negative values of $\nabla^2\Pi$ in the slow regime will behave as a metal since its least tightly bound electrons have laminar dynamics. On the other hand, a substance that has very positive values of $\nabla^2\Pi$ in the slow regime will behave as an insulator since its least tightly bound electrons have turbulent dynamics. Semiconductors are anticipated to be substances whose slow electron dynamics transition from turbulent to laminar ($\nabla^2\Pi$ at low values of p goes from positive to negative) in the presence of an external electric field.

The original computational support for the physical validity of the postulate, and the resulting relationship between the topology of $\nabla^2\Pi$ and the electron transport properties of molecules and matter, was heuristic [13]. The values of $\nabla^2\Pi$ at the origin in momentum-space, electrons with zero momentum, can be computed from MacLauren expansions of the spherically averaged total electron momentum density that have been tabulated by Thakkar et al. for ground state atoms of elements from H to U [16]. It was found that, with few exceptions, atoms of metallic elements had laminar slow electron dynamics, sometimes highly so, whereas atoms of nonmetallic elements had slightly turbulent slow electron dynamics. It is interesting to note that, of all the elements, an atom of silicon had slow electron dynamics that were nonlaminar, but the least so on a per electron basis (i.e., $\nabla^2\Pi/\Pi$ at the origin in momentum-space). In other words, assuming the validity of the postulate, the semiconducting nature of silicon's valence electrons is an intrinsic property of the element [13].

An essential property of semiconductor devices is that their electron transport properties be switchable, preferably with the application of a modest electric field.

So-called Tour wires are small conjugated organic molecules that can be functionalized and assembled into larger electronic devices that may eventually constitute carbon-based molecular electronics [17, 18]. The veracity of the postulate, and the utility of the Laplacian analysis for predicting electronic transport properties of molecules and matter in conditions relevant to electronic applications, has been successfully tested for molecules that have been proposed as molecular diodes, with both rectifying and resonant tunneling behavior [18]. Tolane, which contains two phenyl rings linked by an alkyne group, is the prototypical Tour wire. At around 0.5 au of momentum, there are several local concentrations of momentum density, but the slowest electrons (nearest the origin) are turbulent. This topology of $\nabla^2\Pi$ is consistent with the expectations of a semiconductor. When the molecule is embedded in a uniform external electric field of 0.05 au (2.6×10^{10} V/m), the moderately fast momentum concentrations shrink in size but remain. However, there is a topological transition in the slow regime. The momentum depletion at the origin dramatically switches to a pronounced momentum concentration. In other words, our model predicts that the slow electron dynamics of the tolane molecule suddenly switches from semiconducting to metallic behavior in the presence of an external electric field.

In addition to switches, electronic devices must also possess logic gates, through which the current can be induced to flow in one way only. By replacing only one of the para hydrogens in tolane with sulfur, an internal bias is introduced in the molecule. Relative to tolane, the tolane thiolate anion has several conserved features, as well as a few additional momentum concentrations, in the moderately fast regime. However, it also has turbulent slow electron dynamics, as a semiconductor would. As for tolane, the application of a weak electric field of 0.01 au, in either direction, causes no noticeable change in the topology of $\nabla^2\Pi$. However, when the field strength in the reverse bias direction is increased to a moderate 0.025 au, the slow-regime transition observed in tolane is not seen. The slow electron dynamics remain turbulent. Amazingly, when a moderate external field is directed with a forward bias, the slow electrons in tolane thiolate again undergo a sudden transition to laminar! [18]

10.4 STRUCTURE OF MOTION, TRANSFERABILITY, AND ANISOTROPY

A cornerstone of the quantum theory of atoms in molecules, according to its primary architect, is that "The constancy in the properties of an atom of theory, including its contribution to the total energy of a system, is observed to be directly determined by the corresponding constancy in its distribution of charge" [19]. When atomicity is lost, is transferability absent as well?

Figures 10.2 and 10.3 illustrate what we refer to as the structure of motion (electronic) in models of important examples of carbon-based electronic materials. Just as key characteristic aspects of the electron dynamics of bulk silicon were evident in the Laplacian of the momentum density of a single atom of silicon, we have probed the slow electron dynamics of a single molecule of naphthalene as a simple model of single-walled carbon nanotubes. Parameterized computational methods that separate and model only the π electrons in conjugated organic molecules, such as the Hückel molecular orbital theory, are surprisingly good at predicting chemical reactivity trends. Yet, when ab initio methods are used, and properties of the total electron density are investigated, a strong interdependence has been found between the σ and π electron subsets [20]. In addition, when topological properties of either ρ or $\nabla^2\rho$ are investigated, there is nothing that by any stretch of the imagination could correspond to the tori of π density that are depicted above and below aromatic rings in most organic chemistry textbooks [21]. Fundamentally, all methods that put symmetry labels on electrons are severely limiting, first because (aside from spin) electrons are indistinguishable, and second, because most systems with practical interest have little or no symmetry. Nevertheless, comparison of the images in Figure 10.2 makes clear that while any anticipated σ/π features were blurred together in the observable (total) density in position-space, there appear to be separately identifiable topological features in $\nabla^2\Pi$. Naphthalene and azulene, the blue hydrocarbon, have different σ-bond frameworks (two fused six-member rings in the former, fused seven- and five-member rings in the latter), whereas the aromaticity of both, and the anomalously large dipole moment of the latter, are among the physical properties that lead chemists to assign hextets of π electrons to each ring in both molecules. The dipole in azulene results from the transfer of one π electron from the seven-member ring to the five-member ring and is responsible for its intense blue color. As for tolane and tolane thiolate (in the absence of an electric field), the slow electron dynamics of naphthalene and azulene are nonlaminar ($\nabla^2\Pi > 0$ at the origin). The momentum concentrations depicted in Figure 10.2 are in the moderately fast valence electron regime. We have not yet determined if the topology of $\nabla^2\Pi$ is stable or susceptible to external electric fields, although it would be very interesting to do so.

It is clear that one set of features in Figure 10.2 is not conserved, while another set is highly conserved. The "snouts" that appear to be almost perfectly transferable between images (there are two in each image, one coming toward the viewer and another on the opposite side) correspond to local concentrations of electron momentum density that we interpret as characteristic of the structure of motion of the π electrons in these fused aromatic rings. The concentric rings of charge concentration, which have different forms in the two images, are in the plane in momentum-space corresponding to electrons whose momenta are in, or parallel to, the plane containing the nuclei, in other words, the σ plane (Fig. 10.2). The snouts in Figure 10.2 are not hexagonal because of the lowered symmetry of the fused rings. (In benzene, the corresponding features must have hexagonal symmetry, but presumably they would otherwise be similar in form and their relative location in momentum-space.) The symmetry of fused aromatic rings in nanotubes is further reduced when planar graphene is rolled into tubes [22]. We have modeled this distortion by forcing an interplane angle of 20° between the rings in naphthalene and azulene (which models adjacent ring-size defects that

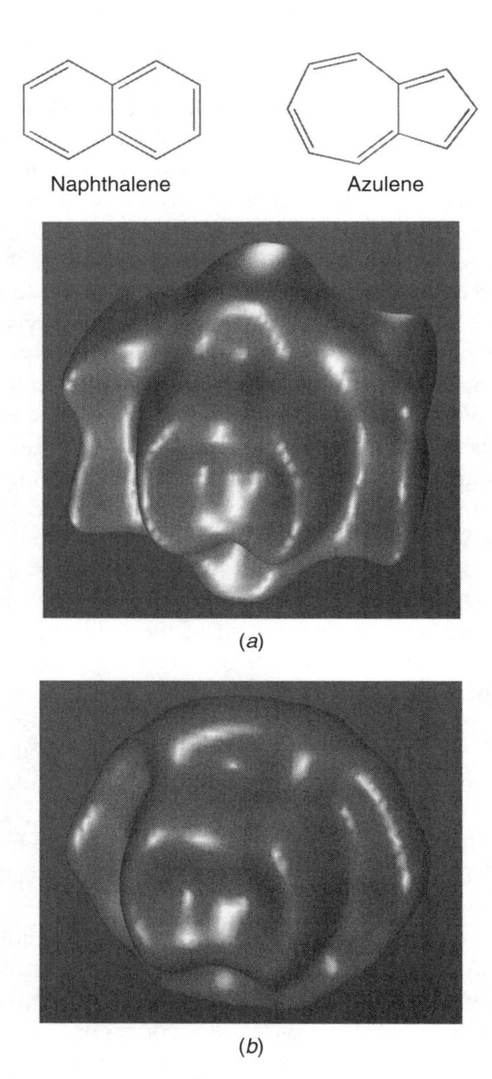

Figure 10.2 Isovalue surfaces for $\nabla^2\Pi = -0.015$ au for naphthalene (a) and azulene (b). All momenta coordinates inside the envelopes are momenta for which the electron dynamics are locally laminar ($\nabla^2\Pi < 0$). The vertical axis in both images corresponds to the component of electron momentum parallel to the direction of the C–C bond that is shared by both rings. The axis that appears to be coming toward the viewer is the component of electron momentum that is perpendicular to the plane containing the nuclei in both molecules. The momentum coordinates in both images extend out to ± 1.0 au. (*See insert for color representation of the figure*.)

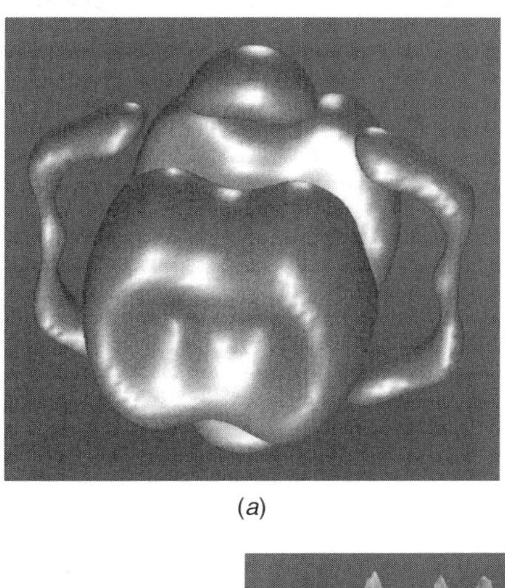

(a)

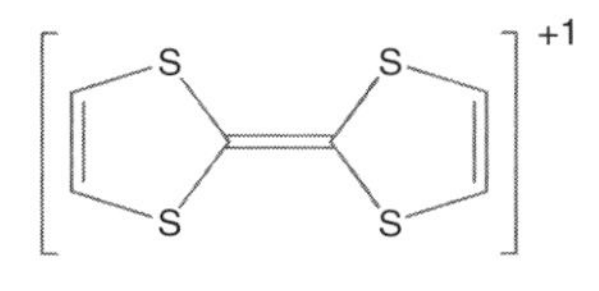

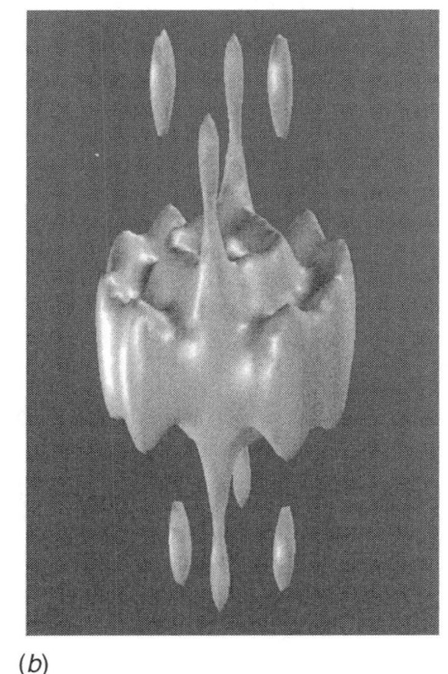

(b)

Figure 10.3 (a) An isovalue surface for $\nabla^2\Pi = -0.030$ au for naphthalene, with the same axis system shown in Figure 10.2. All momenta inside the envelope, therefore, have locally more laminar electron dynamics than points on the surface shown in Figure 10.2. (b) Isovalue surface for which $\nabla^2\Pi$ is marginally less than zero, computed for the TTF radical cation. The vertical axis corresponds to the component of electron momentum that is perpendicular to the plane containing the nuclei. The axis that appears to be coming toward the viewer corresponds to the component of electron momentum that is parallel to the central C–C bond. The momentum coordinates in both images extend out to ± 1.0 au. (*See insert for color representation of the figure*.)

are frequently found in nanotubes), and in both cases, the topological properties of $\nabla^2\Pi$ are unchanged. The snouts of π electron dynamics are imperceptibly altered (not shown), even though σ/π separability is formally absent.

The electronic transport properties of single-walled carbon nanotubes depend sensitively on the angle between the axial direction and the C–C bond vector, as well as the tube diameter [22]. This angle is zero for the so-called armchair nanotubes, which are always metallic. When the angle is between 0° and 30°, the nanotube is chiral and may be metallic or semiconducting. Figure 10.3a shows an interesting and perhaps related anisotropy. For electronic motion in the valence regime, that is within or parallel to the plane of the aromatic rings in naphthalene (which would also be required for ballistic electron transport within graphene or a nanotube), the direction with the highest laminarity is also parallel to the fused C–C bond vector. To illustrate this property of the structure of motion in naphthalene, we have simply chosen a higher magnitude contour of $\nabla^2\Pi$ than that shown in Figure 10.2. The electron momentum concentrations that most closely correspond to the conducting electrons in armchair nanotubes are the kidney-bean-shaped features (one is hidden) above and below the snouts in Figure 10.3. The adjacent depletions indicate the sensitivity of the valence electron dynamics to the angle between the direction of electron motion and the fused C–C bond vector.

The topological properties of $\nabla^2\Pi$ also appear to reflect the unexpected electron transport anisotropy that is observed for salts of the organosulfur compound tetrathiafulvalene (TTF) [23]. In these compounds, the direction of highest electrical conductivity is not parallel to the planes of π conjugation, but rather more closely perpendicular to it, in the direction of π–π stacking interactions. The tall spikes of momentum concentration that are seen in Figure 10.3b correspond to laminar valence electron dynamics that have a fixed direction relative to the π plane of the central C–C bond (motion is parallel to it), but a varying angle relative to the σ plane of a single molecule of the TTF radical cation. If the spikes had been along the center axis of the image, it would indicate that motion exactly perpendicular to the σ plane was the most laminar. The noncentral location of the spikes may be related to the herringbone arrangement (instead of like pancakes) that is observed for the ion stacking in these synthetic metals [23].

10.5 CONCLUSION

The structure of motion for the heterocyclic system in Figure 10.3 is more complex than that seen in Figures 10.1 and 10.2, and we do not doubt that the topology of $\nabla^2\Pi$ for systems that include transition metals will add further complexity. Regardless of whether or not our postulate is valid, further investigation of the topological properties of the Laplacian of the electron momentum density, both computationally and experimentally, will yield interesting insight into the structure of electronic motion in molecules and matter. If our postulate is strengthened, empirically and/or theoretically, such studies will add significantly to our understanding of electron transport properties of matter. Most importantly, this understanding will come without *a priori* assumption of a particular independent-particle model and all the known, as well as hidden, limitations that come with making that choice.

ACKNOWLEDGMENTS

This chapter was dedicated to the late R. F. W. Bader on the occasion of his eightieth birthday. PJM thanks the Office of Science in the US Department of Energy for the funding (DE-SC00005094).

REFERENCES

1. McWeeny R. Rev Mod Phys 1960;32:335.
2. (a) Husimi K. Proc Phys Math Soc Jpn 1940;22:264; (b) Harriman JE. J Chem Phys 1988;88:6399; (c) Schmider H. J Chem Phys 1996;105:3627; (d) Schmider H, Hô M. J Phys Chem 1996;100:17807.
3. Bader RFW. Chem Rev 1991;91:893, references therein.
4. Wiberg KB, Bader RFW, Lau CDH. J Am Chem Soc 1987;109:1001.
5. Matta CF. In: Matta CF, editor. *Quantum Biochemistry—Electronic Structure and Biological Activity*. Weinheim: Wiley-VCH; 2010. p 423–472.
6. (a) Bader RFW, Lau CDH, MacDougall PJ. J Am Chem Soc 1984;106:1594; (b) MacDougall PJ. The Laplacian of the electronic charge distribution [PhD dissertation]. McMaster University; 1989.
7. (a) Bader RFW, Gillespie RJ, MacDougall PJ. J Am Chem Soc 1988;110:7329; (b) Bader RFW, Heard GL. J Chem Phys 1999;111:8789.
8. Lewis GN. J Am Chem Soc 1916;38:762.
9. Gillespie RJ. *Molecular Geometry*. London: Van Nostrand Reinhold; 1972.
10. (a) Bader RFW, MacDougall PJ. J Am Chem Soc 1985;107:6788; (b) MacDougall PJ, Henze CE. Theor Chem Acc 2001;105:345.
11. Frisch MJ, Trucks GW, Schlegel HB, Gill PMW, Johnson BG, Robb MA, Cheeseman JR, Keith TA, Petersen GA, Montgomery JA, Raghavachari K, Al-Laham MA, Zakrzewski VG, Ortiz JV, Foresman JB, Peng CY, Ayala PY, Wong MW, Andres JL, Replogle ES, Gomperts R, Martin RL, Fox DJ, Binkley JS, Defrees DJ, Baker J, Stewart JJP, Head-Gordon M, Gonzales C, Pople JA. *Gaussian 94 (Revision D.3)*. Pittsburgh (PA): Gaussian, Inc.; 1995.
12. FAST: Watson V, Merritt F, Plessel T, McCabe RK, Castegnera K, Sandstrom T, West J, Baronia R, Schmitz D, Kelaita P, Semans J, Bancroft G. Moffett Field (CA): NASA Ames Research Center, NAS Division. Information about FAST is at http://www.nas.nasa.gov/publications/sw_descriptions.html#FAST. Accessed 2012 May.
13. MacDougall PJ. Can J Phys 1991;69:1423.
14. Maxwell JC. Volume I, *A Treatise on Electricity and Magnetism*. New York: Dover; 1954. p 31.

15. Sagar RP, Ku ACT, Smith VH Jr., Simas AM. J Chem Phys 1989;90:6520.
16. Thakkar AJ, Wonfor AL, Pedersen WA. J Chem Phys 1987;87:1212.
17. (a) Tour JM, Wu R, Schumm JS. J Am Chem Soc 1991;113:7064; (b) Goldhaber-Gordon D, Montemerlo MS, Love JC, Opiteck GJ, Ellenbogen JC. Proc IEEE 1997;85:521.
18. MacDougall PJ, Levit MC. In: Dadmun MD, Van Hook WA, Melnichenko YB, Sumpter BG, editors. *Computational Studies, Nanotechnology, and Solution Thermodynamics of Polymer Systems*. New York: Kluwer Academic/Plenum; 2000. pp. 139–150.
19. Bader RFW. *Atoms in Molecules: A Quantum Theory*. Oxford University Press; 1990. p 3.
20. (a) Slee TS. J Am Chem Soc 1986;108:606; (b) Slee TS, MacDougall PJ. Can J Chem 1988;66:2961.
21. Bader RFW, Chang C. J Phys Chem 1988;93:2946.
22. Dresselhaus MS, Dresselhaus G, Jorio A. Annu Rev Mater Res 2004;34:247.
23. Bendikov M, Wudl F, Perepichka DF. Chem Rev 2004;104:4891.

11

APPLICATIONS OF MODERN DENSITY FUNCTIONAL THEORY TO SURFACES AND INTERFACES

G. PILANIA, H. ZHU, AND R. RAMPRASAD

11.1 INTRODUCTION

A surface or an interface provides a doorway through which any solid contacts and interacts with the external atmosphere or a second solid phase. At a fundamental level, surfaces and interfaces present model systems in which physics in two dimensions can be investigated and chemistry of bond breaking and bond formation between dissimilar systems can be studied [1, 2]. Understanding of surfaces and interfaces has not only extended our knowledge of basic physical and chemical sciences but also played key roles in the successful realization of many industrial processes [3–5]. For instance, surfaces form the basis of heterogeneous catalysis without which the present day chemical industry would not exist (at least the way we know it). Surface and interface science phenomena are pervasive in situations involving superlattices, crystal growth control, corrosion abatement, and nanostructured systems (where the surface or interface to volume ratio is large). Owing to its interdisciplinary nature, surface/interface science derives frequent contributions from physical, chemical, and materials sciences.

Density functional theory (DFT) based computations have been used for more than two decades in the arena of surface/interface science [6]. Several success stories of DFT in surface science are already well documented and hence are not repeated here. Famous examples such as the correct prediction of the

A Matter of Density: Exploring the Electron Density Concept in the Chemical, Biological, and Materials Sciences, First Edition. Edited by N. Sukumar.

Si(001)-(2 × 1) [7–12] and Si(111)-(7 × 7) [13, 14] reconstructed surfaces are now part of standard textbooks in the field [15]. DFT-based methods have not only evolved into an important tool for analyzing surface geometries and surface phases at various temperature, pressure, and chemical environments but also been applied successfully to model real-life heterogeneous surface catalysis [16, 17]. There are numerous examples in which DFT calculations have preceded experimental observations (see e.g., Reference 18). A more recent development that has contributed to the bridging of the experiment-theory gap is first principles thermodynamics (FPT) [19–21] which, as we discuss in detail later, involves a seamless combination of zero-temperature DFT results with statistical mechanics to provide pressure- and temperature-dependent observable properties that can be directly compared with experiments.

This chapter attempts to provide examples of some recent contributions made in the arena of surface/interface science using state-of-the-art DFT-based computations. While this is by no means a comprehensive account (and is highly colored by the authors' work and perspectives), materials systems and methodologies spanning several application areas including catalysis, crystal growth, and electronics are explored. Methodological details are provided where appropriate, and several references are provided when the scope and length of this presentation precludes a lengthy exposition of basic concepts.

11.2 THE PREDICTIVE CAPABILITY OF DFT

Among all modern electronic structure methods, DFT [22–26] is seen to offer the best trade-off between computational cost and accuracy. These methods are also referred to as "first-principles" or "ab initio" techniques to emphasize that there are no system-specific fitted parameters utilized during the course of such calculations. DFT has developed into a popular approach for predicting various structural and electronic properties of a wide range of materials systems including molecules, bulk solids, surfaces, and other low-dimensional nanostructures. In this section, we summarize the level of accuracy that one may expect from DFT calculations for some of the more basic properties before we plunge into surface/interface-based discussions.

Within Kohn–Sham DFT [22, 23], the following eigenvalue equation (in atomic units) is solved:

$$[-\nabla^2 + V_{\text{eff}}(r)]\Psi_i(r) = \epsilon_i \Psi_i(r), \tag{11.1}$$

where the first term in brackets represents the electronic kinetic energy (with ∇ being the gradient operator) and the second term, $V_{\text{eff}}(r)$, represents the effective potential energy seen by an electron. $V_{\text{eff}}(r)$ contains all the electron–electron and electron–nuclear interactions, as well as the potential caused by an external electric field. In practice, the quantum mechanical part of the electron–electron interaction is approximated using (semi)local functionals such as the local density approximation (LDA) or generalized gradient approximations (GGA), or nonlocal hybrid functionals. $\Psi_i(r)$ and ϵ_i represent the spectrum of Kohn–Sham orbital wave functions and orbital energies, respectively, indexed by i. We note that for any given set of atomic positions (i.e., for given $V_{\mathrm{eff}}(r)$), the above equation is solved self-consistently to result in converged charge densities (obtained from the wave functions of the occupied states), total energies (obtained from the wave functions and eigenenergies of the occupied states), and atomic forces (obtained from the first derivative of the total energy with respect to the position of any given atom). The atomic coordinates are optimized by the requirement that the total energy of the system is a minimum and that the forces on each atom are close to zero. Once the geometry is converged, several other properties of interest may be computed (as described in the rest of this chapter).

Some comments concerning the expected accuracy of DFT predictions are in order. The greatest strength of DFT is its ability to predict structural details of materials, typically to within 1% of experimental values. Figure 11.1 shows the correlation between DFT predictions of structural properties and experimental data for several classes of systems. Vibrational frequencies of molecules, phonon frequencies of solids, elastic constants of solids, and relative energies are predicted to within 2% of experiments by DFT. Figure 11.2a compares DFT predictions of vibrational frequencies of diatomic molecules with experiments, and Figure 11.2b shows a similar comparison for bulk and shear moduli for various solids. Dielectric constants of insulators are typically predicted to within 5% of experiments, as portrayed in Figure 11.2c for both static and optical dielectric constants. The larger discrepancy in this case is primarily caused by the lack of sufficiently accurate single-crystal experimental data.

The greatest deficiency of DFT is its inability to predict band gaps in semiconductors and insulators to the same level of accuracy achievable in the case of the other properties; DFT band gaps are underestimated relative to experimental determinations by up to 50%, as shown in Figure 11.2d. However, the shape and the width of the bands, and trends in changes in the band gap (e.g., due to external pressure), are predicted accurately [35]. The above deficiencies are the consequences of the approximations made within DFT such as the LDA or GGA, which include spurious electron self-interaction effects. Techniques to handle such deficiencies are currently available and include the use of hybrid functionals that are rising in popularity [36] and quasiparticle *GW* (note that GW is not an acronym; here G and W represent the Green function and screened coulomb interaction, respectively) corrections to the electronic energy levels (References 34 and 37). Such treatments, although computationally more expensive than conventional DFT, result in satisfactory agreement of the computed band gaps with experiments as shown in Figure 11.2d for the case of *GW* corrections.

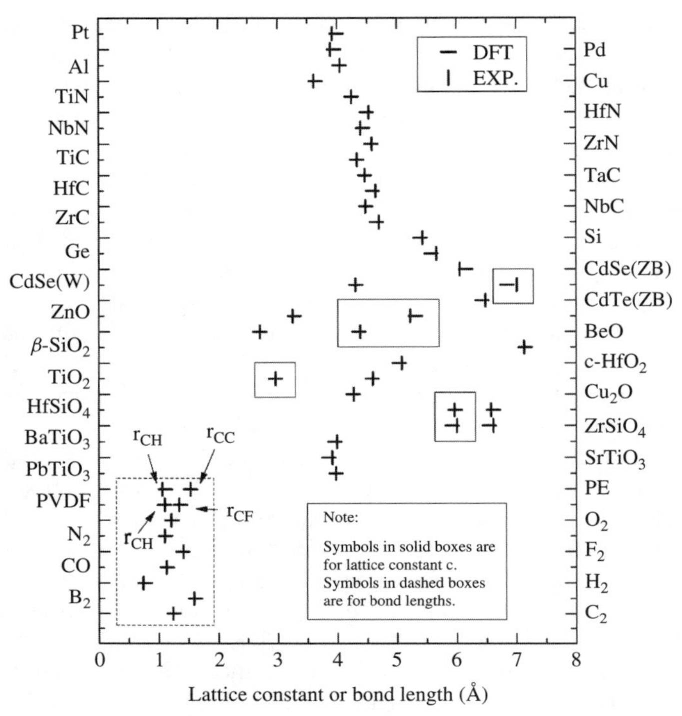

Figure 11.1 Comparison of DFT-computed structural parameters with experiments for a variety of metals, semiconductors, insulators, polymers, and molecules. ZB and W represent the zinc blende and wurtzite crystal structures, respectively. Data are from various sources [27–30]. (Reprinted with permission from R. Ramprasad, N. Shi, and C. Tang. Modeling the physics and chemistry of interfaces in nanodielectrics. In Dielectric Polymer Nanocomposites. J. K. Nelson (Ed.), Springer (2010). Copyright 2010 by Springer.)

11.3 SLAB MODELS USED IN SURFACE/INTERFACE STUDIES

At a surface/interface, periodicity and translational symmetry of a bulk crystal are destroyed. Atoms at these boundaries possess broken bonds and reduced or altered coordination as compared to their bulk counterparts. To simulate a surface/interface, either one has to specifically apply two-dimensional periodic boundary conditions, or one could impose three-dimensional periodic boundary conditions along with incorporation of a vacuum region normal to the surface/interface. The latter approach, referred to as *the slab supercell model*, is more commonly used (as many DFT codes implicitly involve three-dimensional periodic boundary conditions). The supercell, when repeated in all three dimensions, gives rise to a series of stacked slabs of the material separated by vacuum spaces.

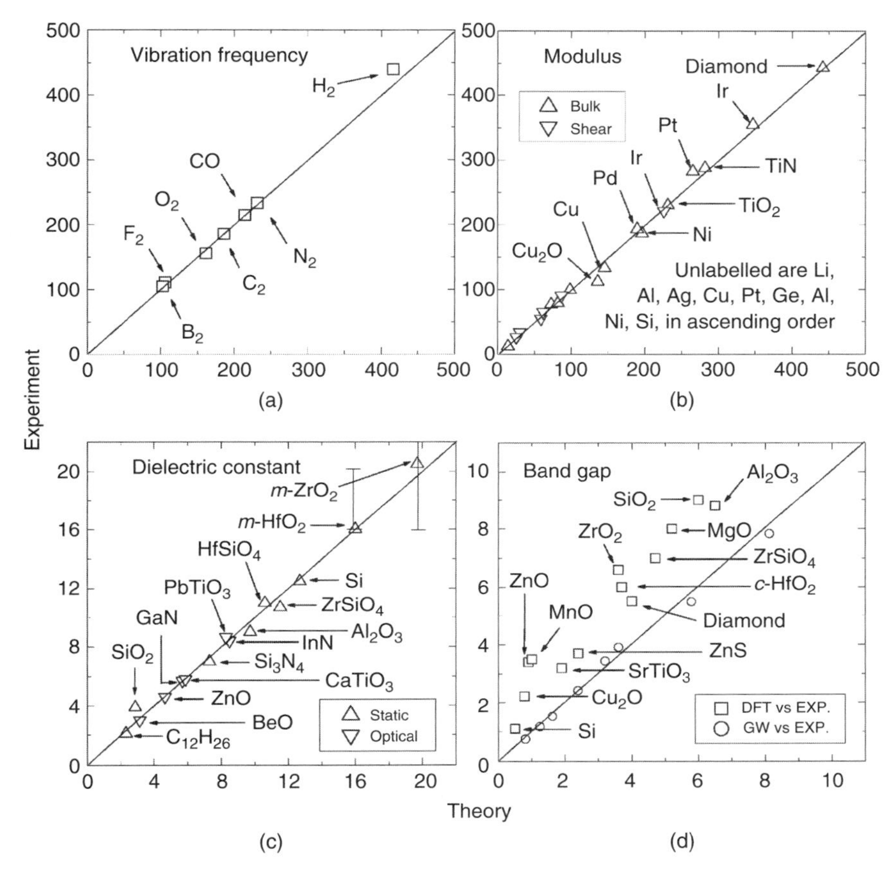

Figure 11.2 Comparison between DFT predictions (horizontal axes) and experimental values (vertical axes) [30]. (Reprinted with permission from R. Ramprasad, N. Shi, and C. Tang. Modeling the physics and chemistry of interfaces in nanodielectrics. In Dielectric Polymer Nanocomposites. J. K. Nelson (Ed.), Springer (2010). Copyright 2010 by Springer). Comparisons for (a) vibrational frequencies of diatomic molecules, (b) bulk and shear moduli of solids, (c) static and optical dielectric constants of insulators, and (d) band gaps of insulators. Frequencies are in cm^{-1}, elastic moduli in gigapascals, and band gaps in electron volts. Data are from various sources including References 28, 29, 31–33. Data for *GW* method in (d) are from Reference 34.

One should always bear in mind that the real surfaces of solids, even when they have no foreign contaminants, are seldom perfect two-dimensional planes. Rather, they contain many imperfections (such as surface vacancies and adatoms), steps, facets, islands, etc., as illustrated schematically in Figure 11.3. To a first approximation, one can assume that when a perfect crystal is sliced along a plane, none of the remaining atoms moves from its original location in the crystal and a perfect crystalline behavior is maintained from the surface throughout the bulk.

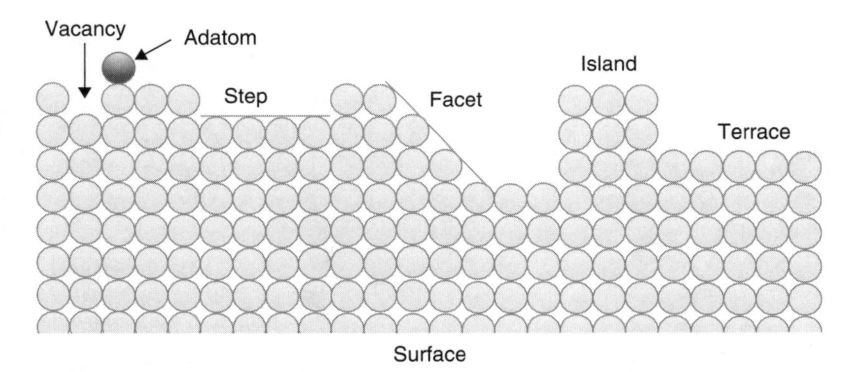

Figure 11.3 A schematic illustration depicting a microscopic view of a surface. In "real-life" situations, a surface of any material may significantly deviate from its idealistic bulk-terminated geometry and often contains zero-, one-, and two-dimensional defects; surface adsorbates; etc. (*See insert for color representation of the figure*.)

However, this idealized bulk-terminated surface assumption is never true in practice. In fact, there is no reason that the material near the surface should retain the same bulk-like interlayer distance, given that the coordination of atoms near the surface is significantly reduced compared to that in the bulk. It is generally observed that the interlayer spacing near the surface is always somewhat different from that in the bulk. This phenomenon is known as *surface relaxation*, which results in only a mild deformation of the crystal at the surface with a slight increase or decrease of the volume per atom near the surface. Relaxation effects generally affect several atomic layers at the surface. A surface layer exposing at least two types of atoms on the surface may also rumple, owing to the different strengths of the surface relaxation for the different atoms. Forces acting at the surface that give rise to surface relaxations and changes in the bonding and interlayer spacing of the surface atoms to different degrees may result in more dramatic effects such as the rearrangement of atoms along the surface or interface plane. Such changes are referred to as *surface reconstruction*. Surfaces of metallic solids generally exhibit a much weaker tendency to reconstruct as reduced coordination on the surface can easily be made up through redistribution of the delocalized electron gas. However, this phenomenon is much more pronounced in the case of covalently bonded semiconductor surfaces (e.g., Si, Ge, CdSe, GaAs, etc.) in which loss of nearest neighbors is rather difficult to compensate for except through passivation of dangling bonds by rehybridization of surface atoms followed by rearrangement of these atoms on the surface.

11.4 THE SURFACE ENERGY AND ISSUES WITH POLAR SURFACES

Any theoretical prediction of stable surface orientation, termination, and reconstruction is made through computation of the Gibb's surface free energy (γ),

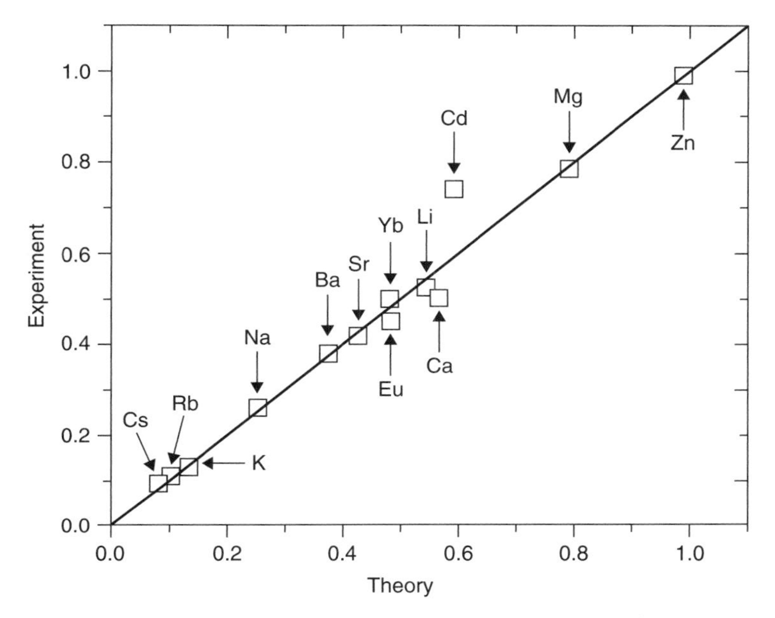

Figure 11.4 Comparison of DFT-computed surface energy (in J/m^{-2}) with experiments for the most stable surface orientations for various materials [38]. (Data collected from L. Vitos, A.V. Ruban, H.L. Skriver, and J. Kollar, Surf. Sci, 1998, 411, 186.)

which can be calculated by subtracting the total Gibbs free energy of a slab from that of the appropriate bulk reference. However, for a solid state system at low temperatures and pressures, one can, to a good approximation, replace the Gibbs free energy with the internal energy of the system, neglecting contributions from configurational and vibrational entropies. The internal energy can be directly computed from the first principles electronic structure calculations. For an elemental system, or for a slab of a compound system containing an integer number of formula units, the surface energy can be defined as

$$\gamma = \frac{1}{A}(E_{\text{slab}} - NE_{\text{bulk}}), \tag{11.2}$$

Here, A is the total area for the top and bottom surfaces of the slab, E_{slab} is the total energy of the supercell containing the surface model in the slab geometry, and N stoichiometric units of the bulk (with energy E_{bulk}) have been used to construct the supercell. This strategy has been used to reliably compute the surface energy for a variety of elemental slab surfaces, as shown in Figure 11.4. [38]

However, one should note that it is not always possible to construct a stoichiometric as well as a symmetric slab with identical terminations at the top and bottom of the slab. Usually, in compound materials (containing more than one constituting element), a stoichiometric slab model may contain nonidentical top and bottom surfaces. For example, in a stoichiometric slab model of a (001) surface in a cubic ABO_3 type perovskite structure, the top and bottom surfaces will have AO- and BO_2-terminations. In such situations, the above expression for the surface energy provides the average surface energy of the top and bottom surfaces.

Another peculiar problem that arises in the case of an asymmetric periodically repeated slab is the appearance of an unphysical electric field in the vacuum region between the slab and its periodic image. To eliminate this artificial electric field in the vacuum region, one may introduce a dipole layer in the midvacuum region that exactly cancels out the electric field created by the normal component of the dipole moment of the slab. Alternatively, one may switch to a symmetric slab geometry with identical top and bottom surfaces. However, in that case, the slab will be nonstoichiometric and will require the introduction of a chemical potential to uniquely define the surface energy of the specific terminating surface. For such a nonstoichiometric slab system composed of n constituents within a supercell containing $N_{\rm i}$ atoms and with a $\mu_{\rm i}$ chemical potential of the $i_{\rm th}$ constituent, the surface energy γ can be written as follows:

$$\gamma = \frac{1}{A}(E_{\rm slab} - \sum_{i}^{n} N_i \mu_i) \tag{11.3}$$

Although the determination of the exact value of μ of each species may be difficult, following simple thermodynamic stability arguments, one can easily derive the allowed range for μ for each of the constituent. Also note that the chemical potential μ of each component depends both on temperature and pressure. We first deal with zero temperature and pressure conditions. The generalizations to the finite temperature and pressure situations is discussed in Section 11.7.

The basic aspects of surface relaxations and issues surrounding surface energy determinations can be understood using a II-VI semiconductor such as CdSe. The surface facets of wurtzite CdSe can mainly be classified as either polar or nonpolar, depending on the stoichiometry of the atoms contained in the surface plane. The nonpolar surfaces are stoichiometric, containing equal numbers of Cd and Se atoms in each surface plane, and carry a net zero dipole moment along the surface normal. The three most stable nonpolar surface facets of CdSe, viz., $(10\bar{1}0)$, $(01\bar{1}0)$, and $(11\bar{2}0)$, are considered for the present illustration. Polar facets, on the other hand, are composed of nonstoichiometric planes of either only Cd or only Se atoms and therefore carry a nonzero component of the surface dipole moment along the slab normal. For a further detailed classification of polar versus nonpolar facets, the readers are referred to a recent review by Goniakowski et al. [39].

The unrelaxed bulk-terminated and the DFT-optimized surface geometries for the polar and nonpolar CdSe facets are shown in Figure 11.5. The relaxation behavior of these facets can, in general, be understood through the electron counting rules for II-VI semiconductor systems [40]. Since Cd and Se have nominal valences of 2 and 6, respectively, sp^3 hybridization in bulk CdSe requires that a Cd atom contributes 1/2 electron to each of its four bonds to Se and an Se atom contributes 3/2 electrons to each of its four bonds to Cd. Atoms at a surface display lower coordination, and hence unshared electrons. By suitable relaxation and reconstruction, a surface attempts to minimize its energy by optimally sharing the electrons at the surface, that is, by rehybridizing, and the extent to which this is accomplished will depend on the nature of the surface and the number of unshared electrons. In the case of the nonpolar $(01\bar{1}0)$, $(10\bar{1}0)$, and $(11\bar{2}0)$ surfaces, significant relaxation is observed. In general, comparison of initial and relaxed structures shows that surface Cd atoms move inward toward the bulk and the surface Se atoms tend to move outward, resulting in a tilting of the surface CdSe bond relative to the horizontal. This tilting of the CdSe bond on relaxation can be understood in terms of the transfer of electrons from the Cd atoms to the more electronegative Se atoms at the surface. We note that, in the case of the $(01\bar{1}0)$ surface, each surface atom displays two dangling bonds, while the $(10\bar{1}0)$ and $(11\bar{2}0)$ surface atoms display one dangling bond each. Thus, Cd atoms at $(10\bar{1}0)$ and $(11\bar{2}0)$ surfaces can donate their unshared 1/2 electron to the surface Se atoms, resulting in a more planar threefold sp^2-type configuration around the surface Cd atom accompanied by the inward movement of surface Cd atoms. The surface Se atoms, on the other hand, possess a doubly filled dangling bond, which is preferentially exposed to any incoming electronegative species. A similar, but more intensified, process occurs at the $(01\bar{1}0)$ surface, as the surface atoms contain two dangling bonds to begin with. Thus, the $(01\bar{1}0)$ surface relaxation is more pronounced. In the case of the polar (0001)Cd, $(000\bar{1})$Cd, (0001)Se, and $(000\bar{1})$Se surfaces—the first two being terminated purely by Cd atoms and the other two by purely Se atoms—no significant relaxation was observed. As these surfaces have only one type of atomic species (either Cd or Se, with one or three dangling bonds), transfer of electrons from the dangling bonds is not possible, and hence there is no clear pathway available for relaxation. Thus, the extent of surface relaxation for polar facets is found to be significantly smaller than those of nonpolar facets.

For a CdSe crystal in thermodynamic equilibrium, the sum of the chemical potential of Cd (μ_{Cd}) and Se (μ_{Se}) atoms should be equal to the chemical potential of the bulk CdSe (μ_{CdSe}). Furthermore, the respective chemical potentials of the constituents (i.e., Cd or Se) at the surface and in the bulk of the crystal have to be the same to avoid any macroscopic mass exchange between the bulk and the surface. Thus, it follows that

$$\begin{aligned} \mu_{\mathrm{Cd}}^{\mathrm{bulk}} &= \mu_{\mathrm{Cd}}^{\mathrm{surface}} = \mu_{\mathrm{Cd}} \\ \mu_{\mathrm{Se}}^{\mathrm{bulk}} &= \mu_{\mathrm{Se}}^{\mathrm{surface}} = \mu_{\mathrm{Se}}. \end{aligned} \tag{11.4}$$

One can further make the important observation that the chemical potential of the Cd and Se atoms in CdSe should always be less than the chemical potential of the condensed phases in their respective elemental form (represented as $\mu_{\mathrm{Cd}}^{\mathrm{Cd,bulk}}$ and $\mu_{\mathrm{Se}}^{\mathrm{Se,bulk}}$ for Cd and Se, respectively), else the CdSe crystal will become

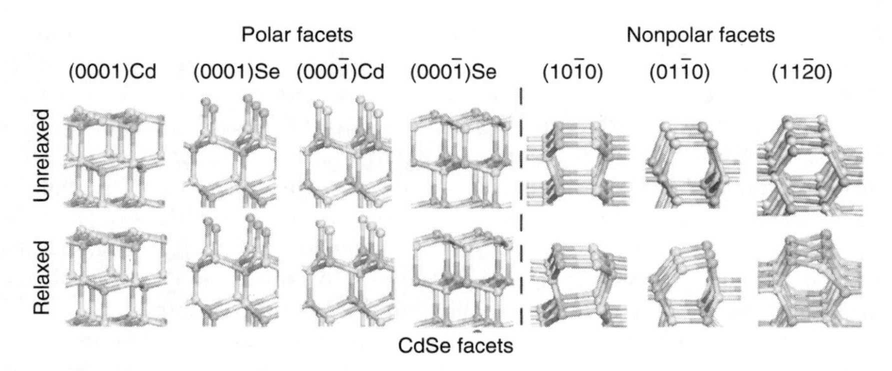

Figure 11.5 Atomistic models of the polar (0001)Cd, (0001)Se, (000$\bar{1}$)Cd, and (000$\bar{1}$)Se and nonpolar (10$\bar{1}$0), (01$\bar{1}$0), and (11$\bar{2}$0) facets of wurtzite CdSe in the bulk-terminated unrelaxed (top) and the relaxed states (bottom).

thermodynamically unstable against its decomposition into respective bulk phases of its atomic reservoirs, that is,

$$\begin{aligned} \mu_{\mathrm{Cd}} &< \mu_{\mathrm{Cd}}^{\mathrm{Cd,bulk}} \\ \mu_{\mathrm{Se}} &< \mu_{\mathrm{Se}}^{\mathrm{Se,bulk}} \end{aligned} \tag{11.5}$$

However, one should also note that the $\mu_{\mathrm{Cd}}^{\mathrm{Cd,bulk}}$ and $\mu_{\mathrm{Se}}^{\mathrm{Se,bulk}}$ are related to the μ_{CdSe} through the heat of formation of CdSe crystal (ΔH_{CdSe}) via

$$\mu_{\mathrm{CdSe}} = \mu_{\mathrm{Cd}}^{\mathrm{Cd,bulk}} + \mu_{\mathrm{Se}}^{\mathrm{Se,bulk}} + \Delta H_{\mathrm{CdSe}} \tag{11.6}$$

Combining Equations (11.5) and (11.6), one can obtain the allowed range of chemical potentials of μ_{Cd} and μ_{Se} in the CdSe crystal as

$$\begin{aligned} \mu_{\mathrm{Cd}}^{\mathrm{Cd,bulk}} + \Delta H_{\mathrm{CdSe}} &< \mu_{\mathrm{Cd}} < \mu_{\mathrm{Cd}}^{\mathrm{Cd,bulk}} \\ \mu_{\mathrm{Se}}^{\mathrm{Se,bulk}} + \Delta H_{\mathrm{CdSe}} &< \mu_{\mathrm{Se}} < \mu_{\mathrm{Se}}^{\mathrm{Se,bulk}} \end{aligned} \tag{11.7}$$

We define the two extreme values for μ_{Cd} as those related to a Cd atom in a "Cd-poor" CdSe crystal (corresponding to minimum μ_{Cd} or maximum μ_{Se}) and in a "Cd-rich" CdSe crystal (corresponding to maximum μ_{Cd} or minimum μ_{Se}). We further make a note that although the allowed range of μ_{Cd} can be properly defined, identification of the value corresponding to specific chemical conditions is nontrivial. For instance, based on the Gibbs–Thompson equation one can easily see that the chemical potential will vary with the size of a nanocrystal [41].

The surface energy of the nonpolar facets can be easily calculated using Equation 11.2. However, determination of the chemical-potential-dependent surface energies of the four polar facets of the wurtzite CdSe is not straightforward because of the absence of inversion symmetry. Note that of the four inequivalent {0001} polar surface facets of wurtzite CdSe, two (i.e., (0001)Cd and (0001)Se) can occur exclusively on one side terminated purely by Cd or by Se atoms and the other two (i.e., $(000\bar{1})$Cd and $(000\bar{1})$Se) on the opposite side in a slab geometry, again terminated purely by Cd or Se atoms. Furthermore, owing to the inherent asymmetry between the positive and negative c-axes, the (0001)Cd and the (0001)Se facets have one and three dangling bonds, respectively. On the other hand, the $(000\bar{1})$Cd and the $(000\bar{1})$Se facets have three and one dangling bonds, respectively. Therefore, it is possible to construct four different types of slab geometries, and, in principle, one can calculate the four unknown surface energies of the polar facets. However, it turns out that such a construction will lead to only three linearly independent equations, which can be further manipulated to give rise to the fourth one. Therefore, owing to the lack of inversion symmetry in the wurtzite crystal structure, the (0001) and $(000\bar{1})$ surface energies cannot be isolated independently using a slab geometry. A similar problem arises for the polar (111) and $(11\bar{1})$ surfaces of the zinc blende structure. For this crystal structure, however, a method of direct calculation of the surface energies has been developed by Zhang and Wei [42] that requires construction of one-dimensional wedge-shaped structures for extracting surface energies. This method has also been applied to CdSe [43, 44] to estimate the surface energy of the polar facets in the wurtzite phase under the plausible assumption that the (0001) and $(000\bar{1})$ facets of the hexagonal wurtzite structure are atomically identical to the (111) and $(11\bar{1})$ facets of the zinc blende structure. Following the technical details described in Reference 44, surface energies of all the four polar facets can be calculated.

The DFT-calculated surface energies of all the relaxed surface facets are plotted in Figure 11.6 over the allowed range of μ_{Cd} values. As already mentioned, the surface energies of nonpolar surfaces do not depend on μ_{Cd} (shown by the dashed lines) and therefore show up as horizontal lines in the plot. Interestingly, the two nonpolar surfaces with one dangling bond per surface atom, $(10\bar{1}0)$ and $(11\bar{2}0)$, have the lowest surface energy, while the surface energy of the $(01\bar{1}0)$ facet, with two dangling bonds per surface atom, is almost double that of the latter ones. The surface energy of the polar facets with one and three dangling bonds is shown as solid and dotted-dashed lines, respectively, in the plot and is a linear function of μ_{Cd}. However, note that the average of surface energies for the two pairs of polar facets, is always constant. Furthermore, the stability of the Cd-terminated polar surfaces increases as we move from Cd-poor to Cd-rich conditions, while the Se-terminated polar surfaces display the opposite behavior. Comparison of the polar surfaces with three dangling bonds per surface atom reveals that the $(000\bar{1})$Cd surface is more stable than the (0001)Se surface throughout the range of chemical potential considered. On the other hand, in the case of polar surfaces with one dangling bond per surface atom, the $(000\bar{1})$Se surface is more stable.

We close this section by indicating that reliable DFT-based schemes are available for computing the surface energies of elemental as well as multicomponent

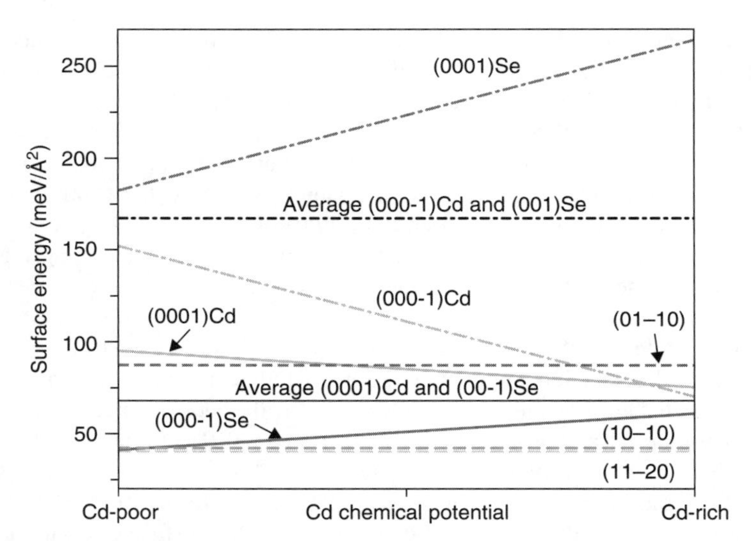

Figure 11.6 Surface energies for polar and nonpolar wurtzite CdSe facets as a function of Cd chemical potential ($\mu_{\text{Cd}}^{\text{CdSe}}$). (*See insert for color representation of the figure.*)

systems. The shape of a crystal is in fact controlled by the anisotropic nature of the surface energies (i.e., because of different surfaces displaying different surface energies). The concept of manipulating the shape of crystallites and controlling the growth of nanostructures through modulating the surface energies, for example, through adsorbates, is discussed in the next section.

11.5 ADSORBATE ON SURFACES—ENERGETICS AND THE WULFF CONSTRUCTION

Surface adsorption is a phenomenon of key importance in surface science and dominates many chemical processes. To obtain a microscopic understanding of the role of adsorbates, it is necessary to know the surface atomic structure. Most of the early structural studies of surface adsorption assumed implicitly that the surface provided a rigid platform of identical adsorption sites, which did not change during the course of adsorption, onto which atoms or molecules were adsorbed. However, it is now well understood that a surface can modify the behavior of adsorbed species in many direct and indirect ways, for instance, through bond breakage, charge transfer, long-range ordering effects and adsorbate–adsorbate lateral interactions. It has also become clear, of course, that the adsorbate also induces changes in the substrate surface. Surface reconstruction, extent of surface relaxation, and surface rumpling are usually strong functions of the type and coverage of surface adatoms or molecules.

Adsorbates also affect the morphology of crystallites through alteration of the surface energies. To understand this, one should note that the equilibrium shape of a crystal is determined by the anisotropic surface energies via the Wulff construction [45, 46]. Determination of the equilibrium shape of crystals that minimizes the total surface free energy proceeds by the following three steps: (i) determination of the surface energies for various facets experimentally or theoretically; (ii) drawing vectors normal to the crystal facets from a common origin, with the length proportional to the surface energies; and (iii) creating at the end of each vector a plane perpendicular to the vector. The shape enclosed by the planes gives the equilibrium shape of the crystal. First principles calculations are playing an increasingly important role in this field, partly due to the capability of such methods to accurately determine surface energies and partly due to the difficulty in quantitatively determining the surface or interface energy experimentally. Some interesting insights have been achieved in the past. For example, using DFT calculations, Shi and Stampfl [47] illustrated that the equilibrium shape of a Au catalyst particle changes with the O_2 atmosphere, with the predominant terminations changing from (111) to (110) with increasing O_2 pressure or decreasing temperature.

In the following sections, we illustrate how the crystal shape and morphology may be controlled through surface energy and surface adsorption using three specific examples: (i) oxygen adsorption on wurtzite CdSe facets, (ii) hydroxyl adsorption on the low-index facets of rocksalt MgO, and (iii) metal adsorption on hexagonal tungsten carbide (WC) surfaces.

11.5.1 CdSe Crystallites

Quantification of the bonding strength of an adsorbate on a substrate is generally done in terms of binding energy (E_b) as

$$E_b = (E_{surf,Ad} - E_{surf,clean} - n_{Ad}\mu_{Ad,gas})/n_{Ad}, \tag{11.8}$$

where n_{Ad} is the number of adsorbate species (atoms or molecules) adsorbed per surface unit cell and $\mu_{Ad,gas}$ is the chemical potential of adsorbate in the reference gas phase (usually taken as the DFT total energy of the gaseous adsorbate molecule). $E_{surf,Ad}$ and $E_{surf,clean}$ are the total energies of the surface models with and without surface adatoms on the surface of interest, respectively.

Knowing the surface energy of the clean surface facet (γ_{clean}) from Equation 11.2 or 11.3, the surface energy of the adatom covered-surface (γ_{Ad}) is obtained by

$$\gamma_{Ad} = \gamma_{clean} + n_{Ad}E_b/A, \tag{11.9}$$

where A is the area of the surface unit cell. It is clear from the above equation that the surface energy for adsorbate-passivated surface facets (which eventually determines equilibrium shape and morphology of the growing crystals) is a function of binding energy, which in turn depends on the temperature, pressure, chemical environment, and surface coverage of adatoms. Therefore, by controlling the growth environment and selecting appropriate surface adsorbates, one can control the equilibrium shape of the crystal. This can be particularly important for oxygen-containing (or water-vapor-containing) environments, where the stability of different surface facets of varying stoichiometry may well be anticipated as a function of oxygen (or hydroxyl groups) in the surrounding gas phase. The surface energy of CdSe facets was discussed in the previous section. CdSe is one of the most technologically important II-VI semiconductors. One-dimensional (1D) nanocrystals of CdSe (i.e., CdSe quantum rods) possess unique optical properties such as linearly polarized emission and higher Stokes shift when compared to zero-dimensional nanocrystals such as quantum dots [48, 49]. From a device standpoint, it is more desirable to have a larger aspect ratio of quantum rods along with the ability to engineer heterojunctions, which renders 1D nanostructures of CdSe more advantageous than the quantum dots [50–54].

An interesting and potentially useful phenomenon observed in wurtzite CdSe nanocrystals is asymmetric anisotropic growth in the presence of oxygen (captured schematically in Figure 11.7a). While anisotropic growth in wurtzite systems refers to preferred growth along one dimension (say, the c-axis) over others, asymmetric anisotropic growth refers to a strong preference to grow along only one of the two complementary anisotropic axes (say, along the positive c-axis rather than along the negative c-axis). As already mentioned, these systems display four inequivalent {0001} surface facets, exclusively occurring in pairs on the either side. Note that for any preferential one-dimensional growth, both the surface facets, that occur in a complementary pair on that side, should have high surface energy. Therefore, growth along one direction will be controlled by one pair of surfaces, while growth along the opposite direction will be controlled by a different pair of surfaces.

Although experimental studies [56] had suggested that oxygen might have a role in directing the growth of CdSe nanocrystals to quantum rods, the underlying mechanism and various factors controlling the asymmetric growth of wurtzite nanostructures were discussed in detail by Pilania et al. [44], based on the results of their ab initio computations. DFT-based surface energy calculations of various polar and nonpolar facets of CdSe showed that both ordering and relative magnitude of these facets change on oxygen adsorption. It was quite crucial to note that the unidirectional and unidimensional growth along the [0001] direction can occur only through successive and interconvertible creation of (0001)Cd and (0001)Se surfaces. Thus, for vigorous growth to occur preferentially along the (0001) direction, the surface energy of both the (0001)Cd and (0001)Se surfaces should necessarily be high relative to that of all other surface facets. Even if one of these two surfaces has a low surface energy, growth along the (0001) direction will be hindered. In fact, the calculated results revealed that both the (0001)Cd and (0001)Se surfaces display large positive values of the surface energy relative to all other surfaces for a large range of allowed Cd chemical potential values (μ_{Cd}, which was used to quantify environmental growth conditions), indicating the possibility of preferential growth along the (0001) direction

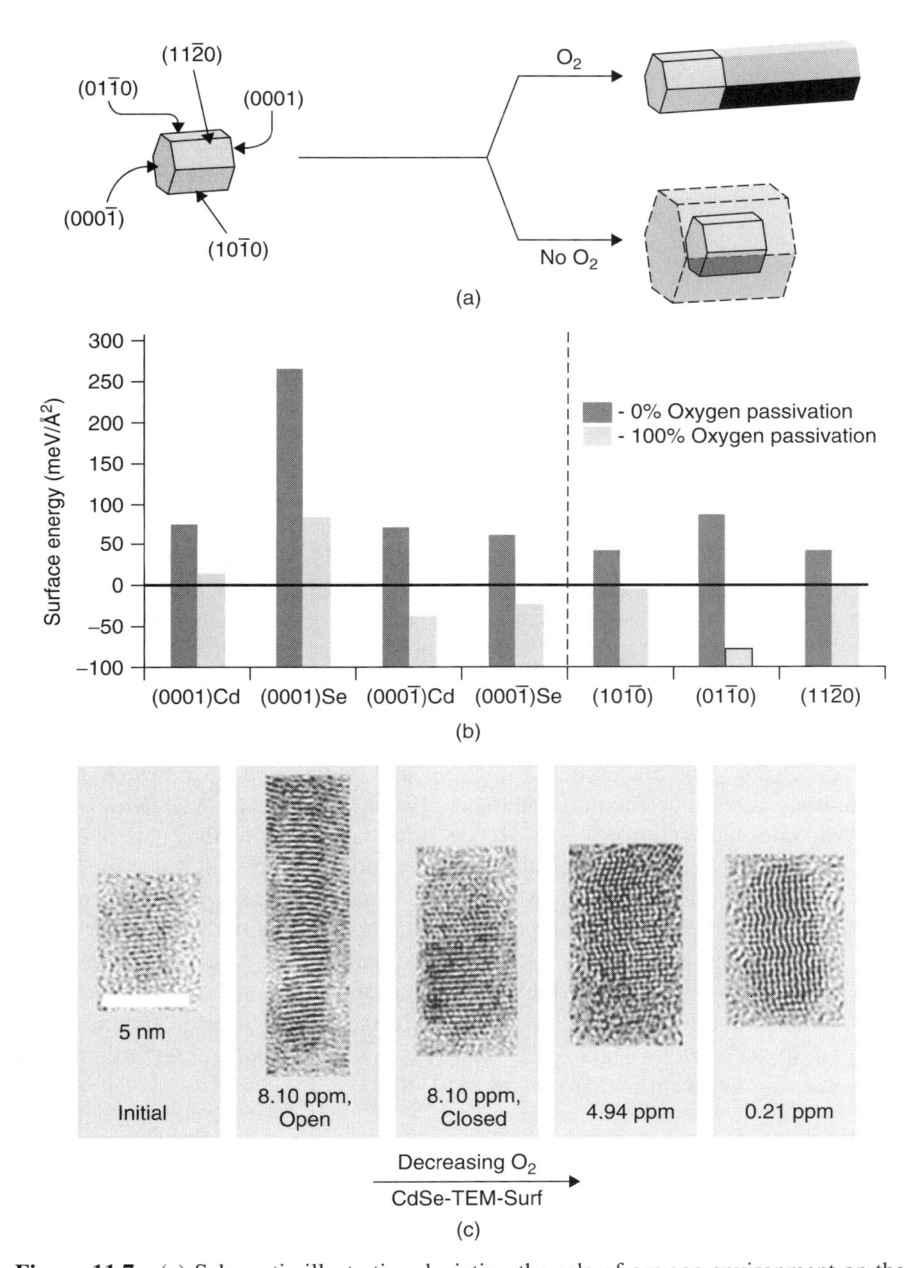

Figure 11.7 (a) Schematic illustration depicting the role of oxygen environment on the growth of wurtzite CdSe nanocrystals. (b) DFT-calculated surface energies for the polar and nonpolar wurtzite CdSe facets before and after oxygen passivation. (c) High resolution transmission electron microscopy images of representative CdSe nanocrystals for various oxygen concentrations. The 5 nm scale bar shown in the leftmost panel applies to all images [55]. (*See insert for color representation of the figure*.) (Reprinted with permission from J. D. Doll, G. Pilania, R. Ramprasad, and F. Papadimitrakopoulos, Nano Lett., 2010, 10, 680. Copyright 2010 American Chemical Society.)

(c.f. Fig. 11.7b). The theoretical predictions were later confirmed through experiments [55]. Figure 11.7c shows high resolution transmission electron microscopy (TEM) images of the CdSe nanocrystals in the starting seed as well as the four grown nanocrystal samples for different levels of oxygen environment (decreasing O_2 from left to right). As compared to the starting CdSe seeds, the extent of the unidimensional growth of the nanocrystals is directly correlated with oxygen concentration.

11.5.2 MgO Crystallites

We now move on to the second example regarding the shape control of MgO crystal through exposure to moisture. MgO crystals grown in dry environmental conditions have cubic shape owing to the stability of (100) crystallographic orientation as opposed to various other possible surface terminations. However, in recent high resolution TEM experiments, it was observed that the shape of highly dispersed MgO nanocrystals (physical dimensions range between 10 and 100 nm) [57, 58] evolves from cubic to octahedral-like when exposed to water. Furthermore, it is also known that the periclase mineral when grown in wet conditions exposes surfaces that mainly have (111) orientation. In the past several years, a consensus has been reached regarding the main features of water adsorption on different terminations of MgO crystals, which results in various surface structures giving rise to environment-dependent morphology of MgO crystals. A mixed molecular and dissociative adsorption of water is predicted for (001) MgO surface facets in both theoretical [59–62] and experimental studies [63], while more open surface orientations such as (110) or (111) facets indicate only dissociative water adsorption [64–76]. It has now become clear that it is due to the change in the pressure- and temperature-dependent relative surface energies of various MgO surface facets that MgO crystals change their shape and morphology on exposure to moisture. A comprehensive theoretical study that summarizes the results for a variety of configurations and compares water adsorption on extended and point defects on various MgO surfaces has recently appeared in the literature [77], while the effects of the choice of exchange and correlation functionals on the stability of various MgO facets and hydroxyl adsorption energy are critically evaluated by Finocchi and Goniakowski [78].

11.5.3 WC crystallites

In this section, we discuss the equilibrium morphology of another important material, WC, under different conditions (i.e., various surface adsorbates as well as carbon concentrations), determined based on DFT results and the Wulff theorem [79]. As important hard metal components, WC-based cermets (WC crystals embedded in a secondary binder phase, such as Co) have been widely used in military and in aerospace, automotive, and marine industries [80]. However, there is strong evidence showing that the size and shape of WC crystals within the

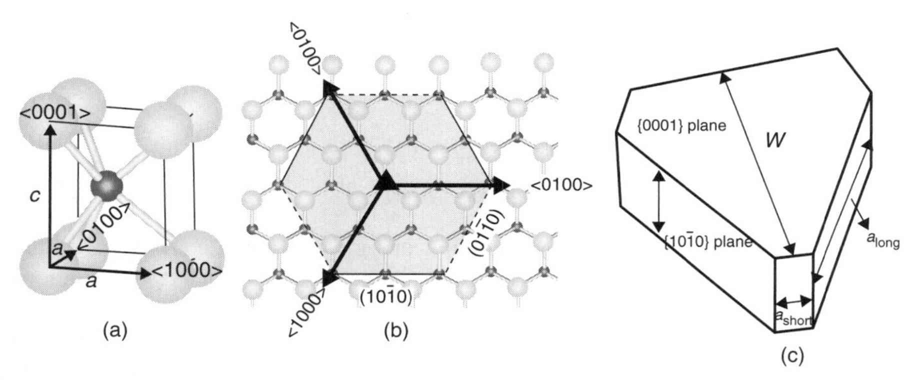

Figure 11.8 (a) The unit cell of WC. (b) Projection along the ⟨0001⟩ direction with the $(10\bar{1}0)$ and $(01\bar{1}0)$ facets represented by solid and dashed lines, respectively. "△" stands for the threefold rotational symmetry about ⟨0001⟩ axis. $(10\bar{1}0)$ surface could be terminated by C with four dangling bonds or W with two dangling bonds. Similarly, $(10\bar{1}0)$ could be terminated by C with two dangling bonds and W with four dangling bonds. (c) Schematic representation of the equilibrium shape for a WC particle. The shape can be characterized by two shape factors, that is, $r = a_{short}/a_{long}$ and $k = t/w$. Green (large) and red (small) spheres represent W and C atoms, respectively. (*See insert for color representation of the figure*.)

cermets affect the mechanical properties [81]. Hence, a better understanding of the growth mechanism and equilibrium shape of WC is of great interest.

Similar to CdSe discussed in Section 11.4, WC has a hexagonal structure, and its unit cell is shown in Figure 11.8a. Because of the threefold rotation symmetry about the ⟨0001⟩ axis, the WC crystal has two sets of three equivalent $\{10\bar{1}0\}$ planes, namely, $(10\bar{1}0)$ and $(01\bar{1}0)$ surfaces, rather than six equivalent $\{10\bar{1}0\}$ planes (Fig. 11.8b) [82]. It is established that the shape of hexagonal WC crystals in cermets is bounded by prismatic $\{10\bar{1}0\}$ surfaces and basal {0001} surfaces, as shown in Figure 11.8c. Two shape factors r and k are normally used to describe the equilibrium shape, where r is the ratio between the lengths of the short (a_{short}) and long (a_{long}) prismatic facets and k is the ratio between the thickness along ⟨0001⟩ direction (t) and the width of the basal plane (w). Along the ⟨0001⟩ and $\langle 10\bar{1}0\rangle$ directions, the stacking sequence of WC consists of alternating W and C planes. As a result, (0001) plane could be either terminated by W or C with three dangling bonds (Fig. 11.8a). On the other hand, as indicated in Figure 11.8b, $(10\bar{1}0)$ surface could be either terminated by C with four dangling bonds or by W with two dangling bonds and $(01\bar{1}0)$ surface could be terminated with C with two dangling bonds or W with four dangling bonds. We should note that although each W (or C) is bonded to six neighboring C (or W) atoms, when projected onto (0001) plane, half the bonds are on top of the others and thus not shown in Figure 11.8b.

The energies of these clean surfaces could be computed based on Equation 11.3 with $\mu_i = \mu_W$ and μ_C by adopting nonstoichiometric supercells. The allowed range of μ_W and μ_C for a stable WC to avoid graphite and W_2C segregation (the decarbonization of WC is in the sequence of WC→ W_2C →W) is indicated by the following inequalities:

$$\mu_C < E_{graphite} \tag{11.10}$$

$$\mu_C + 2\mu_W < E_{W_2C} \tag{11.11}$$

where $E_{graphite}$ and E_{W_2C} represent the DFT energies of graphite and W_2C within its lowest energy form, the ϵ-Fe_2N structure (space group: P-31m) [83]. Equations 11.10 and 11.11 correspond to the so-called "carbon-rich" and "carbon-deficient" conditions. The corresponding surface energy change in the presence of a metal binder phase has been explored by placing a metallic atom on top of these WC surfaces. Co and Ni, as the most common binder materials for WC, have been already explored as discussed below.

The lowest energies of each of the (0001)-, $(10\bar{1}0)$- and $(01\bar{1}0)$-type surfaces along with Wulff theorem determined the equilibrium crystal shape at different conditions in Figure 11.9. Several interesting points are found from this figure. (i) As expected, the crystal shape and its factors vary with the surface conditions (i.e., clean WC surface, Co-adsorbed WC surface, and Ni-adsorbed surface). (ii) At the same surface condition, the chemical potential of carbon affects the equilibrium shape of WC particles. (iii) Co adsorption under the carbon-deficient condition promotes the formation of truncated triangular prisms, whereas Ni adsorption under the carbon-rich condition enhances the formation of near-hexagonal prisms. (iv) The equilibrium shapes of WC crystals under all conditions (especially with Co or Ni adsorption) can be described as "bulky" because their k factors are close to 0.8 or are higher, and thus the thickness of the truncated triangular prism and the width of its basal plane are similar, which is far from the platelet geometry.

Figures 11.10a,b show the SEM images for WC–Co powders in the carbon-rich and carbon-deficient conditions, respectively. Although the shape of large particles may or may not be the equilibrium morphology of WC particles, depending on whether the crystal growth is controlled kinetically or thermodynamically, it is still beneficial to compare the shape of the large particles in the experiments with the simulation results. Note that the truncated triangular prism has appeared in both types of powders. However, r factors for the two powders are different. For the carbon-rich powder (Fig. 11.10a), the short prismatic facets are approximately ~0.4 times the length of the long prismatic facets, which is close to the theoretical prediction shown in Fig. 11.9. For the carbon-deficient powder (Fig. 11.10b), the r factor from the SEM image is in the range of $0.2 \sim 0.3$, which is also in good agreement with the theoretical value, 0.23 (Fig. 11.9). These results reveal that the chemical potential of carbon can affect the morphology of WC particles. However, the ratios of the thickness of the truncated WC prism to the width of the basal plane, that is, the k factor, in both powders are smaller than those predicted from the first principles calculation (Fig. 11.9), which may be caused by the slow growth kinetics along the [0001] direction.

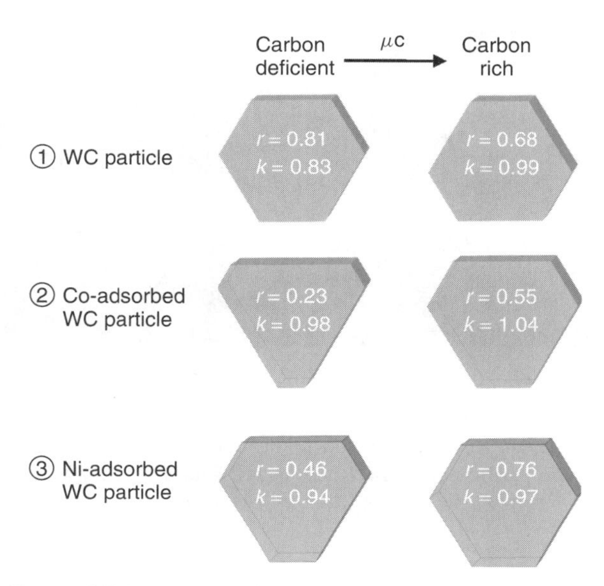

Figure 11.9 The equilibrium shapes for pure WC, Co-adsorbed WC, and Ni-adsorbed WC particles in carbon-rich and carbon-deficient conditions. The two factors, k and r, are also indicated.

In light of the difficulties in experimentally identifying the correlation between morphologies and factors such as surface adsorbates, temperature, pressure, and bulk compositions, etc. DFT-based Wulff construction is becoming a useful tool in this arena.

11.6 ADSORBATES ON SURFACES—ELECTRONIC STRUCTURE

In the previous section, we discussed applications of DFT-based techniques to study the structure and energetics of surfaces at the atomistic level. Now, we discuss the local electronic structure of atomic and molecular adsorbates on the surface. The properties and reactivity of an adsorbed atom or molecule are determined by the nature of the surface chemical bond, which in turn is governed by the newly formed electronic states because of the bonding to the surface. Interaction of an incoming adsorbate with the surface can vary from being very weak (i.e., physisorption) to strong enough such that it actually rearranges the valence levels of the adsorbate (i.e., chemisorption). Discrete and sharp molecular/atomic orbitals in case of an isolated gaseous adsorbate gradually evolve on its interaction with the substrate (and also with other neighboring adsorbates) to produce a new set of electronic states/bands that are usually broadened and energetically shifted with respect to the gas-phase species. In the following discussion, we shed some light on how informative and potentially important local

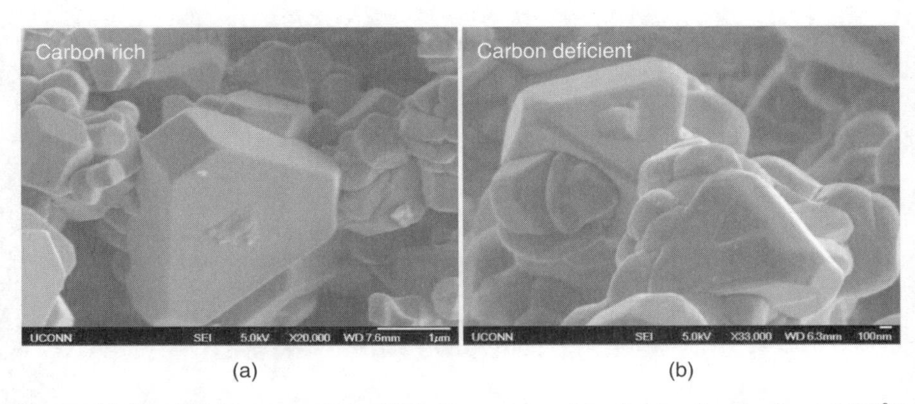

Figure 11.10 SEM images of the WC–Co powders (a) after heating in Ar at 1400°C for 2 h and (b) after H_2 treatment at 800°C and then heating in Ar at 1400°C for 2 h [79]. (Reprinted with permission from Y. Zhong, H. Zhu, L. L. Shaw, and R. Ramprasad, Acta Mater 2011, 59, 3748. Copyright 2011 by Elsevier.)

density of states (LDOS) could be in the case of surface adsorption to probe local electronic structure. For the sake of illustration, we take a specific example of atomic oxygen adsorption on various transition metal surfaces with the aim of understanding trends in oxygen adsorption energies on the metal surfaces.

Figure 11.11a shows DFT-calculated adsorption energies as a function of the distance of the O atom above the surface for various transition metals in a small section of the periodic table. The more negative the binding energy, the stronger is the bond between oxygen adsorbate and the metal surface. The energy per oxygen atom in a gaseous O_2 molecule is shown for comparison. Only metals in which the minimum in the adsorption energy function is below this value will be able to dissociate O_2 exothermally. It can be seen that O binds most strongly to the metals to the left in the transition metal series and more strongly to the 3d metals than to the 4d and 5d metals. It is evident from Figure 11.11a that with O adatoms, Ru bonds much stronger than Pd and Ag. Au is very noble with a bond energy per O atom less than that of O_2, Ag is just able to dissociate O_2 exothermically, and Cu forms quite strong bonds. It is also noted that these DFT-calculated results are in excellent agreement with the experimental findings [1].

However, before we try to understand this trend in the oxygen adsorption energy for various transition metal surfaces through LDOS and local electronic structure, it is important to note that the valence states of the transition metal surface atoms can be divided up into the free-electron-like s electron states and the more localized d electron states. Following the Newns–Anderson model [85, 86], the interaction of the adsorbate atom with the delocalized s electrons leads to a broad resonance, whereas interaction with the narrow band d states leads to distinct new bonding and antibonding levels (cf. Fig. 11.12).

To simplify, one can imagine this complex coupling between the adsorbate atomic level to the metal s and d states to be composed of various discrete steps.

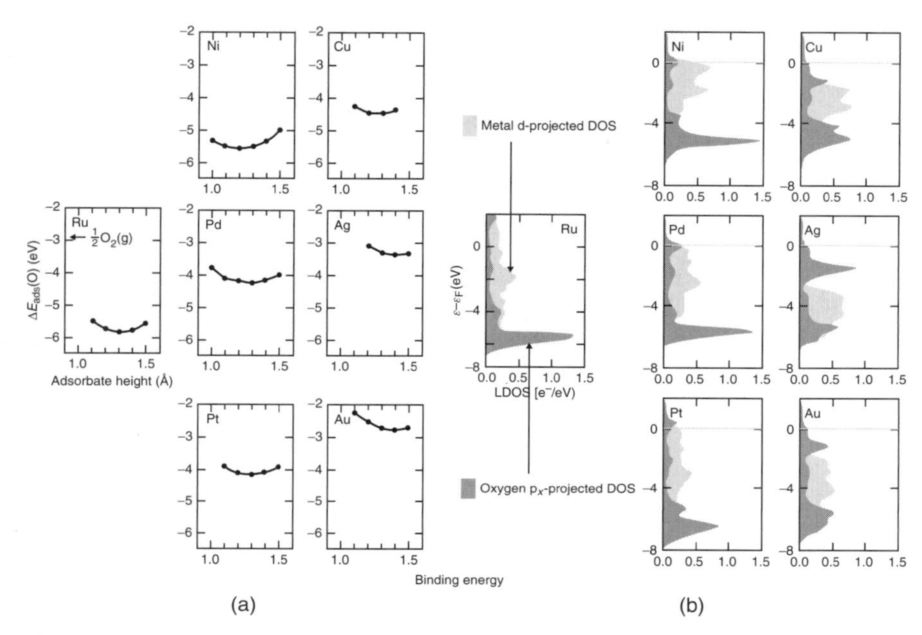

Figure 11.11 (a) DFT-calculated adsorption energy for oxygen adatom as a function of distance of the O adatom above the surface for a range of close-packed transition metal surfaces, ordered according to their position in the periodic table. The energy per O atom in gaseous O_2 is also shown in the leftmost box, showing results for Ru for comparison. (b) The density of states projected onto the d states (gray) of the surface atoms for the surfaces considered in (a). The oxygen $2p^x$ projected density of states (black) for adsorbed O adatom on the same surfaces are also shown. The formation of bonding and antibonding states below and above the metal d states is clearly seen [84]. (Adapted with permission from B. Hammer and J.K. N rskov, Adv. Catal., 2000, 45, 71. Copyright 2000 Elsevier.)

First, let us consider coupling of the adsorbate atomic levels with the metal's s states, and in the next step, we switch on the coupling to the metal d states as well. The coupling to the broad s band in the first step leads to a broadening and shift of the adsorbate state (Fig. 11.13). Usually there are only small differences in this interaction going from one transition metal to the other, owing to the fact that all the transition metals have a half-filled considerably broad s band in the metallic state. Therefore, the differences between the different transition metals must be associated primarily with the d states. The interaction of the adsorbate states with localized d states will give rise to the formation of separate bonding and antibonding states, as shown in the bottom of Figure 11.13. In such a picture, bonding strength of an adsorbates depends on the relative occupancy of bonding and antibonding states. The strength of the bond will be maximum when the bonding states are completely occupied and all the antibonding states are empty, whereas if the antibonding states also start getting filled, the bond gradually becomes weaker.

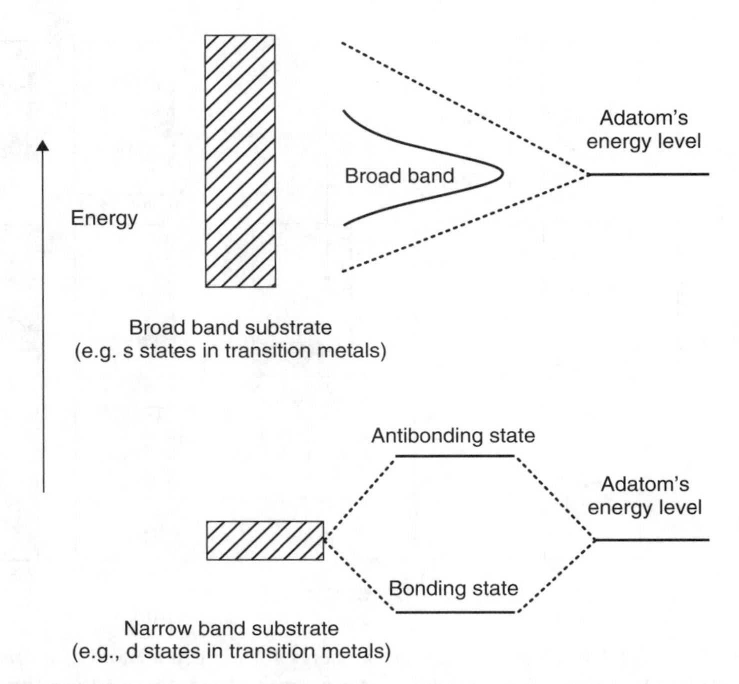

Figure 11.12 Schematic diagram depicting evolution of the projected density of states of an adsorbate as it gets adsorbed onto a surface within the Newns–Anderson model [85, 86] in two limiting cases: when the band width is large (top) and when the band width is small (bottom).

To understand the variations in the binding energy of oxygen adatom, let us look at the variations in the electronic structure of the oxygen adsorbed on the different transition metals (Fig. 11.11b). Since the energy of the d states relative to the Fermi level varies substantially from one metal to the next, the number of antibonding states that are above the Fermi level, and thus empty, will depend on the transition metal under consideration. It can be seen that the antibonding states for oxygen adatom get filled gradually as we move from Ru to Pd to Ag, explaining why the bonding becomes weaker in that order. However, the above picture is not sufficient to understand why 3d metals bond stronger than 4d and 5d metals. More extensive models have been developed to explain these effects and to extend this theory for more complex systems [84, 87].

11.7 SURFACE PHASE DIAGRAMS: FIRST PRINCIPLES THERMODYNAMICS

From the above discussion, it is clear that first principles computational techniques can provide an adequate description (in terms of geometrical and electronic

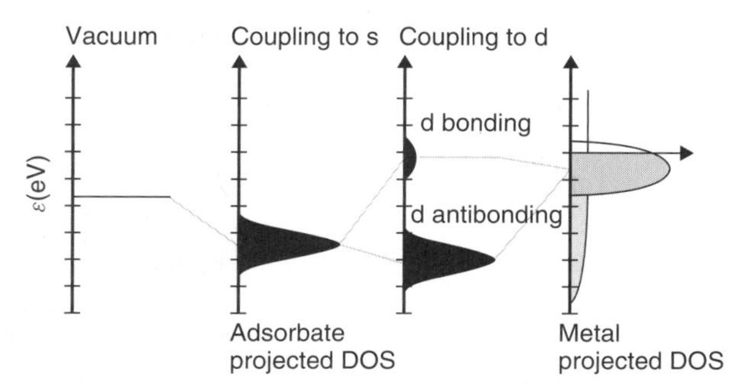

Figure 11.13 Schematic illustration of the formation of a chemical bond between an adsorbate valence level and the s and d states of a transition metal surface [84]. (Adapted with permission from B. Hammer and J.K. N rskov, Adv. Catal., 2000, 45, 71. Copyright 2000 Elsevier.)

structure, energetics, etc.) of various individual microscopic processes taking place on a surface such as adsorption, desorption, surface dissociation, etc. However, this knowledge is not sufficient to predict the behavior of a *real-life* macroscopic-scale system for several reasons. First, the behavior of a macroscopic system of any technological relevance is usually governed by a large number of distinct atomic scale processes. Therefore, in addition to the precise description of elementary processes involved, any predictive modeling of a macroscopic system would also require information regarding statistical interplay of a large number of such possible microscopic processes. In other words, we are interested not only in each individual elementary process taking place in isolation but also in what happens when a large number of such processes are allowed to take place simultaneously. Second, in the microscopic regime, electronic excitations and lattice vibrations in a material take place on time scales of femtoseconds and picoseconds, respectively, while bond breaking and bond formation occur at a length scale of several angstroms. On the other hand, in the macroscopic regime, relevant times are of the order of 10^{-3} to several hundreds of seconds and the relevant length scales can vary from 10^{-6} to several meters. To obtain macroscopically observable system properties and functions from electronic structure theories such as DFT, a gap of several orders of magnitude in *both* time and length scales will have to be overcome. Finally, in the electronic structure calculations, the pressure and temperature effects are not included. Therefore, to actually describe the situations of finite temperatures and pressures, first principles calculations have to be used as an input for further thermodynamic considerations. In the following paragraphs, we discuss how such a description, appropriate for the macroscopic regime, can be achieved by combining the results of ab initio DFT with concepts from statistical mechanics and thermodynamics.

The primary aim of the first principles atomistic thermodynamics approach is to use information pertaining to the potential energy surface calculated at the level of electronic structure theory, for a system in equilibrium (or in a meta stable state), to calculate an appropriate thermodynamic potential function such as the Gibbs free energy G. The Gibbs free energy of any system decomposed in terms of various contributing factors can be given as follows:

$$G(T, P) = E^{\mathrm{DFT}} + F^{v} - TS^{\mathrm{c}} + PV. \tag{11.12}$$

In this equation, the leading term E^{DFT} represents the internal energy of the system and can be obtained directly from electronic structure theory calculations. The second term represents the vibrational free energy, which contains contributions due to *both* the zero point energy E^{ZPE} and the vibrational entropy. The third term is the configurational free energy arising from the configurational entropy S^{c}. T, P, and V are the temperature, pressure, and volume, respectively. Once this pressure- and temperature-dependent Gibbs free energy is known, one is immediately in a position to calculate all the macroscopic properties of the system of interest. Although this concept is quite general and applicable to a broad range of problems, here we find it instructive to illustrate the approach by taking a specific example of a metal surface in equilibrium with the surrounding oxygen environment.

As already mentioned, in an ab initio atomistic thermodynamic approach, we are interested only in the equilibrium state (or metastable state) of the system and not in the information as to how the system has evolved to reach that state. Therefore, the first task of the present thermodynamic approach would naturally be to identify a number of known (based on the experimental information available in the literature) and unknown but possibly relevant (based on the computational screening) oxygen-containing surface structures along with the clean and bulk oxide surface structures. Once the time-consuming electronic structure calculations are done, the next step is to evaluate which of the considered surface structures are most stable and at what environmental conditions (i.e., the temperature and partial pressure of O_2). Therefore, much of the remaining discussion in the present section is based on a nonzero T and P generalization of Equation 11.3.

For a two-components system composed of metal M and oxygen, the (T,P)-dependent surface free energy γ is a function of the chemical potential of the metal (μ_{M}) and oxygen (μ_{O})

$$\gamma(T, P) = \frac{1}{A}[G^{\mathrm{surf}}(T, P, N_{\mathrm{M}}, N_{\mathrm{O}}) - N_{\mathrm{M}}\mu_{\mathrm{M}}(T, P) - N_{\mathrm{O}}\mu_{\mathrm{O}}(T, P)], \tag{11.13}$$

where N_{M} and N_{O} are the number of metal and oxygen atoms in the finite supercell of the system that repeats to a produce 2D infinite slab. The chemical potentials of the metal at the surface and in the bulk of the crystal have to be same to avoid any macroscopic mass exchange between the bulk and the surface, and therefore, for a sufficiently thick metallic slab, it is fixed to the corresponding bulk value, $\mu_{\mathrm{M}} = \mu_{\mathrm{M}}^{\mathrm{bulk}}$. Oxygen chemical potential on the other hand is a variable that can cover a range of values and plays a decisive role in stabilizing a particular surface structure. Under ultrahigh vacuum, when oxygen chemical potential μ_O is very low, the clean metal surface prevails, and this situation is known as O-poor limit. On the other hand, O-rich limit is defined by $\mu_O = 1/2E_{O_2}^{DFT}$, with $E_{O_2}^{DFT}$ being the DFT-calculated total energy of an O_2 molecule in the gas phase. If μ_O exceeds this critical limit, condensation of the O_2 molecule from the gas phase to the surface will take place. Furthermore, when considering a metal surface in contact with O_2, a complete conversion of the metal into a bulk oxide phase is also a thermodynamic possibility that should be taken into account. Therefore, the stability of the corresponding bulk oxide (let us say M_xO_y) has to be evaluated with respect to the various other surface oxide adlayer-containing structures. In the bulk oxide phase, the chemical potential of the metal and oxygen do not remain independent any more and are related to the Gibbs free energy of the bulk oxide ($g_{M_xO_y}^{bulk}$) as

$$x\mu_M(T, P) + y\mu_O(T, P) = g_{M_xO_y}^{bulk}(T, P). \tag{11.14}$$

Furthermore, thermodynamic stability of the oxide phase requires that $\mu_M(T, p) \leq g_M^{bulk}(T, P)$; otherwise, the surface oxide will decompose back into the solid metal and the oxygen gas phases. Using this constraint along with Equation 11.14 and taking $T = 0$ K and $P = 0$ atm limit for the bulk phase energies, we can find the minimum oxygen chemical potential $\min[\mu_O(T, P)]$ at which formation of the bulk oxide phase will indeed become thermodynamically favorable, as

$$\min[\mu_O(T, P)] = \frac{1}{y}[g_{M_xO_y}^{bulk}(0, 0) - xg_M^{bulk}(0, 0)]. \tag{11.15}$$

So far, we have seen how the surface free energy and therefore, the thermodynamic stability of various surface structures varies as a function of oxygen chemical potential for a metallic surface in equilibrium with oxygen reservoir. Furthermore, we have also established the bounds on the valid range of the oxygen chemical potential based on thermodynamic stability arguments. However, a quantitative evaluation of surface free energy requires evaluation of each and every component of the Gibbs free energy, as described in Equation 11.12. As already mentioned, the leading term in the Gibbs free energy expression is the total energy E^{tot}, which has to be evaluated directly from the electronic structure theory calculations. The remaining terms in the expression (viz., the vibrational term, the configurational entropy, and the PV-term) are discussed below.

Calculation of the configurational entropy is quite involved and very much system specific. To explicitly treat the configuration contribution to the total free energy, one has to sample the configurational space of all ordered and disordered structures using modern statistical methods such as Monte Carlo simulations. However, when dealing with highly crystalline structures such as metals or metal oxides with moderately disordered surfaces, the configurational entropic contributions are usually quite small and can be estimated as follows.

Let us assume that the system under consideration has N surface sites with n identical defects or adsorbate sites. Then the total number of ways in which the surface can be configured is given by $\omega = \frac{(N+n)!}{N!n!}$. From statistical mechanics, the configurational entropy of such a system is given by

$$S^{c} = k_{B} \ln \omega = k_{B} \ln \frac{(N+n)!}{N!n!}, \tag{11.16}$$

where k_B is the Boltzmann constant. Now, in addition to using the Sterling approximation $(ln(n!) = nln(n) - n, n \gg 1)$, we estimate the total surface area A by summing up the areas of all surface sites (NA_{site}), assuming only a small concentration of defects. Then the configurational entropic contribution per unit area is given by

$$\frac{TS^{c}}{A} = \frac{k_{B}T}{A_{site}} \left[ln\left(1 + \frac{n}{N}\right) + \frac{n}{N} ln\left(1 + \frac{N}{n}\right) \right]. \tag{11.17}$$

Assuming that for a moderately disordered surface the ratio of (n/N) remains with in 5%, Equation 11.17 gives

$$\frac{TS^{c}}{A} \leq 0.20 \frac{k_{B}T}{A_{site}}. \tag{11.18}$$

For temperatures as high as 1000 K and surface areas per site of about 16 Å^2, the configurational entropy will not contribute more than about 1 meV/Å^2 to the surface free energy, and therefore, to a first approximation, this contribution is often neglected. However, it is important to note that for highly disordered systems as well as for various phase boundaries, this contribution to the Gibbs free energy of the surface will not be negligible, and in such cases, the exact contribution has to be calculated using Monte Carlo simulations.

The vibrational contribution is typically computed at the harmonic level (with the quasiharmonic and anharmonic contributions generally ignored). Even the harmonic contribution is rather expensive to compute for large systems. What is determined is the phonon density of states (DOS) $\sigma(\omega)$ for the bulk and relevant surface structures. Once the harmonic phonon DOS is known, the vibrational contribution to the Gibbs free energy can be written as

$$F^{v} = \int \sigma(\omega) \left[\frac{1}{2}\hbar\omega + k_{B}T \ln(1 - e^{-\hbar\omega/k_{B}T}) \right] d\omega. \tag{11.19}$$

However, it is often useful to first obtain an estimation of magnitude of the vibrational contribution to the surface Gibbs free energy before calculating the cumbersome full phonon DOS. In the calculation of the surface free energy, the term F^{v} contains the change in the vibrational modes of the surface atoms or surface adsorbates with respect to their bulk counterparts. As the vibrational modes of atoms in the interior of the surface are not expected to change much because of the formation of surface or surface adsorption or defect formation, the term F^{v} can usually be safely estimated by considering only the change in the vibrational modes of atoms in the topmost layer of the slab and the surface adsorbates with respect to the bulk within the Einstein model of harmonic approximation.

The contribution to the Gibbs free energy due to the last term (i.e., the PV term) in Equation 11.12 is in general negligible for solids. It is easy to see through a simple dimensional analysis that the term PV/A (atm Å^3/Å^2) $\sim 10^{-3}$ meV/Å^2. Even at very high pressures such as 100 atm, the PV term contributes only 0.1 meV/Å^2, which can be safely neglected in comparison with the total surface free energy term, which is usually of the order of 100 meV/Å^2.

An example plot of the surface energy γ as a function of the chemical potential of the surrounding gas phase conveniently defined in terms of $\Delta\mu_O = \mu_O(T, P) - 1/2E_{O_2}^{DFT}$ is shown in Figure 11.14a, while Figure 11.14b shows the schematics of various stable thermodynamic phases for a metal surface in equilibrium with O_2 reservoir, as the oxygen chemical potential varies from a very low to a very high value within allowed bounds. The surface free energy of a clean metal surface will be independent of the gas-phase chemical potential and shows up as a horizontal line on the plot. As one can easily guess, at the lower end of the oxygen chemical potential, the clean metal surface will be the most favored thermodynamic phase, and therefore have the lowest surface free energy. As the oxygen chemical potential gradually increases, more and more oxygen would get adsorbed on the surface. As the oxygen content of the surface increases, the slope of the γ v/s $\Delta\mu_O$ curve gets more and more negative, leading to the formation of stable surface phases I and II and eventually, at the critical oxygen chemical potential (cf. Eq. 11.15), formation of the bulk oxide phase takes place. Increasing the oxygen chemical potential even further would finally lead to oxygen condensation onto the oxidized metal surface (at $\mu_O = 1/2E_{O_2}^{tot}$). We also note that the width of the thermodynamic stability range of the bulk oxide phase on the oxygen chemical potential axis is given by the heat of formation of the bulk oxide (min[$\mu_O(T, P)$]-1/2 $E_{O_2}^{tot} = \frac{1}{y}\Delta G_{M_xO_y}^{f}(0, 0)$); in other words, the width is equal to the stability of the bulk oxide per oxygen atom.

11.8 INTERFACE PHASE DIAGRAMS: FIRST PRINCIPLES THERMODYNAMICS

The determination of the surface phase diagrams when systems are in equilibrium with appropriate gas phase molecules using FPT constitutes a reasonably mature computational methodology in materials research as discussed in the previous sections. In following paragraphs, we discuss the usefulness of this methodology to predict the interface phase diagram using Si-HfO_2 and Pt-HfO_2 interfaces as examples. These two types of interfaces—-found in the sub-45 nm complementary metal oxide semiconductor (CMOS) technology, based on high dielectric constant (or "high-k") oxides and metal electrodes—present a set of issues related to the interface oxidation, morphology, and defect chemistry [88, 89].

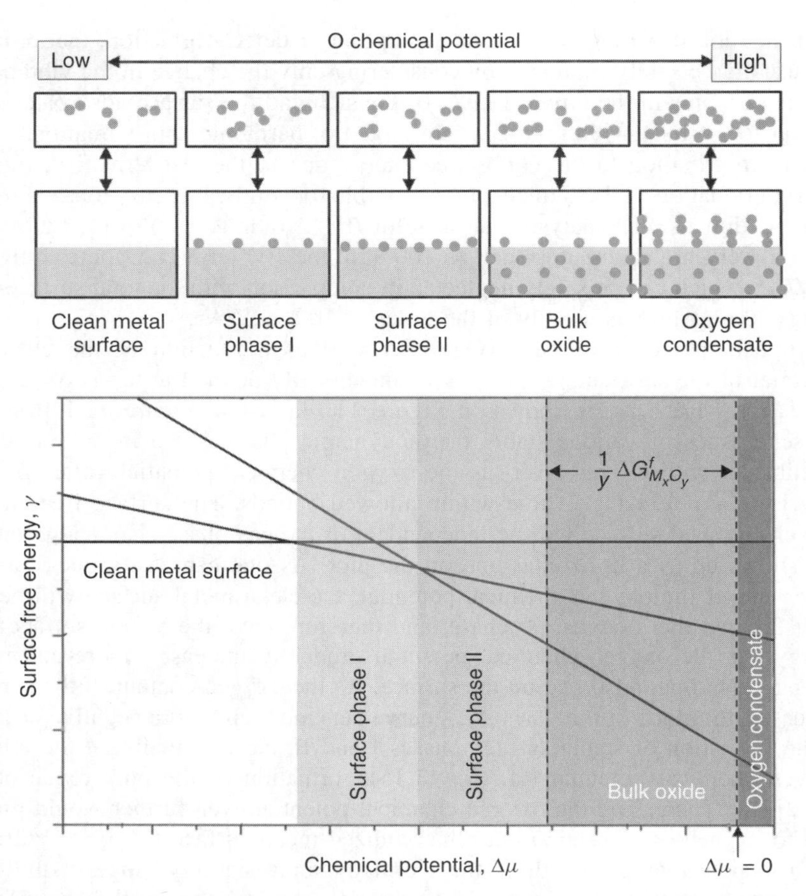

Figure 11.14 Example plot of Gibbs surface free energy (γ) versus oxygen chemical potential ($\Delta\mu_O = \mu_O(T, P) - 1/2E_{O_2}^{\mathrm{DFT}}$) for a surface in equilibrium with a surrounding gas phase. The surface energy of the clean metal surface serves as a zero reference in the plot. Adsorption of an O adatom will become favorable on the metal surface only if the Gibbs free energy of the resultant oxygen adatom-covered surface is lower than that of the clean surface. Also note that as the O content of the surface increases, the slope of the γ versus $\Delta\mu_O$ plot becomes increasingly negative. With increasing gas-phase chemical potential, first the clean metal phase is stable, then the first and second adsorbate phases become stable followed by an oxide phase, and finally, oxygen gas phase condensation takes place on the surface. (*See insert for color representation of the figure*.)

In the case of the Si-HfO_2 interface, a low-k SiO_x or metallic silicide phase (both of which are undesirable) could form at Si-HfO_2 interfaces [88, 90, 91]. In fact, reversible formation of SiO_x and silicides at the Si-HfO_2 interface by annealing in oxygen-rich and oxygen-deficient environments, respectively, have been observed previously [92, 93]. On the other hand, the desired metal electrodes should possess appropriate work functions such that the metal Fermi level lines up with either the valence or conduction band edges of the underlying Si substrate. However, the interfacial chemistries at metal-HfO_2 interfaces (in terms of charge transfer, bond formation, defect accumulation, dipole creation, etc.) lead to shifts in the work function value from its true vacuum value [94] and shows a dependence on the processing conditions. Thus, in both cases of Si-HfO_2 and metal-HfO_2 interfaces, it would be valuable to understand the relationship between ambient conditions (e.g., temperature and oxygen pressure) and interface morphologies.

Figure 11.15 shows the standard interface supercells containing an X-HfO_2 (X = Si or Pt) heterostructure and a vacuum region. The interface oxidation could be modeled by varying the concentration of O at the X-HfO_2 interface, which is represented by θ_O in units of a monolayer (ML). In keeping with the stoichiometry of HfO_2, an ML is defined as two times the number of Hf atoms in a layer (1 ML of O has 4 O atoms in the specific interface

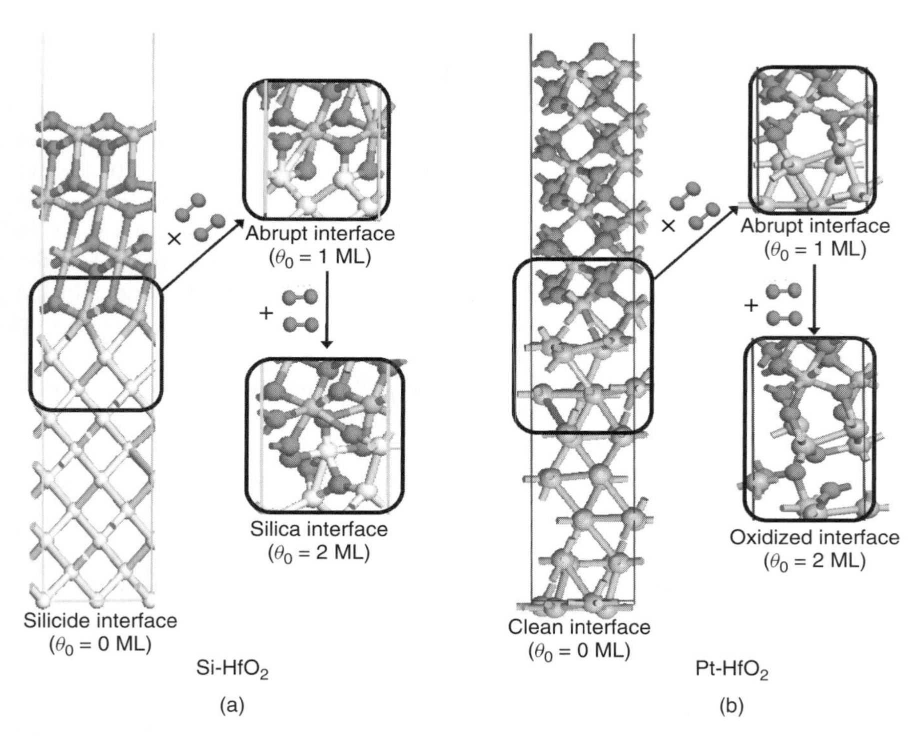

Figure 11.15 The representative atomic structures of the X-HfO_2 heterostructures with X at the bottom: (a) Si-HfO_2 interface and (b) Pt-HfO_2 interface. The Hf and O atoms of the top HfO_2 parts in both (a) and (b) are represented by light blue (gray) and red (dark) spheres, respectively [99]. (*See insert for color representation of the figure*.) (Reprinted with permission from H. Zhu, C. Tang, and R. Ramprasad, Phys Rev, 2010, B 82, 235413 and H. Zhu and R. Ramprasad, Phys. Rev,2011, B 83, 081416(R). Copyright 2010 by the American Physical Society.)

models considered). The corresponding interface configurations are denoted by X:θ_O:HfO_2. Three representative interfaces for Si-HfO_2 and Pt-HfO_2 are indicated in Figure 11.15a,b. For Si-HfO_2, $\theta_O = 0$ corresponds to an interface completely depleted of O and is thus composed of primarily Hf-Si bonds; we refer to this as a "silicide" interface. A well-passivated Si-HfO_2 interface requires a full ML of O, with half of the ML passivating HfO_2 and the other half passivating Si [95]. This situation is referred as an *abrupt interface* ($\theta_O = 1$). $\theta_O > 1$ represents a situation in which O atoms in excess of 1 ML prefer to penetrate the Si side of the heterostructure, with the most favorable sites being between the top two Si layers [96–98]; we refer to this as an *oxidized interface*. When $\theta_O = 2$, the interface is a silica-like interface. Similarly, the three representative interfaces for Pt-HfO_2, Pt:0:HfO_2, Pt:1:HfO_2, and Pt:2:HfO_2 are referred to as *clean, abrupt, and oxidized interfaces*, respectively (see 11.15b).

We now describe the approach used to create the interface phase diagrams based on FPT [99]. The most stable interface is the one with minimum interface free energy (γ_{θ_O}), which can be defined as

$$\gamma_{\theta_O} = (G_{\theta_O} - n_{Hf}\mu_{Hf} - n_O\mu_O - n_X\mu_X)/A - \sigma_{HfO_2} - \sigma_X \tag{11.20}$$

where G_{θ_O} is the Gibbs free energy for the X:θ_O:HfO_2 heterostructure, A is the interface area, and n_{Hf}, n_O, and n_X are, respectively, the numbers of Hf, O, and X atoms in the heterostructure. μ_{Hf} and μ_O are the chemical potentials of Hf and O, respectively, in bulk$m$-HfO_2 and μ_X is the chemical potential of X in its stable elemental form, which is the diamond cubic structure for Si and the face-centered cubic structure for Pt. σ_{HfO_2} and σ_X are the surface energies of the two free surfaces, which are insensitive to the interface structures, and will drop out when relative interface energies are considered (as done below).

Moreover, the chemical potentials in the above equation satisfy the following relationships:

$$\begin{aligned} \mu_{Hf} + 2\mu_O &= G_{HfO_2} \\ \mu_X &= G_X, \end{aligned} \tag{11.21}$$

where G_{HfO_2} and G_X are the Gibbs free energies for bulk m-HfO_2 and bulk X, respectively. Substituting Equation 11.21 into Equation 11.20 results in

$$\begin{aligned} \gamma_{\theta_O} = {} & [G_{\theta_O} - n_{Hf}G_{HfO_2} - (n_O - 2n_{Hf})\mu_O - n_XG_X]/A \\ & - \sigma_{HfO_2} - \sigma_X. \end{aligned} \tag{11.22}$$

When $\theta_O = 0.5$, the whole system is stoichiometric ($n_{Hf} : n_O = 1{:}2$) and the interface energy is

$$\gamma_{0.5} = (G_{0.5} - n_{Hf}G_{HfO_2} - n_XG_X)/A - \sigma_{HfO_2} - \sigma_X. \tag{11.23}$$

By using $\gamma_{0.5}$ as a reference, replacing $n_O - 2n_{Hf}$ by $4 \times (\theta_O - 0.5)$ (four accounts for the fact that 1 ML of O contains four atoms in the interfaces considered here, see Fig. 11.15), substituting Equation 11.12 into Equations 11.22 and 11.23, and assuming that the *PV* terms for X:θ_O:HfO_2 and X:0.5:HfO_2 are roughly equivalent, the relative interface energy can be further defined as

$$\begin{aligned}\Delta\gamma_{\theta_O} &= \gamma_{\theta_O} - \gamma_{0.5} \\ &= [E^{DFT}_{\theta_O} - E^{DFT}_{0.5} + F^{v}_{\theta_O} - F^{v}_{0.5} + F^{c}_{\theta_O} - F^{c}_{0.5} \\ &\quad - 4(\theta_O - 0.5)\mu_O]/A.\end{aligned} \tag{11.24}$$

The X:θ_O:HfO_2 heterostructure reaches its equilibrium state at a certain temperature and oxygen pressure by exchanging O atoms with the surrounding atmosphere. Hence, μ_O in Equation 11.24 could be replaced with half of the chemical potential of an oxygen molecule, μ_{O_2}, which is a function of temperature (T) and oxygen pressure (P_{O_2}), and could be obtained from JANAF tables or ab initio statistical mechanics [100, 101].

$$\mu_{O_2}(T, P_{O_2}) = E^{DFT}_{O_2} + \Delta\mu_{O_2}(T, P_{O_2}), \tag{11.25}$$

where $E^{DFT}_{O_2}$ is the DFT energy of an isolated O_2 molecule and $\Delta\mu_{O_2}(T, P_{O_2})$ contains the zeropoint vibrational energy ($E^{ZPE}_{O_2}$) and the T- and P-dependences of chemical potential. Hence, Equation 11.24 could be written as

$$\begin{aligned}\Delta\gamma_{\theta_O} &= \{E^{DFT}_{\theta_O} - E^{DFT}_{0.5} + F^{v}_{\theta_O} - F^{rmv}_{0.5} + F^{c}_{\theta_O} - F^{c}_{0.5} - (2\theta_O - 1) \\ &\quad [E^{DFT}_{O_2} + \Delta\mu_{O_2}(T, P_{O_2})]\}/A \\ &= [\Delta E^{DFT}_{\theta_O} + \Delta F^{v}_{\theta_O} + \Delta F^{c}_{\theta_O} - (2\theta_O - 1)\Delta\mu_{O_2}(T, P_{O_2})]/A,\end{aligned} \tag{11.26}$$

where $\Delta E^{DFT}_{\theta_O}$ stands for the DFT energy difference between X:θ_O:HfO_2 and X:0.5:HfO_2 + $(2\theta_O - 1)O_2$, that is, $\Delta E^{DFT}_{\theta_O} = E^{DFT}_{\theta_O} - E^{DFT}_{0.5} - (2\theta_O - 1)E^{DFT}_{O_2}$, and may be obtained from normal DFT calculations. Also, $\Delta F^{v}_{\theta_O} = F^{v}_{\theta_O} - F^{v}_{0.5}$ and $\Delta F^{c}_{\theta_O} = F^{c}_{\theta_O} - F^{c}_{0.5}$. Details on how $\Delta F^{v}_{\theta_O}$ and $\Delta F^{c}_{\theta_O}$ are estimated are discussed later (also see Eq. 11.17 and 11.19). Prior FPT works on the surface phase stabilities illustrate that $\Delta F^{v}_{\theta_O}$ and $\Delta F^{c}_{\theta_O}$ will not qualitatively affect the phase diagram, although phase transition regions may be smoothed in the presence of these two terms [102, 103]. Thus, in many treatments, $\Delta F^{v}_{\theta_O}$ and $\Delta F^{c}_{\theta_O}$ are dropped in the computation of the surface phase diagram. In such cases, $\Delta\gamma_{\theta_O}$ is purely a function of $\Delta E^{DFT}_{\theta_O}$ and $\Delta\mu_{O_2}(T, P_{O_2})$ (referred to here as $\Delta\gamma^{DFT}_{\theta_O}$).

$$\Delta\gamma^{DFT}_{\theta_O} = [\Delta E^{DFT}_{\theta_O} - (2\theta_O - 1)\Delta\mu_{O_2}(T, P_{O_2})]/A. \tag{11.27}$$

Equation 11.26 or 11.27 forms the basis for the construction of interface phase diagrams, from which one can compute the interface energy for various interface structures for given (T, P_{O_2}) combinations. The interface phase diagram is then obtained by identifying the most stable (i.e., minimum energy) interface as a function of T and P_{O_2}. In the following paragraphs, we discuss the interface phase diagram obtained using both Equations 11.26 and 11.27 as well as the impact of including $\Delta F^{v}_{\theta_O}$ and $\Delta F^{c}_{\theta_O}$.

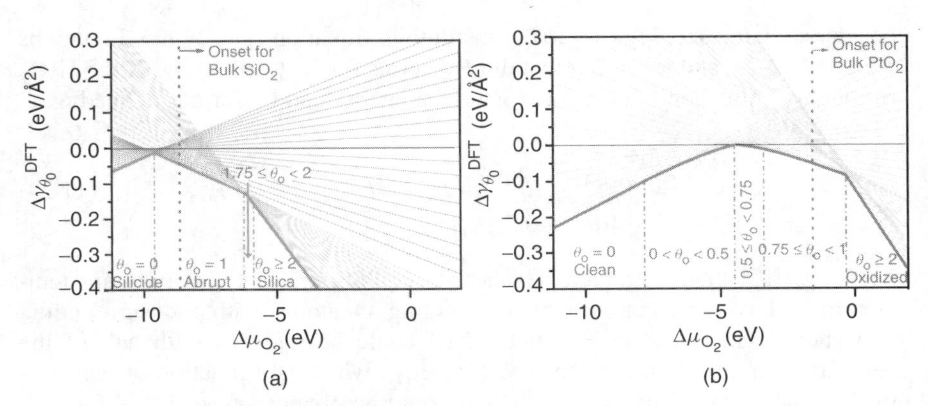

Figure 11.16 The relative interface free energy as a function of $\Delta\mu_{O_2}$ for the (a) Si-HfO_2 and (b) Pt-HfO_2 interfaces. The inner boundaries represent the lowest interface energy at each $\Delta\mu_{O_2}$ value. The corresponding θ_O value is indicated in the figure. The vertical dash-dotted lines are boundaries between two different stable interface structures. The onset $\Delta\mu_{O_2}$ to form bulk SiO_2 and α-PtO_2 is indicated by the vertical dotted line in (a) and (b), respectively [99]. (*See insert for color representation of the figure*.) (Reprinted with permission from H. Zhu, C. Tang, and R. Ramprasad, Phys Rev, 2010, B 82, 235413 and H. Zhu and R. Ramprasad, Phys. Rev,2011, B 83, 081416(R). Copyright 2010 by the American Physical Society.)

11.8.1 Interface Phase Diagrams Not Including Vibrational and Configurational Energies

We consider the interfaces first with the assumption that the contribution to the interface energy due to $\Delta F^{v}_{\theta_O}$ and $\Delta F^{c}_{\theta_O}$ may be ignored, as described by Equation 11.27. The advantage with considering Equation 11.27 is that all T and P_{O_2} dependence of $\Delta\gamma^{DFT}_{\theta_O}$ is contained solely within $\Delta\mu_{O_2}$. Thus, as shown in Figure 11.16, a plot of $\Delta\gamma^{DFT}_{\theta_O}$ versus $\Delta\mu_{O_2}$ yields straight lines, with the slope and intercept depending, respectively, on the interfacial O content and the value of $\Delta E^{DFT}_{\theta_O}$. The lower boundary of $\Delta\mu_{O_2}$ is confined to avoid the segregation of Hf from bulk HfO_2. Lines corresponding to each value of θ_O ranging from 0 to 2 are shown, and the lowest energy interface at each $\Delta\mu_{O_2}$ value is identified along with the corresponding θ_O value. The critical $\Delta\mu_{O_2}$ value, above which formation of the bulk XO_2 oxides occurs, is also indicated. At this critical $\Delta\mu_{O_2}$, bulk X is in equilibrium with bulk XO_2. The stable forms for bulk Hf, HfO_2, Pt, and PtO_2 are the hexagonal, monoclinic, face-centered cubic, and α-PtO_2 structures, respectively.

As shown in Figure 11.16a, the stable interface type in Si:θ_O:HfO_2 heterostructure changes abruptly from $\theta_O = 0$ to $\theta_O = 1$ and then to $\theta_O = 1.75$. Nevertheless, when $2 > \theta_O > 1.75$, the interface changes smoothly. Moreover, bulk SiO_2 is favored at smaller $\Delta\mu_{O_2}$ values than interfacial silica. This is probably due to the fact that the oxidation of interfacial silicon is accompanied by strain and hence requires more energy. The Pt:θ_O:HfO_2 interface (shown in Fig. 11.16b), however, changes more smoothly for $\theta_O \leq 1$, and then abruptly changes from $\theta_O \sim 1$ to $\theta_O = 2$. Similar to the Si-HfO_2 case (and presumably for the same reason), the critical $\Delta\mu_{O_2}$ for an oxidized Pt-HfO_2 interface is larger than the one required to form bulk PtO_2.

Since $\Delta\mu_{O_2}$ is a function of T and P_{O_2}, the information contained in Figure 11.16 may be used to create (T, P_{O_2}) phase diagrams in which each $\Delta\mu_{O_2}$ "turning point" of Figure 11.16 becomes a curve in a T versus P_{O_2} plot, demarcating boundaries between two different phases. Such phase diagrams are presented in Figures 11.17a and Fig. 11.18a for Si-HfO_2 and Pt-HfO_2 interfaces, respectively, under the assumption that $\Delta F^{\mathrm{v}}_{\theta_O}$ and $\Delta F^{\mathrm{c}}_{\theta_O}$ may be ignored.

From Figure 11.17a, we find that (not surprisingly) the interfacial silica phase prefers high oxygen pressure and low temperature, while the silicide phase is stable at low oxygen pressure and high temperature. It is however interesting to note that interfacial silica can occur even at ultrahigh vacuum conditions ($P_{O_2} < 10^{-12}$ atm) in a wide temperature range, which explains why the interfacial silica phase is widely observed. The open and solid circles in Figure 11.17a, respectively, represent experimental conditions at which SiO_2 and SiO at Si-HfO_2 interfaces are known to occur [92, 93]. The gray filled circle stands for the critical point for interfacial SiO_2 to decompose to SiO [93]. The (T, P_{O_2}) boundaries predicted in Figure 11.17a for the decomposition of interfacial silica is consistent with experiments.

Compared to Si-HfO_2, the Pt-HfO_2 interface displays a smoother transition from one level of O coverage to another, especially for $\theta_O < 1$ (Fig. 11.18a). Also, this interface can be oxidized only at a very high oxygen pressure and low temperature. In the ultrahigh vacuum environment, the stable interfacial O coverage between Pt and HfO_2 is 0.5–1 ML. In view of the fact that experimental data for the Pt-HfO_2 interface morphologies is sparse, we compared our phase diagram with experimental data for Pt surface oxidation. In order to facilitate such a comparison, we make the following observation. Since 0.5 ML O at the Pt-HfO_2 interface passivates the dangling bonds on the HfO_2 side of the interface (based on the charge counting notions) and has little interaction with Pt, the net interfacial O strongly bonded to Pt is actually $\theta_O - 0.5$ ML. Within this context, the interfacial Pt in Pt:0.5:HfO_2 behaves like a clean Pt surface. The open and solid squares in Figure 11.18a stand for the (T, P_{O_2}) conditions at which 0.25 ML O-adsorbed (111) Pt and clean (111) Pt surfaces are observed in experiment, respectively, consistent with our $\theta_O - 0.5$ ML values of ~0.25 and 0, respectively, under those same conditions. Another interesting finding of this work is that the oxidation of Pt at the interface is similar to that of a free (111) Pt surface. The saturation coverage of the chemically adsorbed O on (111) Pt surface is 0.25–0.3 ML, after which a layer of PtO_2 forms immediately on the surface [104]. Here, we find that the corresponding $\theta_O - 0.5$ ML value beyond which interfacial PtO_2 is formed is 0.25–0.5.

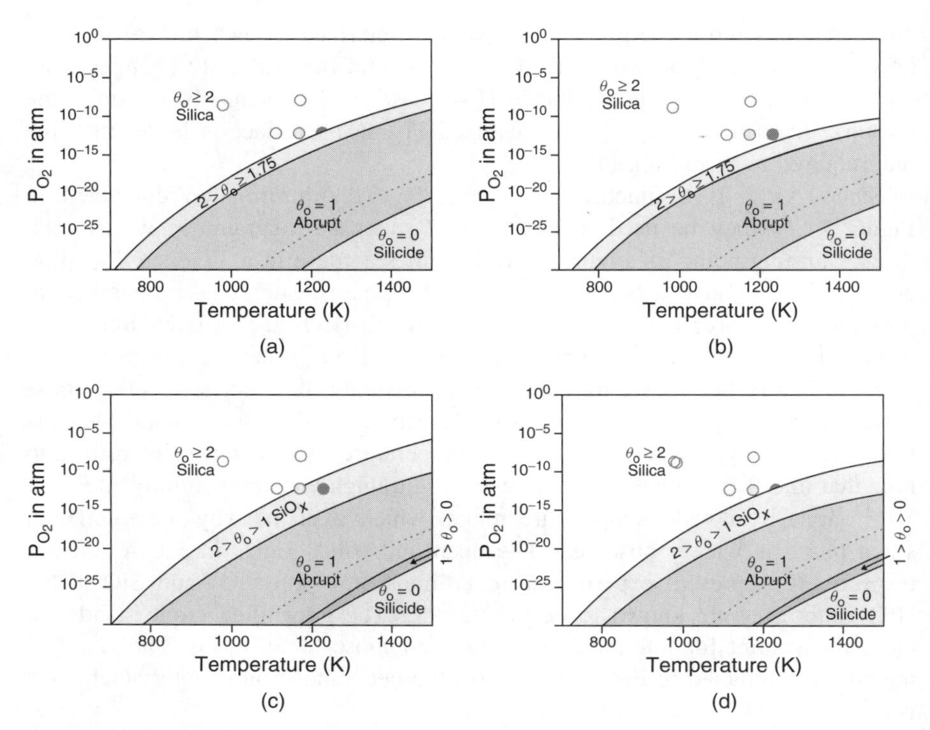

Figure 11.17 (a) The interface phase diagrams for the Si-HfO_2 interface, determined using Equation 11.27), *i.e.*, after neglecting the vibrational and configurational entropic contributions of the condensed phases to the relative interface free energy. (b) Same as (a), but with the vibrational entropic contribution ($\Delta F^{\mathrm{v}}_{\theta_{\mathrm{O}}}$) included. (c) Same as (a), but with the configurational entropic contribution ($\Delta F^{\mathrm{c}}_{\theta_{\mathrm{O}}}$) included. (d) The interface phase diagrams determined using Equation 11.26; this is the same as (a), but with both the vibrational and configurational entropic contributions included. The solid curves indicate the interface phase boundaries, and the dotted curves represent the onset of formation of bulk SiO_2. The open and solid circles stand for the condition to form SiO_2 and SiO at the Si-HfO_2 interfaces in experiments, respectively [92, 93]. The gray filled circle stands for the critical point for interfacial SiO_2 to decompose to SiO [93, 99]. (*See insert for color representation of the figure*.) (Reprinted with permission from H. Zhu, C. Tang, and R. Ramprasad, Phys Rev, 2010, B 82, 235413 and H. Zhu and R. Ramprasad, Phys. Rev,2011, B 83, 081416(R). Copyright 2010 by the American Physical Society.)

11.8.2 Impact of Other Factors (For Example, the Vibrational and Configurational Energy Contributions) on Interface Phase Diagrams

Next, we address theoretical aspects left unexplored in the treatment above, for example, the neglect of $\Delta F^{\mathrm{v}}_{\theta_{\mathrm{O}}}$ and $\Delta F^{\mathrm{c}}_{\theta_{\mathrm{O}}}$ (which allowed us to simplify Equation 11.26 to Equation 11.27]. To estimate $\Delta F^{\mathrm{v}}_{\theta_{\mathrm{O}}}$, the vibrational DOS, $\sigma_{\theta_{\mathrm{O}}}(w)$, for an interface with a certain O coverage and frequency w have been determined

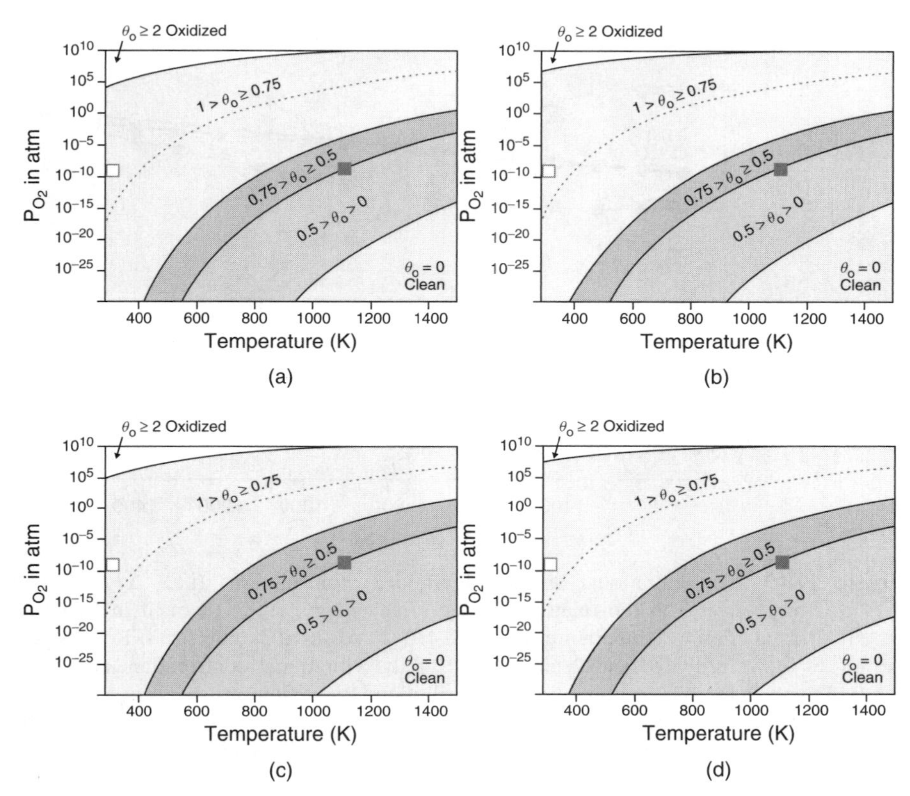

Figure 11.18 (a) The interface phase diagrams for the Pt-HfO_2 interface, determined using Equation 11.27, that is, after neglecting the vibrational and configurational entropic contributions of the condensed phases to the relative interface free energy. (b) Same as (a), but with the vibrational entropic contribution ($\Delta F^{v}_{\theta_O}$) included. (c) Same as (a), but with the configurational entropic contribution ($\Delta F^{c}_{\theta_O}$) included. (d) The interface phase diagrams determined using Equation 11.26; this is the same as (a), but with both the vibrational and configurational entropic contributions included. The solid curves indicate the interface phase boundaries, and the dotted curves represent the onset of formation of bulk PtO_2. The open and solid squares are the T and P_{O_2} when a 0.25 ML O-adsorbed (111) Pt surface and a clean (111) Pt surface were observed, respectively [99, 104]. (*See insert for color representation of the figure*.) (Reprinted with permission from H. Zhu, C. Tang, and R. Ramprasad, Phys Rev., 2010, B 82, 235413 and H. Zhu and R. Ramprasad, Phys. Rev., 2011, B 83, 081416(R). Copyright 2010 by the American Physical Society.)

through harmonic normal mode analysis by allowing only the interfacial O, Si, Pt, or Hf atoms, as appropriate, to vibrate. The vibrational contribution to the interface free energy (to be used in Equation 11.26), can be calculated by Equation 11.19. The vibration of atoms away from the interface for θ_O and 0.5 are expected to be roughly equivalent and are assumed to cancel out in the computation of $\Delta F^{v}_{\theta_O}$. For clarity, we show $\Delta F^{v}_{\theta_O}/A$ for θ_O values of only 0 and 2 in

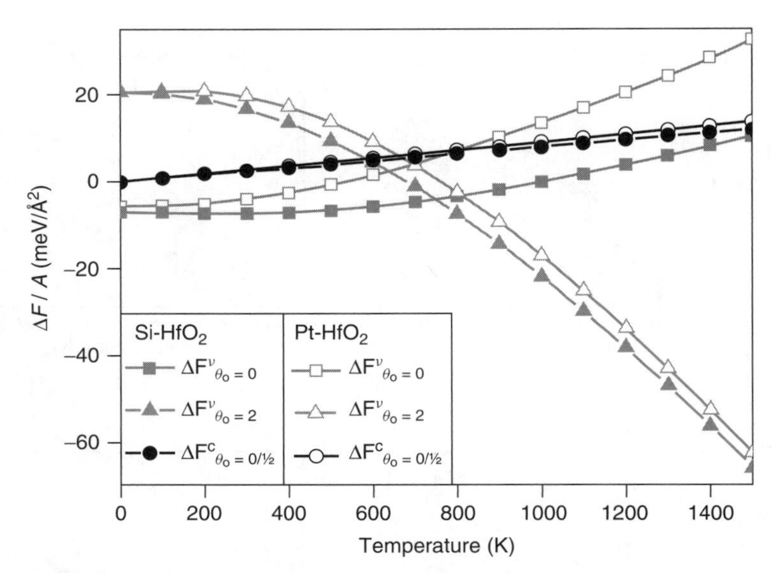

Figure 11.19 The vibrational and configurational contributions (i.e., $\Delta F^{\upsilon}_{\theta_O}/A$ and $\Delta F^{c}_{\theta_O}/A$, respectively) to the relative interface free energy, $\Delta\gamma_{\theta_O}$ (defined in Equation 11.26], as a function of temperature for the Si-HfO_2 and Pt-HfO_2 interfaces. Results pertaining to θ_O values of 0 and 2 are shown for the vibrational contribution, and those for 0, 1, and 2 for the configurational contribution. By definition, the $\Delta F^{c}_{\theta_O}/A$ values are identical for the three coverages. The slight difference in the $\Delta F^{c}_{\theta_O}/A$ values for the Si-HfO_2 and Pt-HfO_2 interfaces are due to the difference in the interface area A for these two cases.

Figure 11.19 as these two extreme coverages provide an idea of the magnitude of the vibrational energy contribution. The roughly flat region at low temperatures is due to zero point vibrations (first term of Equation 11.19]. The temperature-dependent (second) term of Equation 11.19 contributes to a decrease (increase) of $\Delta F^{v}_{\theta_O}$ with increasing temperature for θ_O values larger (smaller) than 0.5, and in all cases going through zero above 500 K. This has the implication that lower (higher) values of the O coverages are favored at lower (higher) temperatures, with the vibrational entropic contribution having negligible impact in an intermediate temperature range when $\Delta F^{v}_{\theta_O}$ goes through zero. Features that reflect these expectations indeed manifest in the phase diagrams of Figures 11.17b and 11.18b, which have been created for the Si-HfO_2 and Pt-HfO_2 interfaces with the explicit inclusion of $\Delta F^{v}_{\theta_O}$ in the determination of the interface free energy (that is, Eq. 11.26 with $\Delta F^{c}_{\theta_O} = 0$). It can be seen that the impact of including $\Delta F^{v}_{\theta_O}$ results in rather minor changes to the features of the phase diagrams (most notably, the enlargement of the $\theta_O \geq 2$ phase region at high temperatures in the Si-HfO_2 interface phase diagram, and a shrinking of this region in the Pt-HfO_2 interface phase diagram at low temperatures). While this analysis provides a justification for the neglect of the vibrational contribution in prior treatments, we note that systems involving lighter atoms, for example, H, may require explicit inclusion of the vibrational contribution to the free energy.

Finally, we estimate the impact of $\Delta F^{c}_{\theta_O}$ on the interface phase diagram by assuming that the O atom sites for $\theta_O \in [0,1]$ are equivalent and O atom sites penetrating X layers are energetically identical. We note that this assumption constitutes an upper bound to the configurational entropy contribution and is justified by the lack of dependence of the binding energy per O atom at the studied interfaces [99]. Thus, using Stirling's approximation, $F^{c}_{\theta_O}$ may be defined as follows. When $0 < \theta_O < 1$, $F^{c}_{\theta_O} = 4k_BT \times [\theta_O ln\theta_O + (1-\theta_O)ln(1-\theta_O)]$. When $1 < \theta_O < 2$, $F^{c}_{\theta_O} = 4k_BT \times [(\theta_O - 1)ln(\theta_O - 1) + (2-\theta_O)ln(2-\theta_O)]$. When $\theta_O = 0$, 1 and 2, $F^{c}_{\theta_O} = 0$. Fig. 11.19 also shows the $\Delta F^{c}_{\theta_O}$ contribution for $\theta_O = 0$, 1 and 2. Unlike $\Delta F^{v}_{\theta_O}$, $\Delta F^{c}_{\theta_O}$ displays a steady rise with temperature. Moreover, as the variation of $\Delta F^{c}_{\theta_O}$ with θ_O is much smoother, inclusion of this contribution is expected to eliminate abrupt transitions between phases. Figures 11.17c and 11.18c show the Si-HfO_2 and Pt-HfO_2 interface phase diagrams when $\Delta F^{c}_{\theta_O}$ is explicitly included (but $\Delta F^{v}_{\theta_O}$ is not included) in the determination of the interface free energy (i.e., Eq. 11.26 with $\Delta F^{v}_{\theta_O} = 0$). As expected, the inclusion of the configurational energy makes the transitions less abrupt and makes some intermediate interfacial phases appear in the phase diagram, for example, the $0 < \theta_O < 1$ and $1 < \theta_O < 1.75$ regions. We refer to $1 < \theta_O < 2$ as a "SiO_x" suboxide region, and note that the presence of this phase region in the phase diagram brings our predictions for the formation and decomposition of interfacial silica more in line with experimental observations. In contrast to the Si-HfO_2 case, the interface phase diagrams with and without $\Delta F^{c}_{\theta_O}$ for Pt-HfO_2 are very similar to each other. The inclusion of $\Delta F^{c}_{\theta_O}$ causes only slight shifts to the phase boundaries, and no new phase regions appear.

Figures 11.17d and 11.18d display the corresponding phase diagrams when both the vibrational and configurational energies are included in the treatment (i.e., when Equation 11.26 is used]. Not surprisingly, simultaneous inclusion of these two contributions also does not result in significant differences with respect to the phase diagrams of Figures 11.17a and 11.18a. Nevertheless, these calculations provide an estimate of errors that may be introduced due to the neglect of such contributions.

We note that the obtained phase diagrams are in agreement with available experimental data. While such an agreement with experiments may be viewed as fortuitous, this may indicate that all the dominant contributions to the interface free energy have been included in this treatment (even at the level of Equation 11.27). The unaddressed issues of deficiencies inherent to approximations within DFT are presumably unimportant in the class of systems studied here (or may have participated in a fortuitous cancelation of errors). We do note that FPT studies may benefit from (the relatively inexpensive) explicit inclusion of configurational entropic contributions to the free energy, especially when a lack of such inclusion results in abrupt transitions between phases.

11.9 OUTLOOK AND CONCLUDING THOUGHTS

This article focused on the practical applications of modern DFT-based computations in the area of surface and interface science. After a general survey of the current state-of-the-art of conventional DFT computations, applications of such methods to the computations of surface energies, optimization of nanostructure shapes, and the determinations of surface and interface phase diagrams were explored. Methods for combining traditional zero-temperature DFT results with statistical mechanics to obtain free energies of processes were also described.

The future prospects for DFT-based computations remains exciting. The spectrum of problems that may be addressed using DFT methods is rapidly increasing. Moreover, methods to deal with some of the remaining fundamental deficiencies of DFT—such as predictions of band gaps and gap levels in insulators [105] and the treatment of secondary bonding interactions (e.g., dispersive van der Waals interactions) [106]—are becoming available and practical. As we go to more extreme conditions, such as high temperatures and pressures, the free energy of the system has to include terms beyond the harmonic approximation (for the vibrational part). Robust schemes for the proper treatment of such factors are also beginning to mature [107]. The technologically important topic of surface/interface science will benefit from these recent and anticipated developments.

ACKNOWLEDGMENTS

The authors acknowledge support of their surface/interface research by the National Science Foundation, the Office of Naval Research, the ACS Petroleum Research Fund, the Department of Energy, and the Alexander von Humboldt Foundation.

REFERENCES

1. Somorjai GA. *Introduction to Surface Chemistry and Catalysis*. New York: John Wiley and Sons, Inc.; 1994.
2. Woodruff DP, Delchar TA. *Modern Techniques of Surface Science*. 2nd ed. Cambridge: Cambridge University Press; 1994.
3. Noguera C. *Physics and Chemistry of Oxide Surfaces*. New York: Cambridge University Press; 1996.
4. Groß A. *Theoretical Surface Science A Microscopic Perspective*. Berlin: Springer; 2002.
5. Kolasinski KW. *Surface Science: Foundations of Catalysis and Nanoscience*. England: John Wiley and Sons, Ltd; 2008. Yip S, editor. *Handbook of Materials Modeling*. Berlin: Springer; 2005.
6. Horn K, Scheffler M, editors. Volume 2, *Handbook of Surface Science: Electronic Structure*. Amsterdam: Elsevier; 2000.

7. Chadi DJ. Phys Rev Lett 1979;43:43.
8. Brink RS, Verwoerd WS. Surf Sci 1981;154:L203.
9. Ihm J, Cohen ML, Chadi DJ. Phys Rev B 1980;21:4592.
10. Yin MT, Cohen ML. Phys Rev B 1981;24:2303.
11. Verwoerd WS. Surf Sci 1981;99:581.
12. Krüger E, Pollmann J. Phys Rev B 1988;38:10578.
13. Brommer KD, Needels M, Larson BE, Joannopoulos JD. Phys Rev Lett 1992; 68:1355.
14. Stich I, Payne MC, King-Smith RD, Lin J-S. Phys Rev Lett 1992;68:1351.
15. Sholl DS, Steckel JA. *Density Functional Theory: A Practical Introduction*. New Jersey: John Wiley and Sons, Inc.; 2009.
16. Honkala K, Hellman A, Remediakis IN, Logadottir A, Carlsson A, Dahl S, Christensen CH, Nøskov JK. Science 2005;307:555.
17. Rieger M, Rogal J, Reuter K. Phys Rev Lett 2008;100:016105. C Stampfl, M Schemer, Surf. Sci., 319, L23 (1994).
18. Burchhardt J, Nielsen MM, Adams DL, Lundgren E, Andersen JN, Stampfl C, Scheffler M, Schmalz A, Aminpirooz S, Haase J. Phys Rev Lett 1995;74:1617.
19. Stampfl C, Kreuzer HJ, Payne SH, Pfnür H, Scheffler M. Phys Rev Lett 1999; 83:2993.
20. Stampfl C, Kreuzer HJ, Payne SH, Scheffler M. Appl Phys A 1999;69:471.
21. Reuter K, Stampfl C, Scheffler M. In: Yip S, editor. Volume 1, *Handbook of Materials Modeling*. Berlin: Springer; 2005. p 149–194.
22. Hohenberg P, Kohn W. Phys Rev 1964;136:B864.
23. Kohn W, Sham L. Phys Rev 1965;140:A1133.
24. Parr RG, Yang W. *Density Functional Theory of Atoms and Molecules*. New York: Oxford University Press; 1989.
25. Dreizler RM, Gross EKU. *Density Functional Theory*. Berlin: Springer; 1990.
26. Martin R. *Electronic Structure: Basic Theory and Practical Methods*. New York: Cambridge University Press; 2004.
27. Kobayyashi K. Surf Sci 2001;493:665.
28. Kamran S, Chen K, Chen L. Phys Rev B 2009;79:024106.
29. Sun G, Kurti J, Rajczy P, Kertesz M, Hafner J, Kresse G. J Mol Struct (Theochem) 2003;624:37.
30. Ramprasad R, Shi N, Tang C. Modeling the physics and chemistry of interfaces in nanodielectrics. In: Nelson JK, editor. *Dielectric Polymer Nanocomposites*. New York: Springer; 2010.
31. Bernardini F, Fiorentini V. Phys Rev B 1998;58:15292.
32. Beche AD. Phys Rev A 1986;33:2756.
33. Robertson J, Xiong K, Clark SJ. Thin Solid Films 2006;496:1.
34. Shishkin M, Marsman M, Kresse G. Phys Rev Lett 2007;99:246403.
35. Ramprasad R, Classford KM, Adams JB, Masel RI. Surf Sci 1996;360:31.
36. Krukau AV, Vydrov OA, Izmaylov AF, Scuseria GE. J Chem Phys 2006;125: 224106.
37. Mitas L, Martin RM. Phys Rev Lett 1994;72:2438.

38. Vitos L, Ruban AV, Skriver HL, Kollar J. Surf Sci 1998;411:186.
39. Goniakowski J, Finocchi F, Noguera C. Rep Prog Phys 2008;71:016501.
40. Pashley MD. Phys Rev B 1989;40:10481.
41. Peng ZA, Peng XJ. J Am Chem Soc 2002;124:3343.
42. Zhang SB, Wei SH. Phys Rev Lett 2004;92(8):086102.
43. Manna L, Wang LW, Cingolani R, Alivisatos AP. J Phys Chem B 2005;109(13):6183.
44. Pilania G, Sadowski T, Ramprasad R. J Phys Chem C 2009;113:1863.
45. Wulff G. Z Kristallogr 1901;34:449.
46. Herring C. Phys Rev 1951;82:87.
47. Shi H, Stampfl C. Phys Rev B 2008;77:094127.
48. Hu J, Li L, Yang W, Manna L, Wang L, Alivisatos AP. Science 2001;292:2060.
49. Chen X, Nazzal A, Goorskey D, Xiao M, Peng ZA, Peng X. Phys Rev B 2001;64:245304.
50. Huynh WU, Dittmer JJ, Alivsatos AP. Science 2002;295:2425.
51. Carbone L, Nobile C, De Giorgo M, Sala FD, Morello G, Pompa P, Hytch M, Snoeck E, Fiore A, Franchini IR, Nadasan M, Silvestre AF, Chiodo L, Kudera S, Cingolani R, Krahne R, Manna L. Nano Lett 2007;7(10):2942.
52. Talapin DV, Koeppe R, Götzínger S, Kornowski A, Lupton JM, Rogach AL, Benson O, Feldmann J, Weller H. Nano Lett 2003;3(12):1677.
53. Mokari T, Rothenberg E, Popov I, Costi R, Banin U. Science 2004;304:1787.
54. Sheldon MT, Trudeau PE, Mokari T, Wang LW, Alivisatos AP. Nano Lett 2009;9:3676.
55. Doll JD, Pilania G, Ramprasad R, Papadimitrakopoulos F. Nano Lett 2010;10:680.
56. Li R, Luo Z, Papadimitrakopoulos FJ. J Am Chem Soc 2006;128:6280.
57. Anpo M, Che M. Adv Catal 1999;44:119.
58. Hacquart R, Krafft JM, Costentin G, Jupille J. Surf Sci 2005;595:172.
59. Giordano L, Goniakowski J, Suzanne J. Phys Rev Lett 1998;81:1271.
60. Odelius M. Phys Rev Lett 1999;82:3919.
61. Delle Site L, Alavi A, Lynden-Bell RM. J Chem Phys 2000;113:3344.
62. Giordano L, Goniakowski J, Suzanne J. Phys Rev B 2000;62:15406.
63. Kim YD, Stultz J, Goodman DW. J Phys Chem B 2002;106:1515.
64. Stirniman MJ, Huang C, Smith RS, Joyce SA, Kay BD. J Chem Phys 1996;105:1295.
65. Liu P, Kendelewicz T, Brown GJr., ParksGA. Surf Sci 1998;412:287.
66. Liu P, Kendelewicz T, Brown GJr., ParksGA. Surf Sci 1998;412:315.
67. Abriou D, Jupille J. Surf Sci 1999;430:L527.
68. Ahmed SI, Perry S, El-Bjeirami O. J Phys Chem B 2000;104:3343.
69. Scamehorn CA, Harrison NM, McCarthy MI. J Chem Phys 1994;101:1547.
70. Goniakowski J, Noguera C. Surf Sci 1995;330:337.
71. Langel W, Parrinello M. J Chem Phys 1995;103:8.
72. Refson K, Wogelius RA, Fraser DG, Payne MC, Lee MH, Milman V. Phys Rev B 1995;52:10823.

73. de Leeuw NH, Watson GW, Parker SC. J Chem Phys 1995;99:17219.
74. Mejias JA, Berry AJ, Refson K, Fraser DG. Chem Phys Lett 1999;314:558.
75. Finocchi F, Goniakowski J. Phys Rev B 2001;64:125426.
76. Ealet B, Goniakowski J, Finocchi F. Phys Rev B 2004;69:054419.
77. Costa D, Chizallet C, Ealet B, Goniakowski J, Finocchi F. J Chem Phys 2006;125:054702.
78. Finocchi F, Goniakowski J. Surf Sci 2007;601:4144.
79. Zhong Y, Zhu H, Shaw LL, Ramprasad R. Acta Mater 2011;59:3748–3757.
80. Jack DH. In: Schwartz MM, editor. *Engineering Application of Ceramic Materials: Source Book*. Materials Park (OH): American Society for Metals; 1985. p 147–153.
81. Exner HE. Int Met Rev 1979;24:149.
82. Kim CS, Rohrer GS. Interface Sci 2004;12:19.
83. Suetin DV, Shein IR, Kurlov AS, Gusev AI, Ivanovski AL. Phys Solid State 2008;50:1420.
84. Hammer B, Nørskov JK. Adv Catal 2000;45:71.
85. Newns DM. Phys Rev 1969;178:1123.
86. Anderson PW. Phys Rev 1961;124:41.
87. Hammer B, Nørskov JK. In: Lambert R, Pacchioni G, editors. *Theory of Adsorption and Surface Reactions*, NATO ASI Series E 331. Dordrecht: Kluwer Academic Publishers; 1997. K Reuter, M Scheffler, Phys. Rev. Lett., 90, 046103 (2003). RB Getman, WF Schneider, AD Smeltz, WN Delgass, FH Ribeiro, Phys. Rev. Lett., 102, 076101 (2009). R Grau-Crespo, KC Smith, TS Fisher, NH de Leeuw, UV Waghmare, Phys. Rev. B, 80, 174117 (2009). CG Van de Walle, J Neugebauer, Phys. Rev. Lett., 88, 066103 (2002).
88. Wallace RM, Wilk GD. Crit Rev Solid State Mater Sci 2003;28:231.
89. Wu M, Alivov YI, Morkoc H. J Mater Sci Mater Electron 2008;19:915.
90. Qiu XY, Liu HW, Fang F, Ha MJ, Liu JM. Appl Phys Lett 2006;88:072906.
91. Preisler EJ, Guha S, Copel M, Bojarczuk NA, Reuter MC, Gusev E. Appl Phys Lett 2004;85:6230.
92. (a) Wang SJ, Lim PC, Huan ACH, Liu CL, Chai JW, Chow SY, Pan JS. Appl Phys Lett 2003;82:2047; (b) Copel M, Reuter MC, Jamison P. Appl Phys Lett 2004;85:458.
93. Miyata N, Nabatame T, Horikawa T, Ichikawa M, Toriumi A. Appl Phys Lett 2003;82:3880.
94. Cho D-Y, Park K-S, Choi B-H, Oh S-J, Chang YJ, Kim DH, Noh TW, Jung R, Lee J-C. Appl Phys Lett 2005;86:041913.
95. Peacock PW, Robertson J. Phys Rev Lett 2004;92:057601.
96. Tang C, Tuttle B, Ramprasad R. Phy Rev B 2007;76:073306.
97. Tang C, Ramprasad R. Phy Rev B 2007;75:241302(R).
98. Tang C, Ramprasad R. Appl Phys Lett 2008;92:182908.
99. (a) Zhu H, Tang C, Ramprasad R. Phys Rev B 2010;82:235413; (b) Zhu H, Ramprasad R. Phys Rev B 2011;83:081416(R).
100. Stull DR, Prophet H. *JANAF Thermochemical Tables*. 2nd ed. Washington (DC): U.S. National Bureau of Standards; 1971.

101. (a) Hill TL. *Introduction to Statistical Thermodynamics*. New York: Dover; 1986; (b) Nash LK. *Elements of Statistical Thermodynamics*. Reading (MA): Addison-Wesley; 1972.
102. Reuter K, Scheffler M. Phys Rev B 2001;65:035406.
103. Reuter K, Scheffler M. Phys Rev B 2003;68:045407.
104. Engstrom U, Ryberg R. Phys Rev Lett 1999;82:2741.
105. Lambrecht WRL. Phys Status Solidi B, 1–12 (2010).
106. Grimme S. Wiley Interdisciplinary Rev: Comput Mol Sci 2011;1:211.
107. Grabowski B, Ismer L, Hickel T, Neugebauer J. Phys Rev B 2009;79:134106.

INDEX

A Matter of Density: Exploring the Electron Density Concept in the Chemical, Biological, and Materials Sciences, First Edition. Edited by N. Sukumar.